Alien Life and Human Purpose

Alien Life and Human Purpose

A Rhetorical Examination through History

Joseph Packer

LEXINGTON BOOKS
Lanham • Boulder • New York • London

Published by Lexington Books
An imprint of The Rowman & Littlefield Publishing Group, Inc.
4501 Forbes Boulevard, Suite 200, Lanham, Maryland 20706
www.rowman.com

Unit A, Whitacre Mews, 26-34 Stannary Street, London SE11 4AB

British Library Cataloguing in Publication Information Available

Library of Congress Cataloging-in-Publication Data

Packer, Joseph, 1983-
Alien life and human purpose : a rhetorical examination through history / Joseph Packer.
p. cm.
Includes bibliographical references and index.
ISBN 978-1-4985-1301-2 (cloth : alk. paper) -- ISBN 978-1-4985-1303-6 (pbk. : alk. paper) -- ISBN 978-1-4985-1302-9 (electronic)
1. Cosmology--History. 2. Ethics. I. Title.
BD494.P33 2015
111'.82--dc23
2015029302

∞ TM The paper used in this publication meets the minimum requirements of American National Standard for Information Sciences Permanence of Paper for Printed Library Materials, ANSI/NISO Z39.48-1992.

Printed in the United States of America

For my parents

Table of Contents

Acknowledgments

I have many individuals to thank for helping me bring this project to fruition. My interest in academia and aliens began with my participation in college debate. I owe a great debt of gratitude to Tim O'Donnell for teaching me how to make an argument and convincing me not to go to law school.

I began my project at Wake Forest and I would like to thank my Master's Thesis committee of Allan Louden, Eric Carlson, and Ananda Mitra for their invaluable early direction. While I was at Wake I also received important help developing as a writer from Mary Dalton and Michael Hyde.

I have many thanks to offer to individuals who offered me advice despite being busy graduate students, who were pursuing their own research. I engaged in productive conversations about my work with Paul Johnson, Jeff Kurr, and George Gittinger. I would also like to thank Brita Anderson, John Rief, and Michael Vicaro for editing parts of the project for me. Finally, I would like to thank Rebecca Conklin who has read over my book manuscript three times and offered innumerable helpful ideas.

I would like to thank Alison Pavan and Kasey Beduhn from Lexington Books for guiding me through the publication process, and my anonymous reviewers for their countless helpful ideas.

At the University of Pittsburgh I owe a debt to many scholars outside of my committee, who I would like to thank. John Poulakos helped brainstorm with me about my chapter on classical Greece. Paul Scade (despite being out of the country and working at a different university) read several drafts of my chapter on Plato and offered a crucial philosophical perspective. Finally, I owe a great debt of gratitude to Mary Zboray, who acted as an unofficial member of my committee, reading and commenting on countless drafts.

I was lucky to have an amazing committee at the University of Pittsburgh, who shepherded me through the dissertation process and gave feedback that enabled me to transition the dissertation into a book. Philip Smith took a chance on a strange idea and agreed to be my outside reader. William Fusfield represented an enthusiastic voice in support of my work from the get-go. Barbara Warnick graciously agreed to step in for my defense when William was unable to attend due to illness. Brent Malin offered excellent feedback throughout the process. Finally, Ronald Zboray played a crucial role in shaping my book. I literally could not have written this document without him.

Finally, I would like to thank my parents, whose endless sacrifices on my behalf gave me the opportunity to pursue my passions.

Introduction

> "Belief in extraterrestrial life is a worldview in itself, a 'biophysical cosmology,' that—once accepted—forever changes the perception of our place in the universe."[1] —Steven J. Dick, 1999

During the height of the Cold War, when nations scanned the skies terrified of an impending nuclear catastrophe, scientists in the Soviet Union also searched the heavens for signs of extraterrestrial life. Important Soviets claimed that dialectical materialism acts as a universal principle, and they expected to discover advanced life forms, perhaps fighting their own class struggles on other planets.[2] Over two thousand years earlier, the ancient Greek atomists postulated an infinite cosmos filled with every imaginable being.[3] In 1277, Etienne Tempier, the Bishop of Paris, began a debate about how the gospel applies to extraterrestrials and in the nineteenth and twentieth centuries, scientists like William Herschel and Percival Lowell pioneered telescope technology for the purpose of seeking alien life.[4] The last three centuries have birthed dozens of religions, from Mormonism to Scientology, based in whole or in part on the idea of inhabited alien planets. In the 1970s, Erich von Däniken sold millions of books worldwide by making the argument that rather than communing with God, Moses and other religious figures were speaking to alien visitors.[5] In 1997, a CNN/Time poll reported that fifty-four percent of Americans said they believe that life exists on other planets.[6] A 2005 poll by the Center for Survey and Research Analysis at the University of Connecticut found the number was sixty percent.[7] Belief in the existence of alien life is an important current in the development of Western thought, but it is only half the story.

For every ardent advocate of the existence of a cosmos with other planets inhabited by alien life (plurality), there is one who vehemently opposes it (unity).[8] Plato denied the possibility of extraterrestrial life in the *Timaeus* and said that the Earth's solitude in the universe indicated its connection to perfection.[9] Early Christian thinkers found Plato's ideas in line with their biblical cosmology and incorporated the *Timaeus* into religious doctrine, while ignoring the works of Plato's contemporaries who argued for many inhabited worlds. The result was that Christian theologians merged their own beliefs with a cosmic structure adapted from Plato's *Timaeus*, and this became the standard interpretation of the universe throughout Europe during the Medieval period.[10]

In the year 1600, Giordano Bruno, a mathematician, astronomer, and memory theorist best known for his support of plurality, was put to death by the Papal inquisition, sending a powerful message that unity was the law of the land.[11] Within the same century the Renaissance begat a surge of belief in an inhabited cosmos, with Bernard Le Bovier de Fontenelle's pro-plurality book, *Discourse on the Plurality of Worlds* (1686), selling a massive number of copies and being translated into ten languages.[12] Even at the height of the Age of Reason, however, plenty of unity holdouts remained, like Thomas Baker, an Ashton Fellow at St. John's College, who derisively referred to proponents of plurality as "World-Mongers."[13] Belief in the existence of alien life featured among Emmanuel Kant's and David Hume's many differences of opinion, with Kant in defense of the proposition and Hume in opposition.[14] Just when it again seemed as though an academic consensus had formed around the plurality hypothesis, the mid-1800s brought a significant backlash from William Whewell, a renowned English scientist, philosopher, and Anglican theologian.[15] While many dialectic materialists in the Soviet Union believed in aliens, Hegel, who pioneered the modern dialectic, insisted that life on other planets violated the "spirit" behind the dialectic.[16]

The debate continues to the present, and while the emergent field of astrobiology (the study of alien life) now suggests that scientists have codified belief in the existence of aliens, *Rare Earth* (2000), a book written by two prominent scientists that argues against the existence of intelligent life on other planets, nonetheless became a popular and academic success.[17] Veteran science writer John Gribbin, wrote *Alone in the Universe* (2011) making the case for the absence of aliens, a decade in to the new millennium.[18] While over half of the American population believes the universe contains alien life, between forty to forty-five percent believe we are alone.[19]

This book does not offer an answer to the empirical question of whether inhabited planets exist outside the planet Earth. Instead, it examines the way individuals rhetorically deploy the question of alien life in order to advance other values and beliefs. There exist several bodies of work, for example, that examine the way some individuals use the belief in alien life to advocate particular worldviews.[20] Underexplored in academic research are individuals and groups who deny the possible existence of alien life. This book examines how rhetors deploy the belief that we are alone in the universe to advance ethical, religious, philosophical, and political arguments and look for linguistic styles, argumentative tactics, and deep-seated values, which are repeated in texts that oppose pluralism written centuries apart.

DEFINING UNITY AND PLURALITY

This book traces a constellation of religious, ethical, philosophical, and political beliefs, attached to unity throughout history. This task is made difficult not just because of the evolution of these attached beliefs, but also the evolution of the science behind unity. In 1853, Whewell popularized "unity" as a foil to the more widely used term "plurality," which referred to the existence of other inhabited planets. Originally, however, Plato used "unity" as a value attached to the physical state of the universe, rather than a term to describe the physical state of the universe. The astronomer Timaeus, in Plato's eponymous dialogue, ascribes the attribute of unity to the absence of what is best described in modern terminology as multiple universes, rather than other inhabited planets. The absence or presence of alternative universes defined the earliest unity/plurality debates, even though other elements of most unity cosmologies would preclude the existence of other inhabited worlds. For example, Aristotle rejected the possibility of other universes and provided a scientific framework that made the existence of both multiple universes and other planets in our universe impossible. The Platonic and Aristotelian cosmologies were adapted into Christian Europe and with them the idea that what we now know as the planets were celestial spheres incapable of supporting life.

Unity was defended primarily as a rejection of the existence of other such universes, until the Renaissance, after which it underwent several shifts and clarifications. These clarifications start with the reaction to Giordano Bruno's cosmology in the late 16th century. Bruno accepted the existence of one universe, but he posited an infinite number of inhabited worlds within it. The response to Bruno clarified unity as depending on the absence of both other universes and inhabited worlds within our universe. The science of unity shifts when William Whewell resurrects it as an argument against utilitarianism. Whewell argued that the existence of other inhabited planets was consistent with unity, as long as those planets contained only animals and not intelligent and spiritual life like humanity. Alfred Russel Wallace, writing after the ascendance of belief in evolution, (a theory he co-discovered) treats unity as the absence of any life. In more recent deployments of unity as an argument for philosophical, ethical, religious, and political values, authors like Ward and Brownlee and Simon Conway Morris treat unity as the absence of complex animal life. The complex/simple creature divide becomes the new version of Whewell's spiritual/nonspiritual distinction. Ward and

Brownlee's version of unity even opens the door ever so slightly to the rare possibility of intelligent life on other planets, as is discussed in chapter five.

While many of the authors examined never deploy the term unity, there are strong reasons to associate them under this term. These authors frequently quote each other in their own works in a way that suggests that they view themselves as part of an intellectual tradition, even though they may disagree on the science of what threshold constitutes a meaningful absence of inhabited planets/alien life. Historians of ideas like Crowe and Dick also treat these figures as a part of an intellectual tradition, despite their disagreements of what constitutes meaningful human aloneness. Plato, Whewell, and the other authors' scientific views on the existence and nature of other worlds are also connected by their use of the absence of inhabited planets as argument. Of paramount importance in the works of figures like Morris and Whewell is that the unity cosmology (however defined) has meaning for ethics, politics, philosophy, and religion. Even as they disagree on specific aspects of what constitutes unity, they use the absence of other inhabited planets as an argument against the forms of relativism that exist within their time. This means that while what constitutes an inhabited planet changes, it changes on the basis of what the rhetors deploying unity as argument and their audiences see as threatening to the sense of human specialness and aloneness in the cosmos. This provides a shared thread that connects these scientific worldviews under the label unity. Self-identification, scholarly identification, and similar deployment of the lack of inhabited planets as argument justifies grouping the cosmologies of Whewell, Wallace, Plato, Aquinas, etc. under the label of unity.

Because this study focuses on the rhetorical deployment of the scientific idea of unity, it addresses plurality primarily as a foil. The term loosely means the scientific belief in the existence of celestial abodes for life (typically intelligent life) and had its most prominent early defenders in the Greek Atomists, who posited an infinite number of universes that contain all possible forms of life. Like unity, the scientific definition of plurality has shifted throughout time. For example, up until 1250 CE the term was more or less synonymous with the prospect of an infinite number of other universes, whereas it now primarily refers to the existence of inhabited planets within our universe.[21] Despite these shifts in definition, the scientific defenders of unity and plurality have found little common ground in that those that defend unity limit meaningful life to Earth, while those defenders of plurality tend to view life as a relatively common phenomenon.

Also like unity, many individuals throughout history have attributed values to the idea that plurality exists. While this book does not examine the values associated with plurality in depth, it is important to understand that individuals are making connections between plurality and various values concurrently with many of the case studies examined. This shows that finding values in cosmology is a common rhetorical trope, rather than an underhanded tactic deployed solely by nefarious teleologists. The next section offers a brief examination of some of these connections, before moving on to a preliminary look at the beliefs associated with unity.

PLURALITY OF WORLDS, PLURALITY OF MEANINGS

Alien life serves as a cosmic Rorschach test, a great unknown factor to which individuals can project their hopes, fears, and desires. George Basalla offers a comprehensive study of the plurality cosmology from Copernicus to the present.[22] He examines scientists, like Carl Sagan and Percival Lowell, writing in defense of plurality and argues that these figures connect plurality to their belief in progress and scientific rationality. For these scientists, super intelligent aliens stand in for God.[23] By making contact humanity will achieve salvation, with aliens providing technological and ethical solutions to humanity's most pressing problems. In the meantime, humanity should seek to emulate these super intelligent aliens by focusing on scientific advancement.[24]

Basalla's excellent study only tells part of the story, however; he focuses on a very particular set of alien life proponents: modern, secular scientists. Vast gulfs exist between the way individuals deploy plurality as argument. Sagan (the prototypical modern, secular scientist) accuses the unidentified flying object (UFO) movement of being a refuge for anti-scientific and irrational thought.[25] In the 1800s, minister and professor of theology Thomas Chalmers gained massive popularity preaching that boundless inhabited planets showcased the omnipotent and beneficent Christian God. Even today the Vatican Observatory remains open to the idea of discovering alien life.[26] While Basalla does identify a clear affinity of plurality among a set of influential modern, secular scientists, it is clear that many foils to this group have also embraced plurality.

Plurality does not always manifest as a trope in traditional debates between Christians and secularists. Mark Harrison traces the development of the modern UFO movement from the ashes of Theosophy.[27] Theosophy developed in the 1870s as a reaction to the conflict between Darwinism and Christian Creation Science. Theosophy mixed racial science, folktales, and superficial knowledge of Eastern traditions into one of the more bizarre movements in history. Theosophy became a haven for Nazis and other racists who rejected the idea that all humans emerged from the same origin (whether God or evolution). Harrison argues that Theosophy initially deified magical Ascended Masters (enlightened individuals), but these figures got replaced in cultural lore with alien visitors as the UFO craze overtook popular imagination. The UFO movement in the West still contains many of the problematic racist beliefs that it inherited from its Theosophical origins. The Theosophists project their racism onto the structure of the universe and came up with a complicated hierarchy of visiting alien races correspondent to the "races" of humans on Earth. The racial hierarchy of aliens served as a justification for the Theosophists' earthly racism. Similarly, the Raelian Church ties its religious and ethical beliefs to aliens. For example, they claim that humans were cloned by an alien species, which provides empirical and ethical justification for cloning here on Earth.[28]

Then, of course, there are aliens in popular media, serving as useful metaphors for whatever the author chooses. Hostile aliens often represent contemporary political threats, like Communism during the Cold War and Terrorism following September 11th.[29] Klaatu came to save Earth from nuclear conflict in *The Day the Earth Stood Still* (1951), during a time when nuclear conflict pervaded the international conscious. The aliens of *Star Trek* serve as reflections on diversity within our own world.[30] C. S. Lewis wrote a science fiction trilogy that uses aliens as a device to explore humanity's spirituality, whereas H. P. Lovecraft uses aliens as a meditation on cosmic dread.[31] Everywhere the plurality of worlds is infused with a plurality of meanings.

UNITY OF THE WORLD, UNITY OF MEANING

While individuals connect plurality with a host of divergent ethical, religious, and moral programs, in Western culture unity has intertwined itself, perhaps inextricably, with one particular set of ideas. From Plato to major figures of today, individuals associate the idea of unity with the idea that our species has special value. This view takes many forms. Some argue that aloneness demonstrates that objective and immutable values and ethical codes exist, rather than the rules of proper behavior being social constructs or determined by utilitarian calculation.[32] Others suggest that our solitude indicates that humanity exists for a special purpose (teleology), whether it is explicitly religious or simply spreading consciousness throughout space.[33] Others still connect the harm of human extinction not to the fact that humanity as a species would be lost, but the idea that humanity as the only intelligent life in the universe would be lost.[34] Many important figures describe unity as a supplement to or replacement for (depending on when they write) humanity's physical centrality in the universe.[35] This centrality carries with it notions of value that extend back to humanity's earliest civilizations.[36] All of these examples converge on the general connection between a cosmology where humanity is alone and the idea that this aloneness gives value and purpose to our species.

While scholars have devoted relatively little attention to studying unity's role in shaping religious, ethical, political, and philosophical beliefs, this has not been the case for unity's close relative, geocentrism. The consensus of scholars agrees that the Aristotelian cosmology, including geocentrism, played an enormous role in shaping the society of medieval Europe.[37] The Copernican revolution and shift to heliocentrism helped undermine the Medieval order. The rationalists used the heliocentric paradigm as a justification to rewrite the academic disciplines from the ground up. An alliance of scientists, philosophers and political thinkers interpreted the finding that the Earth and its inhabitants did not occupy a privileged position in the solar system as a justification for an empirical methodology.[38] New social and political institutions found space to develop when the prevailing social assembly lost its connection to the eternal structure of the universe.

Scholars have focused their attention almost exclusively on the rise and dramatic collapse of heliocentrism, without attention to how heliocentrism was but one of a constellation of cosmological features that medieval Europeans used as a justification for their social order. Unity may not necessitate a physically central location, but, as the only life, it gives humanity a metaphorically central location in the universe. It is no accident that both the unity and geocentrism of the Medieval world come from *Timaeus*, where Plato specifically connects these cosmological ideas with his ethical philosophy. Unity carries on the rhetorical work of geocentrism after the Copernican revolution, offering a cosmic justification for universal values and movement towards a particular *telos*.[39] The Copernican revolution had such an impact in part because of the tight connection between a set of political, philosophical, religious, and ethical ideas and the scientific question of heliocentrism. The next section details the ways that connections between science and extra-scientific values serve as a particularly effective form of argumentation.

COSMIC MYTH VS. COSMIC FACT

The Copernican revolution helped build support for an empirical methodology, but criticisms of empiricism still exist in Europe and elsewhere. Some Soviets claimed that dialectic materialism requires the existence of alien lives leading their own class struggles, but one can still support Marxism even in an empty universe. The facts of cosmology, unity vs. plurality, and geocentrism vs. heliocentrism, exist independently from their interpretations. John D. Caputo, for example, takes a very different view of the implications of unity:

> We pass our days on the surface of a little star which drifts aimlessly through endless skies, inventing such fictions as we require to make it through the day and to persuade ourselves of our meaning and significance. Until at last, weary

> of its peculiar little local experiment, the cosmos draws another breath and moves on. Then we disappear without a trace. "Knowledge," "obligation," "justice,"—these are so many obsolete inventions of little animals, now useless vapors dissipating in interstellar space.[40]

For Caputo, unity does not give humanity a privileged position, instead humanity is a mistake—a "local experiment" that will inevitably be destroyed by an uncaring universe. Without humanity all of our supposedly transcendent values become nothing more than "useless vapors." The connection of unity with teleology, universal and immutable values, and divine protection is a rhetorical move, more precisely a myth.

It is relatively uncontroversial to suggest that myths, understood as fables that explain some natural or social phenomenon and typically involve supernatural elements, shaped the political, religious, scientific, and cultural existence of early humankind.[41] Barthes argues, however, that myths are better understood as forms of language that hide the ideological nature of particular terms and thus cover over the inevitable ambiguities that arise from language.[42] The result is that myths act as propositions that need no support, such that society simply takes myth for granted, whether it's the prevailing belief that everything can and should be understood in terms of monetary value or the medieval European practice of justifying monarchy on the basis of religion.[43]

Science provides a particularly fertile ground for Barthes form of myth, because it replaces the explanatory function once filled by supernatural elements of traditional myths. Ancient Greeks attributed lightning to the god Zeus, in the way we now attribute it to atmospheric electricity. The Greek gods like Zeus also served functions beyond explanation of the physical world, for example, grounding ethics and proper behavior. One could answer the question of why an action was right or wrong on the basis of the gods. Stephen Toulmin argues that this non-explanatory aspect of the gods represents myth. He claims that similar myths occur in modern society, derived from our current method of explanation: science. Just as the ancient Greeks, for example, connected their method of explanation (god) with ethics, many people today imbue science with ethical implications. These "scientific myths" bring scientific data to bear on nonscientific questions. Toulmin gives examples of people using evolutionary theory to advocate for particular ethical values and individuals that feel nihilistic in the face of evidence that heat will eventually distribute itself evenly throughout the universe making life impossible (heat death). While Toulmin allows that these outlooks may be justifiable, he argues that "physics does not *oblige* us to adopt one, or indeed any, of these attitudes."[44] Just because humans evolved in a particular manner does not mean they must conform their ethical beliefs to an evolutionary test. For example, one can argue that a willingness to sacrifice your life for others may be ethical, despite it preventing you from spreading your genetic code. The way that rhetors deploy scientific myth blurs science and ethics into one indivisible idea so that the ethics naturally follow from the science, just as falling naturally derives from gravity.

Toulmin's scientific myth depends on delineation between scientific modes of inquiry and all other disciplines. The Oxford English dictionary defines science as, "The intellectual and practical activity encompassing the systematic study of the structure and behavior of the physical and natural world through observation and experiment."[45] Whether one can draw a definitive boundary between science and other disciplines is up for debate, but as a practical matter science in the modern era seems to have etched out its own domain.

This study argues for a longstanding scientific myth that connects unity with teleology, immutable values, and centrality and extends in various forms from ancient Greece to the present day. This raises the concern of whether the conception of scientific myth in ancient Greece or Medieval Europe is too anachronistic to offer any insight. Ancient Greece did not have science in the modern sense or even a clear and consistent word to designate scientific like practices. Using scientific myth as a lens into arguments of the past raises two primary concerns. First is whether something approximating science existed for all of the case studies in this book. For example, while the Oxford English dictionary defines experimentation as a key element of science, Aristotle and many others viewed experiments as unreliable because they involved humans intervening in nature.[46] If we apply the modern understanding of science that undergirds Toulmin's examples of scientific myths associated with evolution and heat, then there exists no unity science to serve as the foundation of the "scientific" myths of ancient Greece.

If, however, we treat science as a set of evolving practices, then it is possible to use scientific myth to examine arguments in ancient Greece. Lloyd defines ancient Greek practitioners of science as those that describe the world absent religious explanation and have a willingness to engage in debate over their explanations grounded in naturalistic arguments.[47] Lloyd and others view these early figures, sometimes called natural philosophers, as engaging in a chain of ideas that developed into the modern practice of science.[48] So while the specific contours of science change throughout the ages, which merits contextualization, the underlying elements of observation and arguments based on nature rather than metaphysics persist. This persistence allows the framework of scientific myths to operate effectively in all of the case studies in this book.

Having a definition of science does not satisfy the second concern about using the idea of scientific myth to analyze arguments in ancient and Medieval society, whether it is too anachronistic to apply the separation of science from questions of ethics, philosophy, religion, and politics to eras where this distinction was not recognized. The first response to this objection is that scientific myths operate in the present, a time when in theory science should be distinct from these other values. In other words, Toulmin's work acts to separate science from other values precisely where they have become conflated. The second response is that it is not anachronistic to apply the label scientific myth to the attribution of nonscientific beliefs to scientific phenomenon, because the distinction between science and other values can be traced back to ancient Greece. Plato criticized the earliest scientists for not engaging in the practice of scientific myth, as will be explored in the next chapter.

When this book refers to the unity myth, it refers to a scientific myth that connects teleology, immutable values, and centrality to a scientific defense of unity. This is distinct from unity or the unity cosmology, which refers to the absence of other inhabited planets, an empirical question separate from other nonscientific beliefs. While the definition of unity, the scientific defense of unity, and the extra-scientific implications of unity shift to meet the demands and constraints of the rhetors that deploy the unity myth, this book argues that it is productive to view the unity myth as a distinct rhetorical and argumentative strategy. Having established the definition of a scientific myth, the next section examines the stakes of the unity myth by assessing the rhetorical power of scientific myths.

THE POWER OF SCIENTIFIC MYTH

Once science and myth become entwined, it greatly bolsters the rhetorical power of the myth. All of the case studies examined draw on science to defend unity and then unity to ground their philosophical, religious, political, and ethical claims. Scientific rhetoric plays a powerful role in conditioning audiences to accept arguments. Philip Wander argues that the standard of scientific refutation forecloses the avenues of response to all but others versed in the discourse.[49] This means the scientists' ethos stands in for the ability of the audience to understand the specifics of the scientific arguments.[50] The public puts its faith in science because of its steady progress; one can look around and see the fruits of the scientific discipline whether in the form of computers, refrigerators, or any number of other inventions. As the next chapter shows, even the ancient Greeks held practitioners of science in high regard. The public's trust in scientists, despite not necessarily having a full understanding of their work, is important, because despite its claims to objectivity, scientific discourse can easily serve as a conduit to hide metaphysical assumptions.[51] Once framed as a scientific question, the trueness of unity becomes difficult to separate from the ideological implications of the unity myth.

Attaching values to a scientific question is also a means of forcing a definitive, provable conclusion to philosophical debates. Scientific theories tend to achieve a critical mass of followers and then replace older scientific theories.[52] In contrast, philosophy has shown almost no progress toward achieving consensus on any issue, as philosopher Leszek Kolakowski writes:

> For centuries philosophy has asserted its legitimacy by asking and answering questions inherited from the Socratics and Pre-Socratics: how to distinguish the real from the unreal, true from false, good from evil . . . There came a point, however, when philosophers had to confront a simple, painfully undeniable fact: that of the questions which have sustained European philosophy for two and a half millennia, not a single one has been answered to general satisfaction. All of them, if not declared invalid by the decree of philosophers, remain controversial.[53]

By linking philosophical ideas to the empirical, disprovable theory of unity, rhetors create the illusion of an objective scientific foundation for their beliefs; for example, if unity is proven scientifically, so teleology is also proven. This offers the hope of achieving the kind of consensus science routinely achieves on a philosophical question.

Scientific myths can emerge from any area of science, but cosmological scientific myths are often the most powerful. Barthes argues that effective myth makes the contingent appear eternal.[54] The appeal to the eternal calls upon a force viewed as more powerful than any human creation. Various ethical, philosophical, religious, and political ideas have come into and fallen out of fashion; Barthes and others would likely label these ideas as contingent (in that they could be and are rearranged to exist in a different formation by human action). In contrast, the nature of the universe is not contingent on human behavior. Claiming that one finds justification for a particular idea manifest in the very cosmos we inhabit makes the contingent appear eternal. It is no accident that many ancient myths that impart stories for how to live also deal with the structure and creation of the universe. The question of plurality and unity can be made to speak to eternal questions: the purpose and nature of the cosmos.

Cosmology also provides a strong foundation for myth because of its connection to the physical location we inhabit. Michael Hyde identifies the *ethos* of rhetoric as "discourse [that] is used to transform space and time into 'dwelling places' with the hope that 'we might feel more *at home* with others and our surroundings.'"[55] This passage introduces a series of essays on the various ways that language makes sense of the world; absent these discourses everyday human existence would appear chaotic and inhospitable. At the risk of taking the idea of "dwelling places" too literally, one can examine the cosmologies used to describe the universe as performing an *a priori* rhetorical *ethos* function. Before discussions of complex social relations can develop, there needs to exist a sense of what it means to live in the world, which in turn necessitates a way of interpreting the cosmos. Absent the existence of an *ethos* of place, humans could only perceive the universe as terrifying randomness. The unity myth posits a knowable and friendly home (universe) that exists for the purpose of humanity. Within this environment, humanity has the ability to strive toward the goals of perfecting its understanding of ethics, philosophy, religion, and politics. The perfect forms of these disciplines are not contingent creations of humans, but prior existing facts that humans discover.

The appearance of being grounded in objective fact, of being eternal, and its connection to humanity's literal "dwelling place" all make cosmology a particularly powerful myth. Cosmological myths can be so powerful that it is sometimes difficult to accept them as a rhetorical construction, but as Martin Heidegger writes,

> Who the human being is—for philosophy, the answer to this problem is not inscribed somewhere in heaven…the essence of human being is *never* an answer, but is essentially a question…The question cannot be asked adequately within the domain of traditional metaphysics, which essentially remains "physics."[56]

Despite their power to shape events, cosmologies will always tell us more about those who construct and use them then they will about objective reality. Humanity's "being" is not written into the stars, but can become manifest in the way humans position themselves in relation to the stars. This book does not seek to provide insight on the factual accuracy of unity, instead it acts to disentangle and expose the scientific myths associated with the belief that no alien life exists. In order to distinguish between unity and its ethical implications this book refers to the idea that no aliens exist as unity or the unity cosmology, while the ethical implications that Plato and others find in unity are referred to as the unity myth.

SPIRITS, ANIMALS, AND FOREIGNERS: WHY EXTRATERRESTRIALS MATTER

This book focuses exclusively on the question of inhabited planets. Viewed another way, the unity/plurality debate revolves around the question of life alien to humanity. If one views the debate in these terms then it is possible to find plurality absent inhabited planets. Extraterrestrials are not the only possible alien intelligences that could exist in the universe besides humans. Spiritual (non-material beings), animals, and foreigners on Earth all can play the role of plurality and have the potential to threaten one's sense of centrality. These other groups operate differently than the possibility of intelligent extraterrestrial life, however. For starters, these threats were assimilated long ago, such that while they could in theory threaten human centrality, in general practice they do not.

Most important figures separate the issue of spiritual plurality and plurality of worlds. Plato's *Timaeus*, for example, argues Earth (or rather what we recognize as Earth) is the only planet with human-like life, but suggests the wandering stars (what we now know as the planets, sun, and moon) are deities.[57] Plato treated the existence of these deity planets as compatible with the unity of our single world. At various times individuals speculated that the inhabitants of other worlds were not material beings like humans, but either Lucifer and his fallen angels or the souls of saved humans.[58] These figures maintained belief in the unity myth, despite believing in the existence of spiritual beings. The distinction between spiritual and material life maintains the necessary components for one to connect unity with ideologies like anthropocentrism, teleology, and absolute ethics. Spiritual inhabitants of the cosmos are most often not considered "alien," in the sense that religious individuals claim that these beings exist, can be understood through religious texts, interfere in human affairs, and occasionally reveal themselves. In contrast, defenders of plurality can only speculate about the nature of life forms on other planets and, for most of history, did not claim to have any interaction with them. In the Medieval period this distinction is codified in terms of the Great Chain of Being. The picture demonstrates the spiritual value of various life forms with God at the top and inanimate objects at the bottom. While largely invisible to humanity, the angels occupy a particular and known place in the structure of the cosmos.

The Great Chain of Being also shows that despite the existence of spiritual beings, humanity maintains a central position between God and inanimate objects. While the spiritual beings (angels) exist above humanity in terms of purity, the battle for the soul of humanity serves as the focus of the cosmic drama at play with animals, plants, and inanimate objects serving as humanity's tools. While humanity may exist in a spiritually degraded position, it is the center of the action. In the Christian religion, God manifests as a human, rather than an animal or angel, to interact with and save humanity. A plurality of worlds denies human centrality in a way the existence of spiritual beings does not. It divides God's attention and suggests that humanity may exist at the periphery, rather than the center, of existence.

Similarly, this book does not examine reactions to various human cultures on Earth that were/are treated as alien, except in so far as these reactions may shed light on attitudes about extraterrestrials. The "discovery" of the Americas, for example, represents a period in history that for many individuals at the time (European and indigenous) can be equated to an encounter with extraterrestrial life. At first, serious debate occurred in Europe over whether the inhabitants of America were human.[59] Insofar as the European and indig-

A depiction of the Great Chain of Being found in *Rhetorica Christiana* (1579).

enous peoples viewed each other as alien, this encounter serves as a useful case study for what is at stake in the debate over pluralism.[60] This book does not examine the treatment of human groups as alien as a case study, however.

The literature on the question of pluralism clearly delineates itself from debate over the nature of various human groups. Defenders of the unity myth from Plato onward view the confines of the planet as a meaningful boundary to demarcate the discussion of intelligent alien life. The possibility of humanoid intelligence on Earth may undermine the narrative of cultural importance (for example, European cultural importance), but it is not nearly as problematic to the ideologies associated with unity as the potential for a universe full of intelligent life. The planet provides a natural boundary, which serves to make Earth a complete unit and thus a model for perfection, an idea consistently found in the unity writings. Also important is that the view of other humans as separate species has always been much more ephemeral than the unity cosmology. Kupper argues that the belief that the indigenous people in the Americas were not human quickly faded from popular European sentiment, because, among other things, the Christian understanding of *Genesis* allowed most Europeans to assimilate the idea of variations of species as representative of the same species (something not necessarily applicable to alien life).[61]

Also those who do view subsets of humanity as nonhuman often argue that the existence of these quasi or subhumans make "normal" humans more, not less, special/central, which is in direct contrast to the unity cosmology. The works of both Mark Harrison and Christopher Roth trace the birth of the modern UFO movement and suggest that groups that believe in biological racial distinctions between humans are more likely to be pluralists.[62] This is not to say that pluralism is strongly correlated to racist beliefs, only that pluralists are more likely to believe in racial distinctions than believers in unity. In contrast, the major defenders of the unity mythology typically argue that all of humanity is a singular species. In the *Timaeus*, Plato creates a clear category distinction between humans and animals.[63] The idea that all humans belong to the same species runs throughout Plato's dialogues.[64] Whewell, while he still maintains many of the culturally biased beliefs of his time, repeatedly and clearly states that all humans are fundamentally the same.[65] These thinkers argue that all human racial groups have the same biological capacity to achieve knowledge, for example.

This is not to say that parallels do not exist between the unity cosmology and the discovery of new civilizations of humans. Michael Zimmerman touches on the issue briefly in an article about UFO abductions. He argues that cosmologies that exclude the possibility of alien life are part of a long history of cultural myths that allow people to feel more important by narrowly drawing the boundaries of which beings are valued or, in this case, exist.[66] Zimmerman uses the analogy of the European exploration of the Americas to show how dehumanization and denial of existence serve the same mythic function. He describes how the Europeans "discovered" two continents of people blissfully ignorant of Europe, much less ideas Europeans took for granted like Christianity and capitalism. Zimmerman claims the realization that Europe did not exist as a central conduit for all human activity on the planet provided an existential shock to the inhabitants of Europe. He argues that in order to minimize the significance of the discovery, the Europeans first dehumanized the inhabitants of "the new world," then embarked on a plan to eliminate them. The first contact between Europeans and the indigenous peoples of America shows how dehumanization and denial serve the similar rhetorical function of building up a sense of importance.[67] Zimmerman suggests that the same cultural mythology that recoiled at the possibility of a civilization of people on the other side of the world could be at work when individuals deny the possibility of alien life.[68] Despite the parallels, however, the unity of the world cosmology is distinct enough to warrant an independent examination.

Animals are a third "alien" intelligence that could threaten to disrupt the idea of human centrality at work in the unity myth. If one keeps an eye on the news, she will find a steady stream of articles highlighting the intelligence of animals: many animals experience grief, crows can be taught to use money, octopi exhibit playful behaviors, gorillas can learn thousands of words in sign language.[69] In this way, humanity has made contact with alien intelligence long ago in the form of our animal relatives. Why the concern for aliens on other worlds while our own world is full of aliens?

One reason returns us to the Great Chain of Being. Western culture largely treats animals as fundamentally below humanity and highlights a series of barriers between animals and humans: speech, intelligence, the soul, use of technology, etc. Even as researchers prove that some animals exhibit these qualities supposedly unique to humans, the larger anthropocentric divide of humans and animals persists. Alexander Wendt and Raymond Duvall argue that UFOs are deeply threatening to humanity's sense of self because they point to a species of unquestioned intelligence. Given the advanced technology reported on these unidentified craft, humans would have difficulty maintaining their superiority over these aliens.[70] Any alien race that exhibits the characteristics of human civilization would threaten to dethrone humans from the spot they claimed as the pinnacle of creation.

Another reason animals do not represent a serious threat to the unity myth deals with the idea of centrality. Animals and humanity have coexisted together; much like Christianity treats humans and spiritual beings as existing together. The prospect of alien life on other planets, inaccessible to humanity for much or all of its history, raises the prospect that humanity does not hold a central place in the story of the cosmos. This centrality is important to the unity myth.

One can view spirits, foreigners, and animals all as alien intelligences. John Durham Peters goes so far as to suggest that all communication occurs between aliens, as wide gulfs exist even between humans of the same culture.[71] A myopic focus on foreignness and alienness does not fully capture the nature of unity. The effectiveness of unity as a rhetorical device depends as much on the notion of centrality as it does on the absence of alien beings. During the 1800s, for example, many predicted that humans almost identical to us physically, mentally, and spiritually would inhabit alien planets, but that did not make plurality any less problematic for advocates of the unity myth like William Whewell. The unity myth claims that plurality destabilizes teleology and absolute values not just because of the possibility for alienness, but because it acts to rob humanity of its centrality in the universe.

PAN-HISTORIOGRAPHY

This book traces the unity mythology throughout Western history from Plato to the modern day, a span of millennia. It uses Debra Hawhee and Christa Olson's method of pan-historiography, which they designed to facilitate broad historical projects.[72] Many recent histories, in rhetoric and other disciplines, tend to limit their scope in order to focus narrowly on a particular, person, time, place, etc. This narrow strategy allows authors to provide a deeply textured picture of their object of study. Pan-historiography complements these focused histories, by examining ideas across time, people, places, and locations. While this more sweeping method may never capture the detail of a tailored history, it allows more attention to "persistence and change" of ideas.[73] One of the most fascinating elements of the unity myth is that it has persisted for such an extended period of time serving, like depictions of native Ecuadorians Olson references as a justification for the process of pan-historiography, as a "sustained and sustaining rhetorical process."[74]

Broad histories, however, risk "homogen[izing] whole eras, places, and controversies into manageable and misleadingly coherent terms."[75] In order to avoid this, this book adopts the pan-historiographic strategy of "time-slicing," which involves the use of "key historical-cultural moments to anchor individual chapters."[76] These time slices allow an in-depth look at key texts and thus reduce the risk of broad characterizations. In this book the time slices focus on one or two individuals living in the same time period that deployed the unity myth. Often times these individuals wrote one primary cosmological text, which is the main object of analysis for a chapter. The case studies were chosen by using authoritative histories of the unity/plurality debate to find the most influential texts promoting the unity myth.[77]

The book approaches the cosmological works from a rhetorical perspective and analyzes the way in which the authors connect unity with teleology, anti-relativism, and other related values. The primary tool of analysis is Toulmin's idea of scientific myth, which deals explicitly with the use of science to justify extra-scientific values, but the chapters also point out a variety of rhetorical constructs, like enthymemes, appeals to authority, and rhetorical questions that are used to bolster the unity myth. In order to understand the case studies' rhetorical exigencies and audiences, the chapters situate the texts within their historical contexts. The context provides a means of gauging how the texts make arguments suited to the particular situations the authors faced. The chapters do not operate independently from each other, but instead work in harmony to portray the evolution of an argument through time.

The in-depth nature of the case studies allows for attention to areas of continuity and change. The case studies draw attention to areas of persistence by pointing out similarities in the unity myth across time and the many ways in which the authors reference back to earlier instantiations of the myth. This diachronic approach to history is mindful of Braziel's idea that constructs "attain and accrue meaning historically."[78] Whewell and Morris may deploy the unity myth in different ways than Plato or Aquinas, but their scientific myths cannot be understood outside the history of the unity argument. Even if the modern authors had not read Plato's *Timaeus*, the ancient unity myth circulates as a societal undercurrent that their arguments, intentionally or not, tap into. The focus on context across time periods allows for the study of change in addition to persistence. The case studies examine the ways that individuals adapt the unity myth to their particular rhetorical ecosystem. For example, the scientific orthodoxy changes radically from the 19th century to the 20th century and Wallace and Tipler's versions of the unity myth reflect these changes. Shifting science helps explains why the definition of what constitutes "unity" changes from author to author. The pan-historiographic approach may not attend to every nuance of each case study, but it is the best available method to trace the presence of a discourse over such a large span of time.

CHAPTER OUTLINE

The first chapter, "Plato's Rhetorical Cosmology: The Unity of the World as Foundational Myth," begins the examination of the unity cosmology. The unity myth does not necessarily begin with Plato, but he provides the best starting point for an examination of the unity mythology. The unity myth he presents, primarily in the *Timaeus* (fourth century BCE), has two important features that appear to be absent in previous eras. Earlier Western cosmologies that claimed no other intelligent life existed in the universe did not encounter serious opposition from pluralists. While myths have the most power when society takes them for granted, they also become very difficult for scholars to study. When myths have become entrenched in a culture they do not need to be specifically referenced in order to be effective. Plato wrote his cosmology in the face of the atomists' rival materialist cosmology, which postulated an infinite number of inhabitable worlds. The conflict between Plato and the atomists provides an opportunity to examine what the aspects of the unity cosmology Plato highlights (including its rhetorical components) in opposition to a pluralist cosmology. Second, Plato's dialogues demonstrate an early understanding of the rhetorical effectiveness of scientific myths and show a progression from traditional to scientific myths. This allows an exploration of the unity myth as an outgrowth of more traditional myths.

In the *Timaeus* Plato connects the structure of the universe to his philosophy by arguing that the cosmos contains messages for how humans should live. Unsurprisingly, Plato suggests that these messages justify the political philosophy that he advocates since his earliest dialogues. Plato describes a cosmos where Earth is literally the center of the universe. Positioning the Earth as the center of the universe was a common theme among early cosmologies. Plato, however, not only locates the Earth physically at the center, but his cosmology obliterates the potential for any form of alien otherness. In Plato's cosmology the universe exists solely for the benefit of humanity; outer space becomes a fixed entity operating in fixed regularity, endlessly transmitting a message of how to live to the human species. This cosmology makes no space for alien lives, literally, because the crystalline universe cannot be inhabited, and rhetorically, because the sophists, atomists, pre-Socratics, and others that disagreed with Plato were acting out of sync with the universe and thus were metaphoric aliens worthy of banishment from Plato's perfect universe.

Plato's cosmology became one of the primary strains of thinking on the universe in ancient Greece, but it had competitors. The atomists' philosophy still survived, as did Aristotle's more scientific version of Plato's ideas. When Christianity became the dominant religion of Europe, however, it was Plato's cosmology that resonated most with the prevailing theology. The second chapter, "The Dominance of the Unity Cosmology: From Plato to Galileo," traces the transition of the unity myth from the ancient to the medieval world. Individuals like Thomas Aquinas, who wrote on unity in *de Potenia* (1265-1266) and the *Summa Theologica* (1265-1274), harmonized Plato's unity cosmology and mythology, the scientific ideas of Aristotle, and Christian theology.[79] This new version of the unity mythology became the dominant, largely unquestioned cosmological framework of Europe well through the 1600s, providing a foundation for theology and ethics throughout Europe.

The second chapter closes with an examination of the Copernican Revolution, which upended much of the stability provided by Plato's cosmic mythology and served as an important catalyst in the rise of modernity. Up until this time, most of the debate about life on other planets focused on whether it was theoretically possible for such a planet to exist, rather than whether plurality in fact existed. These debates typically occurred within a religious context with one side making Plato and Aristotle's arguments for why there could not be a plurality of worlds and the other side saying these explanations incorrectly placed limits on an omnipotent creator. The success of the Copernican cosmology opened up space for a different kind of discussion, which focused not on whether a plurality of worlds was possible, but whether it in fact existed. Bolstered by the adoption of the Copernican cosmology among elite thinkers, the plurality theory became the dominant paradigm in scientific and academic discourse.

Despite the success of the plurality cosmology, vocal detractors, fearful of the ethical and theological implications of plurality, published works that defended the unity cosmology.[80] This minority of scholars fought a losing battle against the modern pluralist cosmology. While Kepler's telescopic observations caused him to proclaim, "Is man the most noble rational creature? Are things made for him?" famed polymath Caspar Bartholin and Jesuit scholar Athanasius Kircher responded definitively in the affirmative.[81] This chapter examines the rise and fall of unity's hegemony in order to provide a bridge between the ancient and modern instantiations of the unity cosmology.

The rise of modernity left some with the impression that unity had disappeared from serious discussion, but upon its publication in 1853, Whewell's *Plurality of Worlds* ignited fierce debate by its suggestion that the plurality thesis had little basis in science or religion.[82] The massive controversy that resulted played itself out in academic rejoinders and a prolonged back and forth in popular magazines. This controversy suggests the victory of the plurality cosmology in the 1600s was much more ephemeral than real. The third chapter, "William Whewell and Alfred Russel Wallace: Unity Cosmology in the Modern Era," examines the unity cosmology of two of the nineteenth century's greatest thinkers. Whewell offers an impressive array of cutting-edge scientific arguments to back his position, but the implications have changed little from Plato's cosmology. Whewell and his supporters argue that unity is crucial to the premise that humanity has a special role in the cosmos. In contrast, the plurality cosmology "repeatedly annihilate[s]" humanity "by the growing magnitude of the known Universe."[83] Whewell used the unity cosmology as an argument against the relativism he saw in the philosophies of utilitarians and others. Alfred Russel Wallace, inspired by Whewell's work, wrote his own defense of the unity cosmology. In *Man's Place in the Universe* (1903) and *The World of Life* (1910) Wallace connects the absence of alien life to a special purpose for humanity.[84] He used unity as an argument against capitalism and social Darwinism, claiming that humanity represented the telos of the universe and thus could escape the survival of the fittest paradigm.

In the early 1900s another wave of belief in extraterrestrial life overtook both the academic and popular imagination. Percival Lowell popularized the idea that Mars was home to an advanced alien civilization, which became a view held by many other prominent scientists. Science fiction and popular science magazines popularized the plurality viewpoint even as Lowell's belief in advanced life on Mars began to fade from mainstream scientific thought. When governments sent probes to nearby planets in the 1970s, the hopes of scientists and the public ran high that at least microbial life would be found. Each new mission of discovery brought more disappointing news for those that were invested in the idea that life was ubiquitous in the universe.

As the chance of finding life in the solar system declined, the time was ripe for another emergence of the unity cosmology. The fourth chapter, "Quantum Unity," traces the recent reemergence of the unity myth in the work *The Anthropic Cosmological Principle*.[85] In the 1980s, professor of Mathematics and Physics at Tulane University, Frank J. Tipler, and Research Professor of Mathematical Sciences at the University of Cambridge, John D. Barrow created an updated version of Plato's unity cosmology that combines the absence of alien life with cutting edge principles from quantum physics to present an anthropocentric picture of the universe. They argue that this anthropocentric universe provides a material foundation to the belief that humans are special and that morality has a foundation in the structure of the cosmos.

The fifth chapter, "Unity in the 21st Century," examines the unity myth in the last few decades. It begins with Peter Ward and Donald Brownlee, science professors at the University of Washington, who wrote a popular best seller arguing the case for unity. In the book they present unity as a response to the idea that humanity is mediocre. This chapter also looks at the works of Cambridge paleontologist Simon Conway Morris. Morris argues that evolution would inevitably result in a creature like humans, because of its tendency to converge on particularly effective solutions. Despite the tendency for evolution to result in humanity, he argues the case for unity. He deploys unity as a warrant against what he sees as the relativistic ideology of Neo-Darwinists, like Richard Dawkins and Daniel Dennett.

CONCLUSION

Individuals throughout history do not make use of the unity myth as some trite metaphor or simple observation of fact. Plato first deployed the unity myth in one of history's greatest culture wars as part of a strategy to undermine his sophist opponents. His victory over the sophists proved a decisive moment in the history of Western philosophical thinking. From then until now, the discussion of alien life frequently intertwines itself with political, religious, and philosophical debates. The unity cosmology's denial of alien life helps its proponents imagine a purposeful, teleological existence for humankind that poses a serious challenge to relativistic viewpoints that eschew seeing history as so purpose-driven. The unity cosmology provides the foundation for an argumentative strategy thousands of years old, but largely unexplored by rhetoricians. This book fills this lacuna in the communication literature by exploring how this argumentative strategy exists throughout history and has evolved through time.

In addition to its role in many ongoing debates the unity mythology acts to construct rhetorically what it means to be human. If society privileges a rhetorical ethos structured around the idea of human perfection, this means extraterrestrial discoveries have the potential to prove highly destabilizing. Much as the "discovery" of the American continents by the Europeans resulted in ideological confusion and ultimately violence, an ethos built on the unity cosmology confines the range of human responses to evidence of alien life to panic or some form of violence (whether physical or discursive).

The constitution of identity in relation to the nonexistence of alien life proves a crucial concern for the Communication discipline. In partial reference to extraterrestrial life, Peters writes,

> "Man" suddenly found many of the distinctions—of species, mechanism, gender and divinity—that had once sustained his status as lord over the earth unprecedentedly permeable. The failure to recognize the paranormal and the inhuman as founding questions for communication theory in our time goes together with the failure to recognize the inhuman when it stares back at you from the mirror. Both are containment strategies, props supporting a dangerously brittle identity.[86]

Peters rightly makes the case that dichotomies and distinctions exist at the heart of communication. The unity cosmology intersects with these dichotomies because it presupposes alien life does not exist and thus represents an *a priori* "no" to the question of whether academics should carve out space for "aliens" within communicative theory.

This book is the first in-depth case study in the way that a cosmology acts as rhetoric. The exploration of the unity cosmology as argument strategy and rhetorical ethos function should present an important addition to scholarship and open the door to rhetorical examination of other cosmological beliefs.

NOTES

1. Steven J. Dick, *The Biological Universe: The Twentieth-Century Extraterrestrial Life Debate and the Limits of Science*, 1st pbk. ed. (Cambridge: Cambridge University Press, 1999), 502.
2. George Basalla, *Civilized Life in the Universe: Scientists on Intelligent Extraterrestrials* (Oxford: Oxford University Press, 2006), 151; I.S. Shklovskii and Carl Sagan, *Intelligent Life in the Universe*, trans. Paula Fern (New York: Dell, 1966), viii; Dick, *The Biological Universe,* 348-49.
3. Steven J. Dick, *Plurality of Worlds: The Origins of the Extraterrestrial Life Debate from Democritus to Kant* (Cambridge: Cambridge University Press, 1982).
4. Ibid., 28; Grinspoon, *Lonely Planets*, 39-40.
5. Michael J. Crowe, "A History of the Extraterrestrial Life Debate," *Zygon* 32, no. 2 (1997); Erich von Däniken, *Chariots of the Gods? Unsolved Mysteries of the Past* (New York: Bantam Books, 1971); Erich Robert Paul, *Science, Religion, and Mormon Cosmology* (Urbana: University of Illinois Press, 1992).

6. "Poll: U.S. Hiding Knowledge of Aliens," *CNN*, http://www.cnn.com/US/9706/15/ufo.poll/.

7. Tariq Malik, "Most Americans Believe Alien Life is Possible, Study Shows," *Space.com*, http://www.space.com/news/050531_alienlife_survey.html.

8. The limitations section below will address what I consider to be "alien" life for the purposes of this discussion.

9. Plato's *Timaeus* responds to the cosmology of the atomists who believed that there existed an infinite number of universes. Plato argued that the universe we exist in is the only universe and that in the universe Earth is the only location habitable by anything other than mortal creatures. Plato, "Timaeus," in *Plato Complete Works*, ed. John M. Cooper (Indianapolis: Hackett, 1997), 31b.

10. Thomas F. Glick, Steven John Livesey, and Faith Wallis, *Medieval Science, Technology, and Medicine: An Encyclopedia* (New York: Routledge, 2005), 412.

11. Scholars debate to this day the precise reason for Bruno's execution. Revisionist accounts are a dime a dozen that claim that it was more likely the Church killed Bruno for his other heresies, like belief in hermetic magic. The whole affair is shrouded in mystery given the destruction of the court documents and the lack of credible accounts of the trial. David Wootton, *Galileo: Watcher of the Skies* (New Haven: Yale University Press, 2010), 53-56. What is certainly true is that Bruno's writing treats plurality as justification for his other an extension and physical manifestation of his other heretical a foundation for his other "more" heretical beliefs. Giordano Bruno, *The Ash Wednesday Supper*, trans. Edwards A. Gosselin and Lawrence S. Lerner (Toronto: University of Toronto, 1995). The Church also told Bruno to abandon his views in plurality prior to his execution. Hans Blumenberg, *The Genesis of the Copernican World*, trans. Robert M. Wallace (Cambridge, MA: MIT Press, 1987), 374. Regardless of why the Church killed Bruno, his death was widely interpreted as a punishment for his advocacy of the Plurality-of-Worlds thesis and created a chilling effect on many intellectuals. Descartes, for example, was hesitant to write about the Plurality of Worlds because of Bruno's execution. Dick, *Plurality of Worlds*, 112.

12. Bernard le Bovier Fontenelle, *A Discourse of the Plurality of Worlds Written in French by the Most Ingenious Author of the Dialogues of the Dead*, trans. W.D. Knight (Dublin: Andr. Crook and Sam Helsham and William Norman, 1687). Michael J. Crowe, *The Extraterrestrial Life Debate, 1750-1900* (Mineola, N.Y.: Dover Publications, 1999), 18, 19.

13. Thomas Baker, *Reflections Upon Learning* (London: A. Bosvile, 1699).

14. David L Clark, "Kant's Aliens: The Anthropology and Its Others," *New Centennial Review* 1, no. 2 (2001): 241-48; Crowe, *Extraterrestrial Life Debate*, 96, 97.

15. Menachem Fisch and Simon Schaffer, eds., *William Whewell: A Composite Portrait* (Oxford: Claredeon, 1991).

16. K. Rosenkranz, "Rosenkranz on Hegel's History of Philosophy," *Journal of Speculative Philosophy* 8, no. 1 (1874); E. T. Winkler, "Religion and Astronomy," *Baptist Quarterly* 5, no. 1 (1871).

17. Scott Holter, "Dead Reckoning," *Columns* 2006 (2003), http://www.washington.edu/alumni/columns/sept03/dead02.html ; Peter Ward and Donald Brownlee, *Rare Earth: Why Complex Life is Uncommon in the Universe*, Paperback ed. (New York: Copernicus 2003).

18. John Gribbin, *Alone in the Universe: Why Our Planet Is Unique* (Hoboken, NJ: John Wiley & Sons, 2011).

19. "Poll: U.S. Hiding Knowledge of Aliens."

20. George Basalla, *Civilized Life in the Universe: Scientists on Extraterrestrials* (Oxford: Oxford University Press, 2006); Christopher F. Roth, "Ufology as Anthropology: Race Extraterrestrials, and the Occult," in *E.T. Culture: Anthropology in Outerspaces*, ed. Debbora Battaglia (Durham, N.C.: Duke University Press, 2005); Debbora Battaglia, ""For Those Who Are Not Afraid of the Future": Raelian Clonehood in the Public Sphere," in *E.T. Culture: Anthropology in Outerspaces*, ed. Debbora Battaglia (Durham, N.C.: Duke University Press, 2005).

21. Grant McColley, "The Seventeenth-Century Doctrine of a Plurality of Worlds," *Annals of Science* 1, no. 4 (1936): 396.

22. Basalla, *Civilized Life in the Universe*.

23. Basalla's argument in regards to Sagan was made earlier by Thomas Lessl. Thomas M. Lessl, "Science and the Sacred Cosmos: The Ideological Rhetoric of Carl Sagan," *Quarterly Journal of Speech* 71 (1985).

24. Basalla, *Civilized Life in the Universe*, 158-60.

25. Carl Sagan, *The Demon Haunted World* (New York: Random House, 1995).

26. Thomas Chalmers, *A Series of Discourses on the Christian Revelation, Viewed in Connection with the Modern Astronomy*, 10 ed. (Glasgow: Robert Chapman, 1822); Guy Consolmagno, *The Heavens Proclaim: Astronomy and the Vatican* (Vatican Observatory, 2009).

27. Mark Harrison, "The Extraterrestrial in US Culture" (PhD diss., University of Pittsburgh, 2006). Roth provides a very similar study from an anthropological perspective. Roth, "Ufology as Anthropology."

28. Battaglia, "For Those Who Are Not Afraid of the Future."

29. Melvin E. Matthews, *Hostile Aliens, Hollywood and Today's News: 1950s Science Fiction Films and 9/11* (New York: Algora, 2007).

30. Elaine L. Graham, *Representations of the Post/Human: Monsters, Aliens and Others in Popular Culture* (New Brunswick, NJ: Rutgers University Press, 2002).

31. C. S. Lewis, *Out of the Silent Planet* (New York Scribner, 2003); C. S. Lewis, *Perelandra* (New York: Scribner, 2003); C. S. Lewis, *That Hideous Strength* (New York: Scribner, 2003); Vivian Ralickas, ""Cosmic Horror" and the Question of the Sublime in Lovecraft," *Journal of the Fantastic in the Arts* 19, no. 3 (2008).

32. Guillermo Gonzalez and Jay W. Richards, *The Privileged Planet: How Our Place in the Cosmos is Designed for Discovery* (Washington, DC: Regnery, 2004); Plato, "Timaeus."; William Whewell, "Of the Plurality of Worlds," in *Of the Plurality of Worlds. 1853 A Facsimile of the First Edition with an Introduction by Michael Ruse*, ed. Michael Ruse (Chicago: University of Chicago Press, 2001).

33. Ward and Brownlee, *Rare Earth*; John D. Barrow and Frank J. Tipler, *The Anthropic Cosmological Principle* (New York: Oxford University Press, 1986).

34. Nick Bostrom, "Existential Risks: Analyzing Human Extinction Scenarios and Related Hazards," *Journal of Evolution and Technology* 9 (2002); Martin Rees, *Our Final Hour: A Scientist's Warning: How Terror, Error, and Environmental Disaster Threaten Humankind's Future In This Century—On Earth and Beyond* (New York: Basic Books, 2003); Richard Posner, *Catastrophe: Risk and Response* (Oxford: Oxford University Press, 2005); John Leslie, *The End of the World: The Science and Ethics of Human Extinction* (New York: Routledge, 1996).

35. Peter Ward and David Brownlee, *Rare Earth: Why Complex Life is Uncommon in the Universe* (New York: Copernicus 2000); Alfred Russel Wallace, *Man's Place in the Universe: A Study of the Results of Scientific Research in Relation to the Unity of Plurality of Worlds* (New York: McClure Phillips & Co., 1903); Simon Conway Morris, *Life's Solution: Inevitable Humans in a Lonely Universe* (Cambridge: Cambridge University Press, 2003).

36. Mircea Eliade, *The Sacred and the Profane: The Nature of Religion* (New York: Harcourt, 1959), 21-30.

37. E. Edson and E. Savage-Smith, *Medieval Views of the Cosmos* (Oxford: Bodleian Library, 2004), 9-20; Stephen Edelston Toulmin, *Cosmopolis: The Hidden Agenda of Modernity* (Chicago: University of Chicago Press, 1990); W. G. L. Randles, *The Unmaking of the Medieval Christian Cosmos, 1500-1760: From Solid Heavens to Boundless Aether* (Brookfield, VT: Ashgate, 1999), X-XI, 3; Norriss S. Hetherington, "Introduction: The Aristotelian World View," in *Cosmology: Historical, Literacy, Philosophical, Religious, and Scientific Perspective*, ed. Norriss S. Hetherington (New York: Garland Publishing, 1993), 177.

38. Toulmin, *Cosmopolis*, 82-84.

39. While most of the case studies I examine would define themselves in opposition to the pluralist position, differences do exist in what exactly they argue. Plato, for example, denies the possibility of any mortal material life outside of what we know as the planet Earth. Whewell argues that intelligent life does not exist elsewhere, but that unintelligent life (animals) may exist (although this is unlikely). Despite the differences in the specifics of their cosmologies, all these authors use similar rhetoric to connect the absence of intelligent life in the universe to the importance of humanity.

40. John D. Caputo, *Against Ethics* (Bloomington: Indiana University Press, 1994).
41. "Myth," http://www.oxforddictionaries.com/us/definition/american_english/myth.
42. Roland Barthes, *Mythologies* (London: Paladin, 1973), 109, 32, 43.
43. Kenneth Burke, *A Grammar of Motives* (Berkley: University of California Press, 1962), 116-17.
44. Toulim, *Metaphysical Beliefs*, 33.
45. "Science," *Oxford Dictionaries*, http://www.oxforddictionaries.com/us/definition/american_english/science.
46. Edward Grant, *A History of Natural Philosophy: From the Ancient World to the Nineteenth Century* (Cambridge: Cambridge University Press, 2007), 49-50.
47. G. E. R. Lloyd, *Early Greek Science* (New York: W. W. Norton & Company, 1970), 8-15.This definition does not preclude a scientist for having religious beliefs and in fact many of these early scientists were religious.
48. William Keith Chambers Guthrie, *A History of Greek Philosophy: The Earlier Presocratics and the Pythagoreans* (Cambridge: Cambridge University Press, 2000), 44-45.
49. Philip C. Wander, "The Rhetoric of Science," *Western Speech Communication* 40, no. 4 (1976): 227.
50. Judy Segal and Alan W. Richardson, "Introduction. Scientific Ethos: Authority, Authorship, and Trust in the Sciences," *Configurations* 11, no. 2 (2003).
51. Lessl, "Science and the Sacred Cosmos."; Thomas M. Lessl, "The Culture of Science and the Rhetoric of Scientism: From Francis Bacon to the Darwin Fish," *Quarterly Journal of Speech* 93, no. 2 (2007).
52. Thomas Kuhn, *The Structure of Scientific Revolutions*, 3rd ed. (Chicago: University of Chicago Press, 1996).
53. Leszek Kolakowski, *Metaphysical Horror* (London: Penguin Group, 2001), 1-2.
54. Barthes, *Mythologies*, 142.
55. Michael J. Hyde, *The Ethos of Rhetoric* (Columbia: University of South Carolina Press, 2004), xiiv.
56. Martin Heidegger, *Introduction to Metaphysics* (New Haven: Yale University Press, 2000), 149.
57. Plato, "Timaeus," 38c-39e.
58. D. B., "Letter 1," *The Royal American Magazine* 1, no. 7 (1774).
59. Lewis Hanke, *All Mankind is One: A Study of the Disputation Between Bartholomé de Las Casas and Juan Ginés de Sepúlveda in 1550 on the Intellectual and Religious Capacity of the American Indians* (DeKalb: Northern Ilinois University Press, 1974), 3-17; David Abulafia, *The Discovery of Mankind: Atlantic Encounters in the Age of Columbus* (New Haven: Yale University Press, 2008). Some of the literature of the time reflects this debate, like Shakespeare's *Tempest* with its quasi-human characters. Lev Manovich, *The Language of New Media* (Boston: MIT Press, 2001), 892-912.
60. Joachim. Küpper, "The Traditional Cosmos and the New World," *Modern Language Notes* 118, no. 2 (2003): 363-92.
61. The suggestion that the belief in sharp racial distinctions proved ephemeral does not suggest that it was unimportant or did not emerge with disastrous consequences at various points throughout history.
62. Roth, "Ufology as Anthropology."; Harrison, "The Extraterrestrial in US Culture".
63. Plato, "Timaeus," 90e-92c.
64. H. C. Baldry, *Unity of Mankind in Greek Thought* (London: Cambridge University Press, 1965), 76.
65. Torben Grodal, "Stories for Eye, Ear, and Muscles: Video Games, Media, and Embodied Experiences," in *The Video Game Theory Reader*, ed. Mark Wolf and Bernard Perron (New York: Routledge, 2003), 43, 94.
66. Michael E. Zimmerman, "Encountering Alien Otherness," in *The Concept of the Foreign*, ed. Rebecca Saunders (Lanham, MD: Lexington Books, 2002).
67. Although Joachim Küpper never makes the connection to life on other planets, he reaches similar conclusions to Zimmerman. Küpper suggests that the Europeans were especially vulnerable to the shock of discovery because a "creationist, monotheistic view of the

world—is liable to be shaken fundamentally by the emergence of 'new' worlds." Again the idea of stability and permanence are juxtaposed with alien life, even when the aliens in question are human. The powerful reactions of many individuals to humans perceived as alien indicate that nonhuman alien intelligence would probably be even more difficult to confront. Ibid., 368-75, 89.

68. Zimmerman, "Encountering Alien Otherness," 157-60.

69. Brendan Borrell, "Are octopuses smart?," *Scientific American*, Feb. 27 2009; Barbara J. King, "When Animals Mourn" *Scientific American* (2013), http://www.scientificamerican.com/article.cfm?id=when-animals-mourn&WT.mc_id=SA_SA_20130618; Joshua Klein, "Joshua Klein: The intelligence of crows," (Ted.com, 2008); Alex Hannaford, "An Audience with Koko the 'Talking' Gorilla," *Telegraph*, Sep. 17, 2011.

70. Alexander Wendt and Raymond Duvall, "Sovereignty and the UFO," *Political Theory* 36 (2008).

71. John Durham Peters, *Speaking into the Air: A History of the Idea of Communication* (Chicago: University of Chicago Press, 1999), 260.

72. Debra Hawhee and Christa Olson, "Pan-Historiography: The Challenges of Writing History across Time and Space," in *Theorizing Historiography in Rhetoric*, ed. Michelle Ballif (Carbondale: Southern Illinois University Press, 2013).

73. Ibid., 92.

74. Ibid., 94.

75. Ibid., 96.

76. Ibid.

77. Crowe, *Extraterrestrial Life Debate*; Dick, *Plurality of Worlds*; Dick, *The Biological Universe*.

78. Jana Evans Braziel, "Being and Time, Non-being and Space: Introductory Notes toward an Ontological Study of Women and Chora," in *Belief, Bodies, and Being: Feminist Reflections on Embodiment*, ed. Deborah Orr (Oxford: Rowman and Littlefield, 2006), 106.

79. Thomas Aquinas, *Quaestiones Disputatae De Potentia Dei*, trans. English Dominican Fathers (The Newman Press, 1952); Thomas Aquinas, *The Summa Theologica of St. Thomas Aquinas. Part I QQ. L.-LXXIV*, trans. Fathers of the English Dominican Province, 2nd Revised ed., 10 vols., vol. 1 (London: Burns Oates and Washbourne, 1920).

80. Robert Jenkin, *The Resonableness and Certainty of the Christian Religion*, 3rd ed., vol. 1 (London: Richard Sare, 1708), 97-98; Baker, *Reflections Upon Learning*.

81. McColley, "The Seventeenth-Century Doctrine of a Plurality of Worlds," 426.

82. William Whewell, *Of the Plurality of Worlds. 1855. A Fascimile of the First Edition with an introduction by Michael Ruse, 1st ed.*, 1st ed. (1855; repr., Chicago: University of Chicago Press, 1855; reprint, 2001).

83. Whewell, "Of the Plurality of Worlds," 63.

84. Alfred Russel Wallace, *The World of Life: A Manifestation of Creative Power, Directive Mind and Ultimate Purpose* (New York: Moffat, Yard and Company, 1911); Wallace, *Man's Place in the Universe*.

85. Barrow and Tipler, *The Anthropic Cosmological Principle*.

86. John Durham Peters, *Speaking into the Air: A History of the Idea of Communication* (Chicago: University of Chicago Press, 2000).

Chapter One

Plato's Rhetorical Cosmology

The Unity of the World as Foundational Myth

> "In modern literature, it has in fact been quite common to investigate Plato's ethics, but much less common to delve into his cosmology. At any rate, these two undertakings have usually been carried out in isolation from each other."
> —Gabriela Roxana Carone, 2005[1]

Plato makes the case for the Good, Beautiful, and True throughout his dialogues. He argues that these eternal, unchanging values should guide ethical behavior for all humans. Plato criticizes rhetoric, a popular practice at the time, for placing eloquence over truth, but as McCoy notes, Plato's characters are not above using rhetoric to persuade the audience. For McCoy, Plato did not oppose rhetoric, but sought to harness it to noble ends.[2] Bertrand Russell offers a much less charitable view and famously wrote,

> [The Platonic Socrates] is dishonest and sophistical in argument, and in his private thinking he uses intellect to prove conclusions that are to him agreeable, rather than in a disinterested search for knowledge... he was not scientific in his thinking, but was determined to prove the universe agreeable to his ethical standards. This is treachery to truth, and the worst of philosophic sins. As a man, we may believe him admitted to the communion of saints; but as a philosopher he needs a long residence in a scientific purgatory.[3]

For Russell, Plato's Socrates would use any tool to persuade his audiences of his philosophy, however ethically suspect or inconsistent with his philosophy. This chapter argues that Plato deploys a scientific myth in his later dialogues as an argument for his philosophy. A central comment of this scientific myth is that the absence of alien life offers proof of the existence of absolute values. Whether a scientific myth is consistent with Plato's philosophy or "dishonest and sophistical" can be discussed elsewhere. This chapter is interested in the way that Plato's dialogues foreshadow and then enact the rhetorical use of the scientific myth of unity.

A rhetorical framework requires a particular strategy for reading Plato's dialogues. Many scholars treat Plato's dialogues as a unified consistent philosophical treatise or as independent self-contained manifestos. Both frames of analysis have value, but each focus deemphasizes the context in which Plato wrote the dialogues. This chapter approaches Plato with an eye toward how they act to persuade given the historical, political, and social environment. With this in mind, the chapter does not limit its analysis to the *Timaeus*, the dialogue that contains most of Plato's cosmology. Instead it examines Plato's cosmology as a rhetorical conduit for Plato's overarching philosophy and the end point of a rhetorical strategy that develops across his dialogues. In order to achieve this goal it begins with the dialogues' earliest mentions of cosmology in the *Phaedo* and ends with one of Plato's last dialogues, *Laws*.

Plato engages in many rhetorical strategies throughout his dialogues in order to persuade audiences. In his early and middle dialogues, Plato uses math, logic, dialectic, and, above all, myth to make the case for the Good, Beautiful, and True. However, in his later dialogues Plato presents, in what has become a decisive moment in the history of Western thought, a scientific myth. His characters ground the Good, Beautiful, and True in a scientific cosmology.[4] Plato presents a vision of the cosmos with orderly celestial objects, a central Earth (geocentrism), and no other inhabited worlds (unity). Plato argues that these features of the universe provide proof of his philosophy. Rather than a distinct and new philosophical insight and/or rhetorical strategy, this chapter situates this cosmological turn as an evolution of Plato's earlier rhetorical strategies brought about by the difficulties of persuading his audiences.[5] Plato moves from a reliance on traditional myth to the development of one of the earliest and most important scientific myths.

This chapter demonstrates that Plato employs his cosmology as a scientific myth, not by investigating his personal beliefs, but by drawing parallels between his use of myth as argument and his later use of cosmology as argument and by providing evidence in the dialogues that Plato understood as the persuasive power of scientific myths. Demonstrating that Plato's *Timaeus* presents a scientific myth does not require that Plato describe a model of the universe he thought was false, nor does it prove that his philosophical predispositions biased the empirical observations he puts forth through the character Timaeus. This chapter does present evidence to suggest that, in fact, it is reasonable to speculate that Plato did not believe in the literal accuracy of his account and that his philosophical predispositions did determine his "observations," rather than his cosmology guiding his philosophical beliefs. Knowing Plato's true beliefs, however, is a futile and ultimately less interesting project than understanding his argumentation strategy.

Before one can analyze Plato's dialogues it is important to have a hermeneutic in place. This chapter begins by offering a framework to interpret Plato's dialogues, which treats them as arguments for fundamental Platonic values, most importantly anti-relativism. It then positions Plato's support for anti-relativism within a particular historical context: his conflict with sophistic thought. This overarching philosophical dispute provides the animating spirit that runs throughout almost all of Plato's dialogues. The second section details the state of cosmology during the era, including the hostile treatment it receives in Plato's early dialogues.[6] This early hostility gives credence to the hypothesis that Plato's later deployment of cosmology represents a rhetorical response to sophistic strategies that defend relativism, rather than a necessary component of Platonic thought. The third section highlights Plato's creation of myths written in defense of his philosophy, in his early and middle dialogues. These writings, based on traditional Greek myths, act as a rhetorical strategy that provides a template for Plato's later use of cosmology as myth. The fourth section examines the way Plato shifts to cosmology as argument for his philosophy and draws clear parallels between Plato's arguments grounded in myth and those grounded in cosmology. The fifth and final section examines the consequences of this rhetorical maneuver for Plato's thought and investigates the way that his attack on relativism becomes intertwined with Plato's cosmology to the point where many thinkers throughout history see attacks on aspects of his own cosmology (including unity) as attacks on absolute values.

READING PLATO'S DIALOGUES

Before moving ahead it is important to have a framework for interpreting Plato's dialogues, because much of what we know about Plato's philosophy comes from these works. Are the ideas in Plato's dialogues Socrates' or is Socrates just a mouthpiece for Plato's original ideas? In order to discuss "Plato's cosmology," we must determine that the cosmology he presents in his dialogues is in fact Plato's. While interpreting texts is always a difficult proposition, it is a unique challenge when addressing the issue of authorship in Plato's dialogues. First, the dialogues are ancient and there exist few documents to contextualize their meaning. Second, they are written as dialogues, which creates ambiguity over when or if Plato speaks his own ideas through a particular character. These difficulties have helped create the conditions for the emergence of a variety of approaches to reading Plato's dialogues. For a long time, scholars sought to demonstrate a unified doctrine across all of Plato's dialogues, treating every argument as harmonious.[7]

More recently, some scholars began to claim a clear break between Plato's early dialogues, which present Socrates' ideas, and his later dialogues, which present Plato's own unique philosophy.[8] Most Platonic scholars read Plato's dialogues in relation to each other, but some argue they should be examined independently.[9] Many scholars take the arguments advanced by the characters in Plato's dialogues at face value, but an important minority claim that Plato hid messages in his texts for the discerning audience member.[10] All of these approaches attempt to read Plato's philosophy in the dialogues, but some scholars question whether this is possible or desirable. Mulhern and Nails argue that because Plato wrote in dialogue form, it is impossible to determine his philosophical intent.[11] Corlett and Griswold claim that the dialogues serve a function totally distinct from philosophical treatise, whether enacting the Socratic method to explore ideas or serving primarily as literature.[12] Given all of the hermeneutic approaches to Plato's dialogues, it is tempting to throw one's hands up in surrender. This chapter, however, argues that Plato does present a cosmology that is his own and that this cosmology acts as an argument for anti-relativism, which is a core Platonic belief that runs throughout the dialogues.

This chapter approaches the dialogues from the premise that Plato had a philosophy and that philosophy manifest itself through the words of certain characters in his dialogues. Typically Plato speaks through Socrates, but there are notable exceptions, like when he speaks through Timaeus in *Timaeus* or the Athenian stranger in *Laws*. The character of Socrates thus provides Plato a consistent mouthpiece to posit his own philosophy and challenge rival philosophies through the dialogues. This holds especially true if, as Aristotle claims, the Athenian stranger is meant to be Socrates, perhaps unnamed because the dialogue takes place in Crete rather than Athens.[13] That Plato would use Timaeus as a mouthpiece to explicate his cosmology instead of Socrates makes sense, given that Socrates was not a cosmologist. Plato's characters Socrates and Timaeus present a variety of arguments against relativism, including the cosmology found in *Timaeus*.

Gerson offers one of the most recent and extensive defensives of this kind of approach.[14] He argues those that view the project of searching for Plato's philosophy in his dialogues as impossible or misguided fail to account for the fact that the dialogues exist within a broader context. Plato founded a school of philosophy (both literally and figuratively) and members of this school are in the best position to interpret Plato's dialogues. Members of Plato's academy, most notably Aristotle but also Speusippus and Xenocrates, all treat Plato's dialogues as reflecting his ideas through the words of key characters from the dialogues.[15] Aristotle and others would not just have encountered Plato's ideas through his dialogues, but through conversations with Plato and other members of the academy. This puts them in better position to interpret the dialogues than revisionist accounts that take place thousands of years after Plato's death. The dialogues even contain evidence that they reflect debates and discussions from within the academy, which highlights the importance of Plato's extra-textual teaching in interpreting the dialogues.[16]

The views of the Platonists also point to some core beliefs that define Plato's dialogues. While Plato and later Platonists, may disagree about things like the forms, Gerson persuasively argues that they are bound by a commitment to anti-materialism, anti-mechanism, anti-nominalism, anti-relativism, and anti-skepticism. Gerson calls these "antis" ur-Platonism and argues that whatever shifts may occur, Plato's dialogues maintain an unwavering commitment to these principles.[17]

This chapter begins with Gerson and many others' premise that anti-relativism represents a core element of Plato's philosophy.[18] Because this is a rhetorical study, it is not interested in whether Plato viewed his dialogues as unified or even whether he personally believed everything he said, like the idea of the forms or the cosmology he presents in the *Timaeus*. Instead it treats the dialogues as means to persuade audiences of the core beliefs of ur-Platonism, specifically anti-relativism. It seeks Plato's rhetorical strategy by examining the arguments made by Plato's spokespersons in the dialogues, but also by looks at historical evidence outside the dialogues to create a fuller picture of how Plato sought to persuade audiences. The next section contextualizes Plato's anti-relativism within the context of his conflict with sophistic ideas.

PLATO VS. THE SOPHISTS

Plato champions universality over relativism in almost all of his dialogues.[19] He provides a strong defense of universal and timeless ideals independent from humanity: the Good, Beautiful, and True. On the basis of these ideas Plato establishes a framework of what the good life would require, including acting justly and placing the group over individual needs.[20] The intellectual zeitgeist favored contingency, as championed by Plato's nemeses, the sophists. In contrast to the Good, Beautiful and True, the sophists typically defended what we today would call moral relativism. Protagoras, one of the most famous and respected sophists, sums up this relativism, "of all things the measure is man, of the things that are, that they are, and of things that are not, that they are not."[21] The passage represents the sophistic view that all values exist only as social agreements; no universal unchanging ethics, morality, philosophy, etc. exist outside of human minds.[22]

This view had serious implications for Plato's efforts to get others to lead Good lives. Under Protagoras's relativistic framework, murder, for example, is not intrinsically wrong, but only bad because society has proclaimed it so. Plato argues that relativism incentivizes selfish and evil behavior, even if the sophists who proclaimed it had no intention of encouraging malevolence. Protagoras's specific maxim "man is the measure" so disturbed Plato that he devoted extensive attention to it, beyond his commonly severe treatment of relativism generally.[23] As Ugo Zilioli writes, "[Protagoras] and his relativism were indeed the most dangerous enemy for [Plato] and the kind of objectivist doctrines he maintained."[24]

The debate between Plato and the sophists took place in the public sphere, with serious consequences for Greek society. The generation prior to Plato saw the sophists grow in prominence, as public speeches, trials, and voting occurred with greater frequency and held more importance. The sophists specialized in teaching public speaking and other skills necessary to navigate the new public sphere. The demand for a sophistic education gave them a platform to spread their relativistic philosophy.[25] The influence of the sophists challenged the traditional ethics and social structure of Athens because their ideas suggest that, rather than being divine and preordained, the organization of society exists as only one of many possible systems. This opened class stratification and many other conventions of ancient Greece up to criticism. The sophists offered an intellectual justification for democracy and taught their students the tools necessary to challenge the status quo.[26] By the time Plato began writing his dialogues, sophistic ways of thinking dominated the public sphere.

While Plato had his own criticisms of Greek society prior to the rise of the sophists, the changes they heralded were anathema to his philosophy. He claims the public-speaking advice of the sophists encourages the use of deception and rhetoric (used pejoratively) over logic and truth. His dialogues oppose democracy and maintain a skeptical attitude towards the ability of the Greek masses to effectively govern themselves. At the broadest level Plato argues that sophistic ideas encourage selfishness and unjust behavior by destroying the possibility for immutable ethical principles. Opposition to sophistic ideas runs throughout Plato's dialogues.

Initially, it may seem difficult to rectify Plato's obsessive battle against sophistic moral relativism with his creation of a cosmology. Since his conflict with the sophists over the nature of values occurred in the here and now, the structure of the universe does not appear to have much relevance for how one should reform the structure of society. An understanding of the practice of cosmology at the time again widens the division between philosophy and cosmology. The presocratic cosmologists, who set the terms for cosmological thinking, largely took an observational, scientific approach that did not involve recourse to traditional myth to answer cosmological questions.[27] Their method removed (if not entirely detached) their cosmologies from ethical concerns. Plato criticized, in his early and middle dialogues, the presocratic practice of cosmology for its failure to draw connections between the physical structure of the universe and morality. Ignoring cosmology except to remark on its problems, Plato's early dialogues focus instead on a variety of other rhetorical strategies to convince others of universal, unchanging ethics.

STATE OF COSMOLOGY AND PLATO'S EARLY DIALOGUES

Before Plato developed his earliest ideas there was a proliferation of cosmologies, which coincided with the development of Greek scientific thought. Thales, Anaximander, and Anaximenes became the first "physicists," a term then used for those who studied the natural world, including the universe. Figures like Heraclitus, Parmenides, Zeno, Melissus, Diogenes of Apollonia, Empedocles, Anaxagoras, Philolaus, and Democritus followed in the initial physicists' footsteps. These early physicists embarked on a project drastically unlike previous thinkers. Guthrie provides a succinct explanation:

> The questions which excited them were of this kind: Can this apparently confused and disordered world be reduced to simpler principles so that our reason can grasp what it is and how it works? What is it made of? How does change take place? Why do things spring up and grow, then decline and die? How can one explain the alternation of day and night, summer and winter? They claim our attention by having been the first to suggest that answers to these questions may be found by taking thought. They abandoned mythological and substituted intellectual solutions. There might or might not be a divine mind at the back of, or permeating, the works of nature (that was a question to which some of them sought an answer), but it was no longer satisfying to say that storms were roused by the wrath of Poseidon, or death caused by the arrows of Apollo or Artemis. A world ruled by anthropomorphic gods of the kind in which their contemporaries believed—gods human in their passions as well as in their outward form—was a world ruled by caprice. Philosophy and science start with the bold confession of faith that not caprice but an inherent orderliness underlies the phenomena, and the explanation of nature is to be sought within nature itself.[28]

The physicists and their intellectual decedents focused much of their thinking on the question of cosmology, which resulted in a variety of competing cosmologies. For example, Thales postulated the Earth floated on a bed of water, whereas Democritus claimed void surrounded the Earth. And yet all the physicists shared a commitment to a nascent scientific method. The physicists and their intellectual heirs broke with the previous cosmologies that relied on religious explanations for the creation and function of the universe, and instead offered cosmologies built on natural explanations derived from logic and observation. The physicists would track the movement and position of celestial objects (astronomy) and study the prevalence and behavior of elements like fire and water in order to make hypotheses about the structure of the universe. In most cases these thinkers isolated their scientific observations from their religious and moral thinking.[29] The gulf between the physicist's methods and ethical principles kept one major intellectual contemporary of the Greek cosmologists, Socrates, from dabbling in the practice.[30] Plato, Socrates' most famous student, appears to take a similar approach in his early dialogues, depicting cosmology as a distraction from philosophy.

In the *Phaedo*, Plato depicts Socrates as initially intrigued by the ideas of Anaxagoras, one of the earliest Greek cosmologists: "I was ready to find out in the same way about the sun and the moon and the other heavenly bodies, about their relative speed, their turnings and whatever else happened to them, how it is best that each should act or be acted upon."[31] Socrates sees potential in astronomy to act as a method for understanding the Good. Upon reading Anaxagoras' writings, however, Socrates realizes that Anaxagoras does not describe his cosmology in terms of what is Good, but instead as the result of a series of scientific processes. Socrates claims Anaxagoras wastes his time because he does not examine the celestial objects through the lens of their "capacity of being in the best place they could possibly be put."[32] Anaxagoras' observational approach displeases Socrates because it fails to provide a model for the Good that can serve a purpose on Earth. Socrates argues that if the stars rotate in the best possible arrangement they would elucidate principles of the Good and provide a guide for proper social behavior. Plato makes clear this critique extends beyond Anaxagoras to all of the cosmologists of the time when he describes Socrates as saying,

> It is what the majority appear to do, like people groping in the dark; they call it a cause, thus giving it a name that does not belong to it. That is why one man surrounds the earth with a vortex to make the heavens keep it in place, another makes the air support it like a wide lid.[33]

Here Plato references ideas from the popular cosmologies of the time in order to dismiss the field as "groping in the dark."

Similarly, in the *Thaetetus*, Plato has Socrates retell an old joke about the astronomer Thales, rumored to have fallen in a well because he was so distracted staring at the stars.[34] Socrates goes on to contrast astronomers like Thales with true philosophers, who concern themselves with issues critical for how to live on Earth. Plato sets up a crucial contrast between astronomy and his own search for the Good, the Beautiful, and the True. Astronomers disconnect themselves from the daily affairs of humanity, preferring to keep their heads in the clouds. In contrast, Plato argues philosophers should focus their attention on how one should best live her or his life.

Plato initially objects to astronomy because astronomers focus on celestial objects to the exclusion of important questions on Earth. One can read this as an indictment of the way certain astronomers practice their craft rather than a broad criticism of the discipline of astronomy. A different astronomy could potentially avoid this problem, for example, if one examined astronomy from the perspective of the Good. Plato, however, provides additional critiques of astronomy in *Rival Lovers* and *Republic* that are more intrinsically connected to the practice of astronomy itself.

In *Rival Lovers* Socrates stumbles upon two boys debating astronomy, or as an onlooker in the dialogue points out, "babbling about things up in the sky and talking philosophical nonsense."[35] Socrates takes offense at the idea that philosophy is nonsense, but later in the dialogue he seems to agree that what the boys were doing was nonsense, because they were debating about astronomy rather than philosophy. One rival lover tries to convince Socrates that philosophy means learning as much as possible in all the subjects. Socrates indicates that the true philosopher must specialize solely in philosophy and not dilute her or his knowledge by learning so many different subjects. Socrates suggests that the rival lovers may have thought they were discussing philosophy, but in reality they were discussing astronomy and thinking it was philosophical. Although the dialogue focuses on academic pluralism in general, the fact that it begins with an argument about cosmology suggests Plato denies cosmology a place as part of the philosophical project. The idea that astronomy detracts from one's ability at philosophy ends the possibility (still an open question in the *Phaedo*) that one could carry out cosmology in a philosophical manner.

In the *Republic*, Plato includes astronomy in the list of subjects that should be studied by the guardians (the enlightened leaders of his ideal city state). This at first appears to represent a major change in his thought, from rejection to embrace of astronomy. Plato, however, makes clear in the dialogue that the astronomy of the guardians should be dramatically reenvisioned from the way astronomers conducted it at the time. He refers to the current astronomers as those who seek to understand higher powers by "studying ornaments on a ceiling."[36] The actual celestial objects move too "slow or fast" for Plato, meaning that their physical movements do not conform to the Good as defined by their relation to "geometrical figures."[37] *The Republic* argues that rather than studying the orbits of celestial objects, astronomy should become a form of four-dimensional geometry. This approach would have students work on theoretical problems involving mathematic ratios and "leave the things in the sky alone."[38] Plato's theoretical astronomy requires students to focus on perfect ratios and other harmonious constructions rather than the messy reality of celestial movement. Plato suggests that looking for higher meaning in celestial motion represents a danger, because the objects move in discordant ways. For Plato the study of discord is not compatible with a search for the Good.

These four dialogues demonstrate Plato's hostility or at best disdain for cosmology during the time of his early and middle dialogues. In the *Phaedo*

Plato argues that cosmology is something different than the philosophical project. In the *Thaetetus* he jokes about how astronomers focus so much attention to the stars that they cannot see what lies directly in front of them. In *Rival Lovers* he expands this critique to say that studying cosmology detracts from one's ability as a philosopher. In the *Republic* he goes further still saying that studying celestial movements actually gives a warped sense of the Good, because the celestial objects move in an unharmonious manner. Given the importance of harmony to Plato's philosophy, this last criticism represents a fairly serious indictment of astronomy.

Despite these seemingly harsh criticisms, however, the cosmologists in Plato's early and middle dialogues play relatively peripheral roles as intellectual antagonists. For the most part Plato does not attack the practice of cosmology directly. Instead, he accuses it of failing to confront the issues of contingency, universality, and ethics at the center of his dialogues. Many of the cosmologists of the time actually shared his values. For example, Democritus proposed the atomist cosmology where random chance and chaos represent the dominant forces in the universe. Yet he agreed with Plato on many important issues of ethics and morality and specifically opposed Protagoras' dictum, "Man is the measure of all things."[39] Despite the agreement over ethics, Plato would not only directly and savagely attack the cosmologists in his later dialogues, but also create his own rival cosmology rooted in his philosophy of the Good, the Beautiful and the True.

In order to understand the shift in strategy that underlies the different treatment of cosmology in Plato's later dialogues, one needs to examine Plato's rhetorical strategies throughout his early and middle dialogues. In his early and middle dialogues Plato relies extensively on traditional myth in order to persuade others of his beliefs. In his later dialogues, Plato begins to rely heavily on cosmological appeals as a new argument strategy to convince others to believe in the Good, Beautiful, and True. These cosmological appeals have many of the elements of traditional myths, but borrow the scientific language of the physicists. In other words, Plato presents his cosmology as a scientific myth in defense of his philosophical values. The next section traces Plato's use of myth in order to provide the background for Plato's cosmological turn.

MYTH AND RHETORIC IN PLATO'S DIALOGUES

Plato's dialogues demonstrate a grasp of the rhetorical power of traditional myth, a force that played a crucial role in shaping the beliefs of Greek citizenry. Plato's philosophical program centered on the Good, Beautiful, and True, but the sophists and others placed these ideas under relentless attack, from the perspective of contingency and materialism. The supernatural elements of his myths allow Plato to ground his conceptions of the Good, Beautiful, and True in a veneer of the eternal, just as Plato argues the Homeric mythology provided a foundation for Greek ethics prior to the sophists.[40] One should not be surprised then that all of Plato's dialogues have "mythical characters and motifs," and starting with the *Protagoras* and *Gorgias* (regarded as the last of the dialogues from Plato's early period), all of Plato's dialogues contain myths (here I mean mythology in the sense of fables, rather than Barthes' conception of myth).[41] Some of the myths are well-known legends altered or combined to help him make his point. Other myths, like the tale of Atlantis, Plato likely made up entirely.[42]

Partenie argues that Plato uses traditional myth to make his philosophy more accessible.[43] She makes a compelling case, but the evidence also suggests that Plato makes use of myth's ability to provide a timeless foundation for his philosophy. This does not represent a simplification of his philosophy useful solely in expanding the audience, but rather draws on the power of myth as a persuasive device.[44] In the *Republic*, Plato argues that individuals

> shape their children's souls with [myths] more than they shape their bodies by handling them....The young can't distinguish what is allegorical from what isn't, and the opinions they absorb at that age are hard to erase and apt to become unalterable.[45]

Plato indicates that teaching children myths could ensure that their opinions become "unalterable," which if true represents a strong rhetorical tactic. Although in this quote he references children, Plato later goes on to discuss the power of myth more generally when he says the poets "persuaded many people to believe" the myth of the Ages of Man.[46] Plato argues that myths so easily persuade the public that a myth with a problematic moral should be hidden from the public even if it is true.[47]

Given Plato's references to the power of myths to shape society it is unsurprising that he made frequent use of them. In the *Timaeus,* Plato has Critias tell the tale of the city of Atlantis. According to the story, Atlantis was a "vast power" that eclipsed all of Athen's accomplishments "in magnitude and excellence."[48] Despite the city's military strength and technical achievements, however, it ultimately collapsed because it lacked justice. The myth serves as a cautionary tale for those that would put the search for material gain ahead of morality. In the *Gorgias*, Plato tells of gods judging the souls of humans in the afterlife as a justification for good behavior.[49] This myth functions very similarly to the myth of Atlantis. In both myths, those that seek material pleasures at the expense of just behavior risk punishment in the future.

Unfortunately for Plato, the use of myths centered on divinities, mystical locales, and ancient peoples posed two distinct problems. The first problem was the preexistence of a long-held set of mythological beliefs passed on through the work of the poets, in which the gods and goddesses behaved in ways unbefitting the Good, Beautiful and True. Plato vehemently attacks the poets, whose mythology he claims promotes values inconsistent with his philosophy.[50] The Gods of Homeric myth could not serve as a justification for universal ideals, because the sophists taught their patrons that one could "persuade" the gods with rhetoric in the form of "prayers and offerings."[51] In other words, rather than upholding universal and timeless values the Gods of Homeric myth were interested in flattery and bribery. Even though one could use myth for noble causes, the degenerate nature of Greek myths made this a difficult venture.

Fixing the problem of a preexisting set of problematic myths was not as simple as creating new myths. In the *Republic*, the character Adeimantus expresses the difficulty with such an endeavor.

> If [the gods] do exist and do concern themselves with us, we've learned all we know about them from the laws and the poets who give their genealogies—nowhere else. But these are the very people who tell us that the gods can be persuaded and influenced by sacrifice, gentle prayers and offerings.[52]

The Greek people already had a set of myths that they had heard since childhood, which supported the sophistic worldview. Later in the dialogue Socrates agrees and says that it would take a sustained effort to entrench a new myth and it would likely take at least a generation before the public believed it.[53] Most of Plato's myths begin by him retelling a legend that he has overheard, which would have no more claim to truth than the old myths that children would have been taught from a young age.

The increase of secularism in Athens represents the second problem for Plato's use of traditional myths. Even if Plato effectively created and disseminated a myth, it may not have done him much good as Athenians at the time began a serious move away from old-fashioned myths centered on Gods, heroes, and monsters. The massive suffering of the plague caused traditional morality to collapse as people struggled for survival.[54] The Peloponnesian league defeated Athens, subjecting the city to rule by tyrants for thirty years, which called into question the idea of divine protection.[55] Critias' play *Sisyphus* had a leader who invented a religion to better control the population, much as Plato literally suggests should be done in the *Republic*. The emergence of materialist cosmologies must have added another argument in favor of abandoning the traditional religious ethics founded in myth. Whether or not materialism actually became the majority opinion among Athenians, it had taken hold of the intellectual climate, and many scholars claim that Plato personally felt that it held enormous sway over popular discourse.[56]

The difficulties facing traditional myth may lead one to question why Plato relied on them throughout his dialogues. Some truth likely lies in Partenie's claim that myth helped elucidate points in ways that made the dialogues more broadly accessible. Brisson argues that passages in the *Protagoras* and *Timaeus* suggest audiences like hearing myths for aesthetic reasons.[57] Beyond these motives, one can imagine Plato hoping to find a way to recapture the previous power of myth to persuade the population. As many problems as Plato had with the Homeric myths, the societal move away from them coincided with the rise of the sophists. These traditional myths, as flawed as they were, had helped create a barrier to relativism that had held for much of Greece's history. In fact, only after the disasters befell Athens and shook public faith did sophists and others really begin to point out the problematic morals of Homeric mythology.

By using traditional myth Plato could appeal to those that had not embraced secularism but were compelled by the arguments made by Pindar and others that Homeric myth's questionable content could not justify moral behavior.[58] Plato's dialogues act to create a new mythology, similar in structure to the traditional myths, but harmonious with his philosophical ideas. Even though the *Republic* recognizes the problems with a strategy of new myth creation, it may be that Plato felt the persuasive potential of myth made the maneuver worthwhile. A new mythology could potentially return the social stability that existed before the sophists.

In light of the problems of new myth creation, Plato's dialogues explore techniques to recapture the power of myth and deploy it to support his philosophy. In the *Republic*, Plato advocates sustained effort to eliminate many of the older traditional myths from Homer and Hesiod.

> We must first of all, it seems, supervise the storytellers. We'll select their stories whenever they are fine or beautiful and reject them when they aren't... Many of the stories they tell now, however, must be thrown out.[59]

Plato goes so far in the Republic as to endorse legal censorship of myths that he believed were problematic.[60] By eliminating the myths that did not support his philosophy, Plato could foster a set of myths that upheld his worldview. The *Republic*'s focus on how to reinvigorate myth as a tool for the Good suggests that myth could shape society for the better if deployed correctly.

Plato's call to censor the myth-telling poets in his middle dialogues represents a radical shift from his early dialogues. Vlastos summarizes the severity of the change between Socrates in Plato's early and middle dialogues when he writes,

> [Early Socrates'] mission 'to live philosophizing, examining [him]self and others' (Ap. 28e), those 'others' being 'anyone of you I happen to meet,... young or old, citizen or alien' (29d-30d). He believes that 'the unexamined life is not worth living by a human being' (Ap. 38a)." By contrast, [Socrates'] project in the *Republic,* where only "an exceptionally gifted, rigorously trained elite" practice philosophy "after they have completed their qualifying mathematical studies," would appear to [early Socrates] to "condemn the great majority of its citizens... to life 'not worth living by a human being.' In the whole history of Western thought no philosophy has been more populist in its outreach than [early Socrates'], none more elitist than Socrates' [middle period].[61]

This chapter contextualizes this shift from populism to elitism in the context of the rhetorical challenges Plato faced in spreading his philosophy.

Plato's own early and middle dialogues indicate that his philosophy, spoken through Socrates, had difficulty making converts. In the *Gorgias*, for example, Socrates says that most people agree with Callicles, the nihilistic, self-interested materialist.[62] In other words, Plato's potential audience was more interested in unsavory ideas than in the Good. Similarly, Plato has his characters lament how easily the public is misled in the *Republic*.[63] Plato's move to restrict the audience of his philosophy may have resulted from practical, rather than philosophical reasons. Rather than suggest that Plato gave up on the philosophical project it seems more likely that Plato's failure to impact social practice acted as an incentive to shift strategies. Even though the citizens of the *Republic* may not have lived examined lives, Plato may well have reasoned that the guardians could trick them into leading Good lives.

Unfortunately for Plato, the model in the *Republic* also faces numerous problems as a method for encouraging people to live a good life. Plato does not even seem very enthusiastic about the chances of an ideal city coming about in the *Republic* dialogue itself. This view may have been further bolstered by a failed attempt by Plato to direct the governance of Syracuse.[64] By the time Plato writes *Timaeus* he appears to completely surrender to the idea of the ideal city. The *Timaeus* begins with the story of Atlantis, a city in the model of Plato's ideal republic that ultimately collapses because of hubris, which suggests that even if an ideal city can be achieved it will only exist ephemerally. The story of Atlantis indicates that Plato has moved away from a political solution to the problems of new myth creation, at the time of the *Timaeus* dialogue.

Given the growing rejection of traditional myths generally, the ubiquity of problematic traditional myths, and the difficulty of imposing new myths, it is unsurprising that Plato would try other strategies to lead individuals to the Good. Among these strategies Plato uses scientific and empirical argument in ways that foreshadow his turn to cosmology. In the *Meno*, for example, Plato depicts an experiment with a slave boy to prove that morality is independent from the human mind.[65] No evidence exists as to whether this experiment convinced others of his philosophy, although its absence from Plato's other dialogues suggests it did not. Evidence does exist for the public reaction to another attempt to offer evidence for the Good, Beautiful, and True. Plato reportedly offered a free lecture on mathematics, which he attempted to connect with his philosophical program. His lecture ended in disaster when much of the audience left in the middle of his speech.[66] Even if this event did not occur, there is more solid evidence to support the notion that the public dismissed Plato's melding of math and ethics. In *Greater Hippias*, Plato has Hippias reference the population's disinterest in math and harmony, the very alternatives Plato offers as justifications for the universal in place of traditional myths.[67]

Plato's philosophical project centers on encouraging individuals to lead moral lives, but historical evidence and his own dialogues suggest that a sophistic value system continued to dominate public thinking. Plato deployed myth, math, and empirical experiments as tools in order to convince his readers and listeners to live justly. If one views Plato's ultimate goal as persuading individuals to behave in a manner befitting the Good, Beautiful, and True, one should not be surprised that Plato altered his rhetorical strategies over time in an effort to best reach the most people. Plato's cosmological turn builds on his previous failures, taking the most effective elements of math, empiricism, and myth and blending them together into one very effective rhetorical strategy.

PLATO'S RECONCILIATION WITH COSMOLOGY

Plato first invokes cosmology in a positive light (and may first invoke cosmology at all, depending on the chronology of the dialogues) in the *Gorgias* when Socrates debates with Callicles. Callicles proves a particularly difficult opponent for Plato because of his shameless disregard for the conventional standards of the Good. Socrates cannot get Callicles to concede points in the same way he could the dialogue's earlier interlocutors, Gorgias and Polus. In the dialogue Socrates defends the Good based on universal and immutable values and Callicles makes case for rule by the strongest by citing natural justice as seen in the cruelty of nature. The debate stretches on until Callicles refuses to answer any more questions, which he argues are irrelevant to the ethical issues at hand.[68] After Callicles cedes the floor, Plato closes the dialogue by referencing both cosmology and the judgment of souls.[69] The idea that souls will be judged does not seem out of place since Plato refers to this myth in his earlier dialogue, *Phaedo*.[70] The reference to cosmology does seem strange, however, given Plato's critique of it in his other early dialogues. Plato has Socrates say,

> Yes, Callicles, wise men claim that partnership and friendship, orderliness, self-control, and justice hold together heaven and earth, and gods and men, and that is why they call this universe a world order, my friend, and not an undisciplined world-disorder. I believe that you don't pay attention to these facts, even though you're a wise man in these matters.[71]

The passage does little to explore the details of what he means by "world order," so much so that without further background knowledge one can easily read it independent of the structure of the universe. The Greek words Plato uses strongly suggest that he references the universe, however.[72] Given this knowledge, one can read this passage to say that the structure of the universe justifies behaving in concordance with universal morality. Callicles' rejection of "partnership and friendship, orderliness, self-control, and justice" means that he operates out of step with the "world order." Contrary to Callicles' belief that nature justified might over right, nature in the form of the universe actually supports just behavior. This quote provides an important milestone in Plato's argumentation, because it connects the question of cosmology to ethics. Plato suggests here that cosmology can serve as a model for order that should guide how humans live their lives.[73]

In the *Gorgias*, Socrates is unsuccessful in his use of conversational dialectic to change the mind of his opponent. The dialogue ends not with dialectic interaction, but instead with Plato lecturing Callicles and referencing both cosmology and religion (the idea that humans have souls that are judged).[74] Charles Kaufman argues that Plato's monologue at the end of *Gorgias* represents a collapse of the dialectic and a desperate turn to a rhetorical appeal.[75] This rhetorical appeal centers on myth—a Platonic standby—and, surprisingly, a cosmological appeal. This early cosmological argument for the Good provides some context for Plato's own later effort to defend a scientific cosmological model that harmonizes with his values.

THE WORLD ORDER AND UNITY

In the *Gorgias* Plato directs Callicles to the cosmos to find direction for how to live a good life. In other dialogues Plato attacks the prevailing approach to astronomy for depicting a universe that cannot provide a strong ethical model. In the *Timaeus*, Plato presents a vision of the cosmos in line with his philosophical beliefs, precisely the kind of cosmos that would act as a justification for Callicles to lead an ethical life. In the dialogue Timaeus, an astronomer from Locris, presents a new cosmology with numerous divergences from the cosmologies of the presocratics, as the next section details in more depth. Most important for this study, *Timaeus* provides a cosmology with three components: orderly celestial movement, one world, and Earth's centrality. Plato argues through the character Timaeus that these features constitute a world order with implications for how to live a good life.

Timaeus defends unity by providing arguments against the existence of multiple universes.

> Have we been correct in speaking of *one* universe, or would it have been more correct to say that there are many, in fact infinitely many universes. There is but one universe, if it is to have been crafted after its model. For that which contains all of the intelligible living things couldn't even be one of a pair, since that would require there to be yet another Living Thing, the one that contained those two, of which they then would be parts, and then it would be more correct to speak of our universe as made in the likeness, now not of those two, but of that other, the one that contains them. So, in order that this living thing should be like the complete Living Thing in respect of uniqueness, the Maker made neither two, nor yet an infinite number of worlds. On the contrary, our universe came to be as the one and only thing of its kind, is so now and will continue to be so in the future.[76]

Although Plato refers to "universe," potentially leaving open the possibility of inhabited life on say Mars, his stance on the nature of celestial objects as ethereal exemplars of behavior, rather than physical locations in the model of Earth, makes it clear that this is not the case in his cosmology.[77] Plato's opposition to the idea of other worlds (οὐρανός) is at its essence a unity claim. Given the prevalence of Plato's cosmological thought, channeled through Aristotle into the medieval world, the plurality vs. unity debate in Western culture largely takes the form of debates over other universes until advances in telescope technology provided evidence that planets share similarities to Earth.

Timaeus does not advance unity as a value neutral proposition, but instead argues that unity reflects some fundamental truths about existence. If the universe was not a complete unit, it would not represent a model of the "Maker," what Plato also refers to as the "Living Thing." Whether or not Plato refers to a literal God or a metaphor for the Good is not important to his rhetorical deployment of the cosmology. The Maker represents perfection; the universe exists in the maker's image by virtue of its unitary nature. If the world did not exist as unity, it could not represent a model of the Good, either literally or metaphorically, which means it could not serve as empirical proof of Plato's philosophy.

Plato returns to the connection between unity and the Good several times throughout the dialogue.[78] Timaeus argues those who ponder the question of infinite worlds are "'unfinished' in things he ought to be 'finished' in."[79] This statement implies that even raising the possibility of plurality suggests a personal failing. Plato ends the dialogue with Timaeus reiterating unity's connection with cherished universal values.

> And so now we may say that our account of the universe has reached its conclusion. This world of ours has received and teems with living perceptible god, image of the intelligible Living Thing, its grandness, goodness, beauty and perfection are unexcelled. Our one universe, indeed the only one of its kind, has come to be.[80]

The dialogue not only closes on an antipluralist note, but also specifically connects this idea to "grandness, goodness, beauty and perfection." The existence of other universes would reflect a lack within our own universe and open the possibility for radically different ways of living that could conflict with Plato's own recommendations. By embodying the Platonic values, our universe provides a blueprint for appropriate behavior.

Timaeus' defense of unity almost certainly references the cosmology of the atomists, who posited an infinite number of worlds. Democritus, one of the founders of atomism, argues that the universe is comprised of an infinite number of atoms (the tiniest component matter can be broken down into). These atoms move randomly and are distributed randomly. The combination of random movement and infinite atoms results in an infinite number of universes. Democritus argues that some of these infinite universes would have no inhabitants, while others would have life, some universes would have a sun, and others would not.[81] The atomist cosmology was the most recent and one of the most widely known at the time of the *Timaeus*.[82] The *Timaeus* attacks the most prominent plurality cosmology of the time and comes down explicitly in defense of unity. The reference to the atomist also signals a rejection of a materialist universe, as is discussed in the next section.

The centrality of the Earth is the next element of the cosmology associated with ethics. Plato places the Earth at the center of the cosmos and associates this centrality with the value of moderation.[83] The move to place Earth at the center of the universe mirrors almost all religious cosmologies (though not all of the Greek materialist cosmologies).[84] Physical centrality implies importance and value.[85] Protagoras argues that morality exists as an arbitrary human creation, but Earth's centrality implies an empirical proof of humanity's connection to a real and nonarbitrary perfection.[86]

The third and most drastic departure from the cosmologies of the time found in the *Timaeus* is that the dialogue describes the celestial objects moving in an orderly fashion.

> For there is a kinship between them, even though our revolutions are disturbed, whereas the universal orbits are undisturbed. So once we have come to know them and to share in the ability to make correct calculations according to

> nature, we should stabilize the straying revolutions within ourselves by imitating the completely unstraying revolutions of god.[87]

The dialogue suggests that anyone can look up to the sky and find a perfect model of how to live a good life in the "unstraying revolutions" of the celestial objects. The ideas of order and harmony have run throughout Plato's dialogues as justifications for absolute values over relativism. To model oneself on the perfect orbits of the celestial objects means conforming to Plato's views on how to live a just life.

Plato's *Timaeus* argues the structure of the cosmos provides justification for a universal ethic bound in order and harmony. In the *Gorgias*, Socrates accuses Callicles of acting out of step with the cosmos. In the *Timaeus* Plato offers an explanation of what precisely this means; unity, Earth's centrality, and orderly celestial objects are the physical embodiment of the world order, which Callicles opposes. Plato's cosmology suggests the very structure of the universe argues against Protagoras' dictum, "man is the measure of all things."

PLATO'S COSMOLOGY: MYTH OR SCIENCE?

The *Timaeus* has generated a long-standing debate in academia over the best hermeneutic to interpret the text, with some scholars reading the dialogue as a scientific argument and others reading it as a traditional myth.[88] The *Timaeus* contains an extended defense of empirical observation, which lends credence to the idea that it is meant as a work of science.

> Let us conclude, then, our discussion of the accompanying auxiliary causes that gave our eyes the power which they now possess. We must next speak of that supremely beneficial function for which the god gave them to us. As my account has it, our sight has indeed proved to be a source of supreme benefit to us, in that none of our present statements about the universe could ever have been made if we had never seen any stars, sun or heaven. As it is, however, our ability to see the periods of day-and-night, of months and of years, of equinoxes and solstices, has led to the invention of number, and has given us the idea of time and opened the path to inquiry into the nature of the universe. These pursuits have given us philosophy. A gift from the gods to the mortal race whose value neither has been nor ever will be surpassed. I'm quite prepared to declare this to be the supreme good our eyesight offers us. Why then should we exalt all the lesser good things, which a non-philosopher would "lament and bewail in vain"? Let us rather declare that the cause and purpose of this supreme good is this: the god invented sight and gave it to us so that we might observe the orbits of intelligence in the universe and apply them to the revolutions of our own understanding.[89]

Given that Timaeus offers a defense of empirical observation for the basis of his cosmology, its reasonable to infer that the cosmology has empirical grounding. Other aspects of the *Timaeus* support the myth interpretation, like the presence of the demiurge, a God-like entity. The intractability of the question of the *Timaeus*'s claim to truth has caused some scholars to take a hybrid approach, treating some elements as literal and others as metaphorical.[90] What is known is that many of Plato's contemporaries interpreted the *Timaeus* literally, including Aristotle.[91]

Whether or not Plato meant Timaeus's cosmology as a myth, he clearly uses the dialogue to comment on many of the prominent cosmologies of the time. Scholars have noted that Plato imitates the style of the presocratic cosmologists and goes so far as to incorporate elements of their cosmologies. His mathematical descriptions of the universe reference late Pythagorean thought.[92] The use of the four elements owes a debt to Empedocles.[93] Plato's articulation of the four elements as the imperceptibly small building blocks of matter sounds very similar to atomism.[94] Taylor goes so far as to suggest that Plato simply merged the cosmologies of Empedocles ("the most promising line in fifth-century science") and Pythagoras ("the one most directly connected with his own developments"), making the *Timaeus* a record of presocratic thought, rather than Plato's own ideas.[95] Even though Taylor's interpretation has gained little traction in the academy, it does highlight the degree to which Plato's dialogue reads as a scientific text quite unlike anything he had yet produced.

The fact that Plato's *Timaeus* adjusts many of the existing cosmologies in ways more in line with his philosophical worldview suggests something at work beyond a simple recounting of existing cosmologies. The random and chaotic atoms of Democritus become instead the structured and orderly forms. Plato's forms provide a much more stable building block for a teleological worldview than the contingent and unpredictable Democritic atoms. Plato also omits the counter-earth, an unseen earth-like planet, from his cosmology. Philolaus most fully theorized the counter-earth as a justification for the preponderance of lunar as opposed to solar eclipses, but Anaxagoras, Diogenes of Apollonia, and Anaximenes all suggest unseen astronomical bodies can help explain astronomical phenomena.[96] A cosmology in line with Plato's philosophy would have no use for hidden celestial objects, because they would not fulfill their teleological purpose of providing exemplars to humanity. The counter-earth in particular would be troubling for Plato's philosophy because it calls into question Earth's singularity and centrality. Plato thus omits mention of the counter-earth despite its ability to account observed astronomical events. Plato also claims the Earth is at the center of the universe, a departure from the popular Pythagorean belief that the Earth rotated like the celestial objects. All of these changes and their corresponding harmony with Platonic ethics suggest that the *Timaeus* does more than recount the cosmologies of others.

All of the changes from the prevailing science of the time have something important to say about the dialogue's relation to empirical observation. Since chaotic atoms, the counter-earth, and the rotating Earth theory did not have much in the way of direct empirical support, Plato has a certain license to remove them from his cosmology in order to make it more compatible with his philosophy. Plato's orderly orbits, however, are inconsistent with the orbits of the celestial objects as they were seen from Earth. The Pythagoreans discovered the existence of wandering celestial objects without a fixed orbit, but in the *Timaeus* Plato explicitly denies the possibility of disorderly celestial objects.[97] This represents the most problematic element of Plato's cos-

Figure 1.1. A depiction of a Plato's cosmology (with the addition of the Greek god Atlas).

mology, because the presocratics had discovered the wandering planets in exactly the empirical fashion that Timaeus recommends early in the dialogue. Nor can Plato claim ignorance of these wandering objects because he has Socrates turn away from astronomy in his earlier dialogues precisely because of the disorderly movement of the celestial objects. In the *Timaeus*, Plato explains that the celestial objects that seemed to move in a disorderly manner actually move in orderly circles if one views them as traveling around the Earth. This thesis has potential merit, and if Plato had offered an outline of how exactly such a scheme fit with the observed movements of the planets, it would have been a promising hypothesis. But Plato's *Timaeus* offers no explanation of how the celestial objects could travel in regular circular orbits when they have been widely observed to occasionally move in a retrograde fashion.[98] Instead Plato simply bends the stars to his will in an act of authorial fiat, ignoring the actual motions of the stars in favor of perfect geometrical figures, just as he recommends in the *Republic*.[99] He leaves the task of creating workable models of his theory to the students at his Academy, who improve on the explanation found in the *Timaeus* but still face difficulties in fitting the phenomena to the theory.[100]

The decision to make the celestial objects move in an orderly fashion represents a prioritization of ethics over observation. This should come as no surprise given Plato's dialogues' previous attacks on astronomy for emphasizing observation over philosophical concerns (Timaeus's defense of observation at the beginning of the dialogue notwithstanding). One should already approach Plato's 180-degree turn to embrace an empirically driven cosmology with great suspicion given his numerous critiques of astronomy in the past. The failure to account for the wandering objects in order to emphasize an orderly and teleological universe points to the central aspects of the dialogue as another example of what Russell would call intellectual dishonesty, rather than a genuine component of Plato's philosophy. In other words, Plato may have turned to cosmology because it could serve as a method to persuade others to accept his philosophy, regardless of the actual state of the universe.

Cornford is nonplussed about Plato's vague description of how the orbits work precisely for this reason. He argues that since the dialogue is primarily a myth these scientific questions are irrelevant.

> The *Timaeus* is a myth of creation, not a treatise on astronomy. The surprising thing is that Plato should have found room for so many details in his broad picture of a rational design in the cosmos, not that he should have simplified by omitting subtleties which would contribute nothing to his main purpose and which might be superseded at any time, as indeed they were soon afterwards.[101]

It is possible, however, to agree with Cornford that the *Timaeus* is a justification for Plato's philosophy, but disagree that the dialogue is not also a treatise on astronomy. Plato outlines the overlap between myth and science in *Sophist*:

> [The presocratic cosmologists] each appear to me to tell us a myth, as if we were children. One tells us that there are three beings, and that sometimes they're somehow at war with each other, while at other times they become friendly, marry, give birth, and bring up their offspring. Another one says that there are two beings, wet and dry or hot and cold. He marries them off and makes them set up house together. And our Eleatic tribe, starting from Xenophanes and even people before him, tell us their myth on the assumption that what they call "all things" are just one. Later on, some Ionian and Sicilian muses both had the idea that it was safer to weave the two views together. They say *that which is* is both many and one, and is bound by both hatred and friendship. According to the terser of these muses, in being taken apart they're brought together. The more relaxed muses, though, allow things to be free from that condition sometimes. They say that all that there is alternates, and that sometimes it's one and friendly under Aphrodite's influence, but at other times it's many and at war with itself because of some kind of strife. It's hard to say whether any of one these thinkers has told us the truth or not, and it wouldn't be appropriate for us to be critical of such renowned and venerable men.[102]

Plato, in effect, predates important concepts of the rhetoric of science by thousands of years. He demonstrates that science often has mythic qualities that audiences accept because of the status of the speakers as scientists—"renowned and venerable men." Rather than use this knowledge to deconstruct science, however, in the *Timaeus* Plato takes full advantage of the ability of science to mask mythic properties. This makes Cornford's dismissal of the idea that the *Timaeus* is not a "treatise on astronomy" problematic, but no more problematic than those that treat the dialogue solely as a treatise on astronomy, rather than a myth. These individuals deprioritize all of the ways that Plato connects the structure of the universe to ethics, despite their central place in the dialogue. This chapter reads the entire dialogue as both myth and science and suggests the two pieces cannot be unlinked, without irreparably altering the nature of the text.

The *Timaeus* calls out for a rhetorical understanding. The dialogue acts to persuade the audience to live good lives. The dialogue presents a myth of the cosmos, but Plato overcomes the problems with new myths that he identifies in *Republic* by adopting the scientific style of the presocratics. The dialogue claims the mantle of "renowned and venerable" physicist in the form of the astronomer Timaeus, lecturing to an audience as if they were "children." With myth wrapped in the veneer of science, the two become indistinguishable: scientific myth.

Whether or not Plato initially meant the *Timaeus* to be read scientifically is not particularly relevant to how the dialogue acts to persuade others. Furthermore, even if Plato initially meant the dialogue as something other than a literal depiction of the universe, Plato later represents the dialogue as scientifically accurate. The academics that suggest one should not read the *Timaeus* as a claim to scientific truth rely primarily on close textual analysis with little contextualization. Historical records and other artifacts suggest that many of Plato's contemporaries read the document as a literal description of the universe grounded in science, which suggests that Plato later represented the dialogue as true.[103] Plato asked his students at the academy to provide models of the universe consistent with what he describes in the dialogue, which shows a vested interest in preserving at the very least the appearance of scientific truth.[104] Plato references the cosmological elements from the *Timaeus* in the *Laws* as a justification for his philosophy, which suggests that by that time he presents his cosmology as true.[105]

A believable model of the universe that harmonized with Plato's philosophical views provided the persuasive benefits of mythology without its problematic aspects. Plato could defend the truth-value of his myth without endorsing the problematic Homeric myths, because the cosmological myth of the *Timaeus* is grounded in science rather than tradition. While traditional myths were losing currency in the era, the popularity of the works of the presocratic philosophers serves as testament to the ascendency of scientific cosmology. Plato's scientific myth overcomes the two main problems he identifies with using traditional myth as a means of persuasion.

Plato's earlier dialogues contain all of the pieces to justify his use of cosmology as argument. In the *Republic*, Plato says that children could be indoctrinated with new myths to create a societal change for the better, a strategy whose shortcomings have already been discussed.[106] In the *Sophist*, Plato says that in the face of the scientific cosmology of the presocratics, adults are like children listening to myths.[107] In effect, cosmology gives Plato the ability to indoctrinate adults with myths in precisely they way he advocated indoctrinating children with myths. Scientific cosmology allows Plato to deploy myth while overcoming problems of growing materialism and the body of contradictory religious myths. As early as the *Phaedo*, Plato suggests that cosmology can serve as an argument for his philosophy. He returns to this idea in the *Gorgias*. The primary impediment was always the specifics of the prevalent cosmologies. Plato's solution: provide a new cosmology.

In his later dialogues Plato deploys cosmology in the way he had previously deployed myth, for the purposes of persuading his audiences to act ethically. In the *Laws*, Plato frequently references cosmology, as if footnoting the *Timaeus,* as a justification/method to persuade people to live good lives.[108] The use of cosmology in one of Plato's final and most important dialogues demonstrates a mature rhetorical strategy, the culmination of a lifetime of different methods designed to direct the audience to the Good.

THE LAWS

In the *Republic*, Plato sought to use myths to control the population, so he banished the poets, who provided a set of myths counter to Plato's ideals. In the later dialogues Plato largely abandoned his critique of the poets. In the *Laws* he turns his ire to atheists, but from Plato's description it is clear that he means rival cosmologists.

> They maintain that fire and water and earth and air all owe their existence to nature and chance, and in no cases to art, and that it is by means of these entirely inanimate substances that the secondary physical bodies—the earth, sun, moon, and stars—have been produced. These substances moved at random, each impelled by virtue of its own inherent properties, which depended on various suitable amalgamations of hot and cold, dry and wet, soft and hard, and all other haphazard combinations that inevitably resulted when the opposites were mixed. This is the process to which all the heavens and everything that is in them owe their birth, and the consequent establishment of the four seasons led to the appearance of all plants and living creatures. The cause of all this, they say, was neither intelligent planning, nor a deity, nor art, but—as we've explained—nature and chance.[109]

Vlastos argues that this philosophical preamble in the *Laws X* singles out materialist cosmologies as the greatest threat to Plato's ideal society. The indictment was so sweeping that even Heraclitus and Diogenes of Apollonia, who believed in an intelligence that provides a foundation to the cosmos, were not excluded.[110] The punishments were also far graver than banishment for poets who told problematic myths in the *Republic*. The most lenient penalty Plato suggests is five years in solitary confinement for reeducation. Atheists (cosmologists) that did not repent their ideas or that proved particularly outspoken were to be put to death.[111] Plato's call to punish rival cosmologists makes him "the first political thinker to propose that errors of opinion be made crimes punishable by law."[112] Not only was Plato the first to do this, but also the law was "without parallel in any surviving code of ancient Greece."[113]

What can explain this sudden interest in incredibly harsh punishments for cosmologists? In the *Republic*, Plato justifies the banishment of the poets, because they spread myths with ethical messages in discordance with the Good. Many cosmologists, including those with materialist cosmologies, shared Plato's values, but their cosmologies were not conducive to persuading others to live good lives.[114] The shift in Plato's rivalry from poets to cosmologists demonstrates a shift in rhetorical strategies from one based in myth to one based in cosmology.

Further evidence that Plato shifted his concern from poets to cosmologists is his opposition to the atomists. Diogenes Laertus writes that Plato sought to collect as many of Democritus' works as he could to burn them, but was stopped by some Pythagoreans who convinced him that the books were already too widely dispersed.[115] Obviously, if this were true it would suggest that Plato had a deep disregard for Democritus. Chitwood suggests that the story of a philosopher wanting to burn another philosopher's books existed as a trope in philosophical history and thus can be written off as mere exaggeration.[116] This may be true in other cases, but there exists evidence that Plato would be more receptive to censorship than the other philosophers Chitwood cites. Remember, in the *Republic* Plato explicitly condones the destruction of books with messages that would mislead the people and advocates the death penalty for atheists in the *Laws*.[117] Democritus' works taught atheistic messages that Plato would have not wanted in his ideal Republic. Chitwood claims that Diogenes Laertius added the story about Plato wanting to burn Democritus' books in order to spice up history, yet Plato often references a desire to censor his ideological opponents. The other examples of historians saying philosophers wanted to burn each other's books lack evidentiary support in the writings of the philosophers in question.[118] Even if Plato never desired to burn all of Democritus' books, Diogenes Laertus' story still likely reflects some element of truth. While Diogenes invented stories about philosophers, these stories typically reflect a facet of reality. The fact that Laertius, of all the historical figures that Plato disagreed with (and there were many), would write that Plato wanted to burn Democritus' works suggests that for Plato atomism was even more dangerous than, for example, the sophists.

The fact that Plato's negative attention shifts from the poets to rival cosmologists indicates that cosmology represents the new conduit for Plato's ideas. When Plato's dialogues deployed traditional myth to persuade people to live a good life, they called to banish rival myths; when they begin to use cosmology to achieve the same effect, they called to banish rival cosmologies. This does not necessarily mean that Plato believed his cosmology was true. Remember in the *Republic* Plato advocates suppression of problematic myths even if they are true and the propagation of socially helpful myths even if they are false. It does suggest Plato eventually came to see that cosmology could serve the function previously held by myth—indoctrinating the masses to live good lives.

ANCIENT RECEPTION

Given the passage of time, it is difficult to measure the effectiveness of Plato's cosmological argument in ancient Greece, but several factors suggest that it proved largely successful. The first argument in favor of this thesis is Plato's shift to the cosmological argument in his later dialogues. Zeyl makes the argument that the cosmological account in the *Timaeus* was probably only a "prelude," to the story of the Atlantean's defeat by ancient Greece.[119] Plato never finished the Atlantean trilogy, a myth in the more traditional sense, and instead wrote the *Laws*. It very well may be that Plato abandoned the unfinished *Critias* (the second dialogue in the trilogy) and never began the third dialogue of the trilogy, because his cosmology proved more effective than his more traditional myth of Atlantis. This would suggest that Plato believed his cosmological arguments proved more effective than traditional myths.

The adoption of a Platonic cosmology by many important Athenians represents the second argument in favor of the cosmology's rhetorical success. Aristotle, one of the most influential Greek thinkers of the time, creates a cosmology based heavily on the *Timaeus* model.[120] The Stoics, a sect of Greek philosophical thought, also built a cosmological model heavily influenced by the *Timaeus*.[121] Like Plato's later dialogues, the Stoics argued that the structure of the universe showed individuals how to live good lives.[122] The Stoics used their cosmological model as an argument against the more relativist philosophy of the Epicureans. This was one of the earliest appropriations of Plato's scientific myth to battle relativism, but it was far from the last.

CONCLUSION

Platonic ideas are so integrated into the western world that they often escape notice. This makes it difficult to imagine a time when this was not the case. Alfred North Whitehead claimed that the European philosophical tradition consists of "a series of footnotes to Plato," but when Plato wrote, he was the outsider who was pushing teleology and anti-relativism to audiences interested in materialism and personal advancement.[123] Sophistic ideas dominated Greek life, and Plato had to make the difficult case that one should choose moral virtues over self-interest. Plato was the intellectual insurgent, not the paragon of the philosophical project.

Many of interpretations of Plato's dialogues treat them as a grand philosophical treatise, but this minimizes the importance of their historical context. Plato did not have a captive audience beyond the confines of his academy. The public walked out on his lecture on the Good. It makes sense to treat

Plato's dialogues as arguments attempting to persuade largely indifferent audiences of his core beliefs. The specific arguments Plato deploys shift throughout the dialogues, but this does not necessarily reflect a major change in his thought. Instead these shifts may reflect a commitment to using the best means of persuasion to advance things like anti-relativism. One of the instances where Plato shifts his arguments considerably is his treatment of cosmology.

In Plato's early and middle dialogues he treats cosmology with disdain. It is something that detracts from the search for the Good, Beautiful, and True. In Plato's later dialogues, however, he has reversed course. Now every person should look to the cosmos in order to harmonize their ethical behavior with the celestial order. This shift seems abrupt, but Plato's dialogues foreshadow it. In *Republic*, Plato writes how an effective system of myths could convince people to live good lives. Unfortunately, many obstacles exist to creating a new system of myths. In *Sophist*, Plato writes that audiences are captivated like children when they hear the cosmologies of the presocratics, cosmologies that Plato treats as little more than myths in new language. In other words, the cosmologies of the presocratics can achieve the effects of myth without the troubles of creating new myths. It is little surprise then that Plato takes the highly respected proto-scientific language of the presocratics and deploys it as another argument for his ethical beliefs.

Through the course of Plato's dialogues he uses myths as a rhetorical device of persuasion. In his later dialogues, he uses the structure of myth and applies it to the proto-scientific cosmologies of the presocratics. The *Timaeus* contains the ur-unity scientific myth, where Plato connects the singular world to "unity." He infuses the question of unity and plurality with ethical meaning. Of all his arguments for the Good, Beautiful, and True this is the one that resonated most powerfully. The Stoics took up his cosmology as a response to the relativism of the Epicureans, but this was just the beginning. Thousands of years later, individuals like Whewell would use the *Timaeus* as a template for their own versions of the unity myth.

NOTES

1. Gabriela Roxana Carone, *Plato's Cosmology and its Ethical Dimensions* (Cambridge: Cambridge University Press, 2005).

2. Marina McCoy claims that Plato often adopted the rhetorical strategies in his arguments with his philosophical opponents and the real difference between the two reflected their end goals. She examines Plato's use of rhetoric through six dialogues, but not the *Timaeus*. Marina McCoy, *Plato on the Rhetoric of Philosophers and Sophists* (Cambridge: Cambridge University Press, 2008).

3. Bertrand Russell, *The History of Western Philosophy: And Its Connection with Political and Social Circumstances from the Earliest Times to the Present Day* (Great Britain: George Allen and Unwin, 1947), 164.

4. Plato still makes use of traditional myths in later dialogues, like the myth of Atlantis in the *Timaeus*. I argue that he adds cosmological scientific myths to his argument repertoire, rather than replaces traditional myths with scientific ones.

5. Scholars have noted the connections between Plato's ethics and his cosmology, but Carone is one of the few scholars that makes an extended effort to situate Plato's cosmology within the context of the arguments he makes in his other dialogues. G. E. R. Lloyd, "Plato as a Natural Scientist," *Journal of Hellenic Studies* 88 (1968): 81-82, 90; T. K. Johansen, *Plato's Natural Philosophy: A Study of the Timaeus-Critias* (Cambridge: Cambridge University Press, 2004), 6; Francis MacDonald Cornford, *Plato's Cosmology: The Timaeus of Plato* (Indianapolis: Hackett, 1997), 6; R. Ferwerda, "Democritus and Plato," *Mnemosyne* 25 (1972): 360; Gabriela Roxana Carone, *Plato's Cosmology and its Ethical Dimensions* (Cambridge: Cambridge University Press, 2005), 3-7.

6. The precise order of the dialogues is not particularly important for my central argument with the exception that the *Timaeus* and *Laws* were among Plato's later dialogues. What matters is the grouping into early, middle, and late periods, and that is a fairly standard practice among modern Platonic scholars. Donald J. Zeyl, *Timaeus* (Indianapolis: Hackett, 2000), xvi-xx; Carone, *Plato's Cosmology*, 22-23; Gail Fine, "Owen's Progress: Logic, Science, and Dialectic: Collected Papers in Greek Philosophy," *Philosophical Review* 97, no. 3 (1988): 373-99. Even one of the most stout defenders of the belief that the *Timaeus* came earlier in the sequence of dialogues has recently recanted his arguments T. M. Robinson, "Mind-Body Dualism in Plato," in *Psyche and Soma: Physicians and Metaphysicians on the Mind-Body Problem from Antiquity to Enlightenment*, ed. John P. Wright and Paul Potter (Oxford: Clarendon, 2000), 47.

7. Paul Shorey, *The Unity of Plato's Thought* (Whitefish, MT: Kessinger Publishing, 2010); Harold Cherniss, "The Philosophical Economy of the Theory of Ideas," *The American Journal of Philology* 57, no. 4 (1936); Charles H. Kahn, *Plato and the Socratic Dialogue: The Philosophical Use of a Literary Form* (New York: Cambridge University Press, 1996).

8. Gregory Vlastos, *Socrates, Ironist and Moral Philosopher* (Ithaca, NY: Cornell University Press, 1991); Richard Kraut, *Socrates and the State* (Princeton: Princeton University Press, 1984); Terence Irwin, *Plato's Moral Theory: The Early and Middle Dialogues* (Oxford: Clarendon Press, 1977); J. E. Raven, *Plato's Thought in the Making* (Cambridge: Cambridge University Press, 1965).

9. Victorino Tejera, *Plato's Dialogues One by One: A Structural Interpretation* (Lanham: University Press of America, 1999); Christopher Gill, "The Platonic Dialogues," in *A Companion to Ancient Philosophy*, ed. Mary Louise Gill and Pierre Pellegrin (Malden, MA: Wiley-Blackwell, 2006).

10. Leo Strauss, *The City and Man* (Chicago: University of Chicago, 1978), 50-138; Allan Bloom, *The Republic Of Plato*, Second ed. (New York: Basic Books, 1991), 305-436.

11. J. J. Mulhern, "Treatises, Dialogues, and Interpretation," *The Monist* 53, no. 4 (1969); Debra Nails, "Mouthpiece Schmouthpiece," in *Who Speaks for Plato?: Studies in Platonic Anonymity*, ed. Gerald Alan Press (Lanham, MD: Rowman & Littlefield, 2000).

12. Charles L. Griswold, "Plato's Metaphilosophy: Why Plato Wrote Dialogues " in *Platonic Writings/Platonic Readings*, ed. Charles L. Griswold (State College, PA: Penn State University Press, 2001); J. Angelo Corlett, "Interpreting Plato's Dialogues," *The Classical Quarterly* 47, no. 2 (1997); Gerald A. Press, *Plato: A Guide for the Perplexed* (London: Continuum International, 2007).

13. Richard Mckeon, *The Basic Works of Aristotle* (New York: Random House, 1941), Pol.2.6.1265a10-13.

14. Lloyd P. Gerson, *From Plato to Platonism* (Ithaca, NY: Cornell University Press, 2013).

15. Ibid., 83-88.

16. Ibid., 81.

17. The arguments Gerson presents for ur-Platonism and against other methods of interpretation literally fill an entire book. Restating them all here would be redundant. There is however another reason to adopt Gerson's views, apart from their correctness. The view that Plato's dialogues reflect a unified philosophy has dominated reception of his works throughout much of history. It is possible that this interpretive frame is incorrect and that Plato did not mean for

his dialogues to make an argument against relativism, but it would not detract from the value of this study. If this chapter does not reveal Plato's personal ideas, it still acts to traces a line of argument. This line of argument has value as an object of study, because it is so often attributed to Plato. In other words, whether or not Plato the historical figure maintained the ideas under discussion here, the philosophy has treated Plato as the origin of these arguments. This is especially the case for many of the later case studies in this book, like Aquinas and Whewell.

18. Plato's anti-relativism is detailed more in the next section in relation to his opposition to sophistic ideas.

19. John Poulakos, *Sophistical Rhetoric in Classical Greece* (Columbia: University of South Carolina Press, 1995), 89-92; Ugo Zilioli, *Protagoras and the Challenge of Relativism: Plato's Subtlest Enemy* (Burlington: Ashgate, 2007), 3-5; Mi-Kyoung Lee, *Epistemology After Protagoras: Responses to Relativism in Plato, Aristotle, and Democritus* (Oxford: Oxford University Press, 2005), 4-5; Sarah Broadie, "The Sophists and Socrates," in *The Cambridge Companion to Greek and Roman Philosophy*, ed. D. N. Sedley (Cambridge: Cambridge University Press, 2003), 76; Cherniss, "The Philosophical Economy of the Theory of Ideas," 446.

20. Plato, "Gorgias."; Plato, *Republic*, 354-425.

21. DK80b1. Pindar makes a similar claim earlier than Protagoras when he argues, "custom is king." Pindar means that rather local customs determine appropriate behavior, rather than universal truths. Baldry, *Unity of Mankind in Greek Thought*, 19.

22. The sophists of history were a diverse group of teacher that individually held many different philosophical opinions. The sophists are painted in this chapter with the broad brush of relativism, because this is how Plato typically engages with the sophists in his dialogues. Many of the other case studies in this book reference back to the conflict between Plato and the sophists as one between relativism and absolute values. The book accepts generalizations of groups like utilitarians, social Darwinists, and postmodernists deployed by unity thinkers as to focus on the rhetorical deployment of the unity myth, rather than provide a detailed history of these schools of thought.

23. Protagoras' relativistic ideas are at the forefront of two dialogues, *Theaetetus* and *Protagoras*. Plato, "Theaetetus."; Plato, "Protagoras."

24. Zilioli, *Protagoras and the Challenge of Relativism*, 14.

25. John Scenters-Zapico, "The Case for the Sophists," *Rhetoric Review* 11 (1993): 355-57; G. B. Kerferd, *The Sophistic Movement* (Cambridge: Cambridge University Press, 1981), 15-17.

26. Edward Schiappa, *Protagoras and Logos: A Study in Greek Philosophy and Rhetoric* (Columbia: University of South Carolina Press, 1991), 119; Mark Lawrence McPhail, *Zen in the Art of Rhetoric: An Inquiry into Coherence* (Albany: State University of New York Press, 1996), 117; Susan Carole Funderburgh Jarratt, *Rereading the Sophists: Classical Rhetoric Refigured* (Carbondale: Southern Illinois University Press, 1991), 98; W. K. C. Guthrie, *The Sophists* (London,: Cambridge University Press, 1971), 3-13; George Grote, *History of Greece*, vol. 8 (New York: Peter Fenelon Collier, 1844); Kerferd, *The Sophistic Movement*, 173-78. Aristophanes' *Clouds* lampoons the influence of the sophists over the youth (although he included Socrates among their number). Aristophanes, "Wasps," in *Aristophanes: Clouds Wasps Birds*, ed. Peter Meineck, trans., Peter Meineck (Indianapolis: Hackett, 1998).

27. Several prominent early works challenged the idea that the presocratic cosmologists operated under a scientific methodology. Nietzsche famously cast doubt on the scientific nature of the presocratics suggesting instead that they often began with metaphysical principles rather than observation or reason Nietzsche Friedrich, *Philosophy in the Tragic Age of the Greeks*, trans. Marianne Cowan (Washington: Regnery, 1998). More recently, Cornford and Jaeger argue that the work of the presocratics resembled theology much more than science. Francis MacDonald Cornford, *From Religion to Philosophy: A Study in the Origins of Western Speculation* (Princeton: Princeton University Press, 1991); Friedrich, *Philosophy in the Tragic Age of the Greeks*; Francis MacDonald Cornford, *Principium Sapientiae* (Cambridge: Cambridge University Press, 1952); Werner Wilhelm Jaeger, *The Theology of the Early Greek Philosophers* (New York: Oxford University Press, 1947). Even at the time of its writing, however, Cornford's thesis stood out as a highly contrarian position. The view of the presocratics as primarily theological or not scientific has largely faded from the academic discourse, although the ques-

tion of how scientific the presocratics were remains open for debate. Even if the presocratics did not endorse all of the elements of modern science, however, they are widely credited as either founding the discipline or among its earliest practitioners. Gregory Vlastos, "Cornford's Principium Sapientiae," in *Studies in Presocratic Philosophy*, ed. David J. Furley and R. E. Allen (New York: Humanities, 1970), 50-55; Walter Burkert, "The Logic of Cosmogony," in *From Myth to Reason?: Studies in the Development of Greek Thought*, ed. Richard Buxton (Oxford: Oxford University Press, 1999), 104-06; Edward Hussey, *The Presocratics* (New York: Charles Scribner's Sons, 1972), 154, 55; Keimpe Algra, "The Beginnings of Cosmology," in *The Cambridge Companion to Early Greek Philosophy*, ed. A. A. Long (Cambridge: Cambridge University Press, 1999), 60-63; M. R. Wright, *Cosmology in Antiquity*, Sciences of antiquity (London: Routledge, 1995), 4, 5; Gregory Vlastos, "Ethics and Physics in Democritus," in *Studies in Presocratic Philosophy*, ed. R. E. Allen and David J. Furley (Atlantic Highlands, NJ: Humanities Press, 1975), 383; Gregory Vlastos, *Plato's Universe* (Seattle: University of Washington Press, 1975), 18-22; Nicholas Rescher, *Cosmos and Cognition: Studies in Greek Philosophy* (Piscataway, NJ: Transaction, 2005), 26; Karl Popper, "Back to the Presocratics," in *Studies in Presocratic Philosophy*, ed. David J. Furley and R. E. Allen (New York: Humanities, 1970); G. S. Kirk, "Popper on Science and the Presocratics," in *Studies in Presocratic Philosophy*, ed. David J. Furley and R. E. Allen (New York: Humanities Press, 1970).

28. Guthrie, *A History of Greek Philosophy*, 44-45.

29. Furley argues that Pythagoras' cosmology may have been the first to merge the new scientific thinking with religious and ethical beliefs. David J. Furley, *The Greek Cosmologists* (Cambridge: Cambridge University Press, 1987), 59. Furley does not provide a response to Walter Burkert's well-argued case against Pythagoras as a scientific thinker. Walter Burkert, *Lore and Science in Ancient Pythagoreanism*, trans. Edwin L. Minar Jr. (Cambridge: Harvard University Press, 1972), 208-17. More recently, Huffman goes back through the fragments from Pythagoras and his followers and demonstrates that many of the scientific views attributed to Pythagoras really belong to his students. Huffman argues, "Pythagoreanism has no room for examination of ideas and philosophical argument but is based on the authority of its founder. The later tradition reports that Pythagoreans felt no need to argue for positions and rested content with the primary assertion that 'he himself said it.'" Pythagoreas' student Empedocles later shows how Pythagorean ideas are compatible with scientific thinking and another follower Philolaus moves away from the metaphysical aspects of Pythagoreanism to develop its scientific and mathematical aspects. Carl A. Huffman, "The Pythagorian Tradition," in *The Cambridge Companion to Early Greek Philosophy*, ed. A. A. Long (Cambridge: Cambridge University Press, 1999), 75, 76-85. Even if Plato is not the first to make the connection between scientific cosmology and ethical values he still deserves study. Coming second to the Pythagoreans, Plato's cosmology would still be among the first uses of a scientific cosmology to justify ethics and more definitively the first of such cosmologies that persist to this day. Later in the chapter I present the divergent scholarly on to what extent Plato's cosmology represented a scientific endeavor.

30. Wright, *Cosmology in Antiquity*, 6.

31. Plato, "Phaedo," in *Plato Complete Works*, ed. John M. Cooper (Indianapolis: Hackett, 1997), 99b.

32. Ibid., 99c.

33. Ibid., 99b.

34. Plato, *Theaetetus*, 174a.

35. Plato, "Rival Lovers," 132b. Some academics have called into question Plato's authorship of *Rival Lovers*. This chapter operates under Julia Annas' belief that the evidence provided for this claim is relatively weak and it is reasonable to assume that Plato did in fact write the dialogue. Julia Annas, "Self-knowledge in Early Plato," in *Platonic Investigations*, ed. Dominic J. O'Meara (Washington, D. C.: Catholic University of America Press, 1985), 111-15. Even if Plato did not write *Rival Lovers* it makes little difference for my overall argument, given the pattern of Plato's hostility to astronomy in his other early dialogues.

36. Plato, *Republic*, VII 529a

37. Ibid., VII 529d.

38. Ibid., VII 530 b-c
39. Vlastos, "Ethics and Physics in Democritus," 382, 96-98; C. C. W. Taylor, "The Atomists," in *The Cambridge Companion to Early Greek Philosophy*, ed. A. A. Long (Cambridge: Cambridge University Press, 1999), 197; Wright, *Cosmology in Antiquity*, 175-77; Hussey, *The Presocratics*, 112-13; James Warren, *Presocratics* (Berkley: University of California Press, 2007), 11.
40. Plato presents the claim that Homeric mythology provides the foundation for Greek ethics in the Republic. *Republic*, 606e-607b.
41. Catalin Partenie, ed. *Plato's Myths* (Cambridge: Cambridge University Press, 2009), 1.
42. Ibid., xiii; Luc Brisson, *Plato the Myth Maker*, trans. Gerard Naddaf (Chicago: University of Chicago, 1998), 11, 14.
43. Partenie, ed. *Plato's Myths*, 6.
44. Brisson, *Plato the Myth Maker*, 77-85.
45. Plato, *Republic*, 377c.
46. Ibid.
47. Ibid., 378a.
48. Plato, *Timaeus*, 25d.
49. Plato, "Gorgias," 523a-27e.
50. Plato, "Ion," in *Plato Complete Works*, ed. John M. Cooper (Indianapolis: Hackett, 1997); Plato, *Republic*, 365a-66a, 414d-15d, 379a-82d, 91a-95b, 599c-606b.
51. The problematic nature of the Homeric gods goes way beyond their susceptibility to persuasion. The myths of the Olympians tell stories of endless bad behavior, from murder to adultery.
52. Plato, *Republic*, 365e.
53. Ibid., 414d-15d.
54. Thucydides, *History*, trans. P. J. Rhodes, vol. 2 (Warminster: Aris & Phillips, 1988), II.48-55.
55. Wright, *Cosmology in Antiquity*, 172; Mircea Eliade, *Myth and Reality* (New York: Harper & Row, 1963), 1, 2, 153-55; Hussey, *The Presocratics*, 110.
56. Ferwerda, "Democritus and Plato," 357, 58; Wright, *Cosmology in Antiquity*, 175; Gregory Vlastos "Cornford's Principum Sapientiae," in *Studies in Presocratic Philosophy*, ed. David J. Furley and R. E. Allen (New York: Humanities, 1970), 23-25. Of course, materialism's rise as a force within ancient Greece is complicated and many elements of mysticism remained long after the presocratics and in some cases peacefully coexisted within the ideas of some of the presocratics.
57. Brisson, *Plato the Myth Maker*, 83-84.
58. Ibid., viii; Plato, "Gorgias," 488b. Euthyphro, from the eponymously named dialogue, fits into the category of one who still believes in the gods, but views their immorality as a justification for impious actions. Plato, "Euthyphro."
59. Plato, *Republic*, 377bc.
60. Ibid., 385bc.
61. Gregory Vlastos, "Socrates," in *Socrates: Critical Assessment*, ed. W. Prior (London: Routledge, 1996), 139.
62. Plato, "Gorgias," 511b
63. Plato, *Republic*, 491b-95a.
64. Plato, "Letter VII." Some scholars have questioned the authorship of all of the thirteen letters Thrasyllus attributed to Plato in his edition of Plato's works. The seventh letter may have been written by someone other than Plato, which would call into question the veracity of the claim that Plato engaged in the politics of Syracuse. John M. Cooper suggests the seventh letter is "the least unlikely to have come from Plato's pen." John M. Cooper, "Letters," in *Plato: Complete Works*, ed. John M. Cooper (Indianapolis: Hackett, 1997), 1635.
65. Plato, "Meno," 92a-100b.
66. Aristoxenus, *Elements of Harmony*, II, 30-1 in John Niemeyer Findlay, *Plato: The Written and Unwritten Doctrines* (New York: Humanities Press, 1974), 413.
67. Plato, *Greater Hippias*, 285b-e.
68. Plato, "Gorgias," 505d.

69. The judgment of the souls is the myth I mentioned previously, when discussing Plato's use of myth in the *Gorgias*.

70. Plato, "Phaedo," in *Plato: Complete Works*, ed. John M. Cooper, trans., G.M.A. Grube (Indianaopolis: Hackett 1997).

71. Plato, "Gorgias," 507e-08a.

72. Carone argues that the etymology of the Greek words suggests that Plato references the physical universe in this passage. She also argues that the use of the same words in the *Timaeus* in the context of cosmology suggest that Plato is literally referencing the universe, rather than social order Carone, *Plato's Cosmology*, 7-8; John M. Cooper, ed. *Plato Complete Works* (Indianapolis: Hackett, 1997), 1237 footnote 13.

73. Plato shied away from this cosmological argument in his other early and middle dialogues, because astronomers' depictions of the universe did not support the idea of a world order, but instead of a world disorder.

74. Plato, "Gorgias," 501a-01d.

75. Charles Kaufman, "Argument as Enactment in the Gorgias," *Philosophy and Rhetoric* 12, no. 2 (1979).

76. Plato, "Timaeus," 31 a,b.

77. For Timaeus, the celestial objects represent being (fixity, perfection, order, ethical) and Earth and its inhabitants represent becoming (chaos, disorder, unethical). The celestial objects could not contain creatures like humans (perhaps spiritual/godlike creatures) without losing their connection to being. The celestial objects also are comprised of their own distinct form of unchanging matter (not the Earth, Fire, Water, and Air that comprise our planets and inhabitants). Second, the idea that the celestial objects had inhabitants was relatively rare at the time. Plutarch suggested that the moon may have inhabitants, but that was a minority opinion. Dick, *Plurality of Worlds*, 19-22. Whether Philolaus claimed the counter-earth was inhabited is also up for debate given the ambiguity of the fragment (DK A 17). The lack of belief in inhabited planets makes it unnecessary for Plato to explicitly denounce the possibility; he attacks plurality in its most popular form of the time.

78. Plato, "Timaeus," 33a, 55d, 92c.

79. Ibid., 55c-d.

80. Ibid., 92c.

81. Furley, *The Greek Cosmologists*, 136-40. Hippolytus, *Refutations* 1.13.2 = DK 6840, Aristotle, *Physics* 80a. 250b, 203b 24-7, d. Simplicius, *Commentary on Aristotle's Physics*, 80b. 1121.5, c. Philoponus, *Commentary Aristotle's Physics*, 405.23-7

82. Wright, *Cosmology in Antiquity*, 24.

83. Plato, "Timaeus," 62d.

84. Eliade, *Sacred and the Profane*, 36-47.

85. Earth's physical centrality overlaps with the question of plurality of worlds; if multiple worlds existed then some would lack a central place. These planets' potential inhabitants would lack a natural connection to moderation. The existence of other universes would do even more to undermine Earth's claim to physical centrality than the lack of physical centrality in the solar system, because multiple universes call into the question the idea that any one location is central.

86. Eliade, *Sacred and the Profane*, 36-47.

87. Plato, "Timaeus," 47c.

88. For scholars treating the *Timaeus* as myth: A. E. Taylor, *A Commentary on Plato's Timaeus* (New York: Garland, 1967), 66-70; Harold F. Cherniss, *Aristotle's Criticism of Plato and the Academy* (New York,: Russell & Russell, 1962), 392-457; Wright, *Cosmology in Antiquity*, 175-78. For those who treat it as science: Vlastos, *Plato's Universe*, 30-65; Richard Sorabji, *Time Creation and the Continuum* (Ithaca, N.Y.: Cornell University, 1983), 272-75; R. Hackforth, "Plato's Cosmogony," *Classical Quarterly* 27 (1959); Cornford, *Plato's Cosmology*, 29; Johansen, *Plato's Natural Philosophy: A Study of the Timaeus-Critias*, 52-53 ; Andrew Gregory, *Plato's Philosophy of Science* (London: Duckworth, 2000), 8-10.

89. Plato, "Timaeus," 47a-c.

90. Carone, *Plato's Cosmology*.

91. Cornford, *Plato's Cosmology*, viii-ix; Zeyl, *Timaeus*, xxi.

92. Charles H. Kahn, *Pythagoras and the Pythagoreans: A Brief History* (Indianapolis: Hackett, 2001), 56.

93. Edward Hussey, *The Presocratics* (New York: Scribner, 1995), 130; Wright, *Cosmology in Antiquity*, 53, 100.

94. Cornford, *Plato's Cosmology*, 210; Vlastos, *Plato's Universe*, 66, 67.

95. Taylor, *Commentary on Plato's Timaeus*, 18-19. Cornford writes a detailed response to this theory Cornford, *Plato's Cosmology*, v-x.

96. Anaxagoras A42 6, 9, A77 Diogenes of Apollonia A 12, Anaximenes A7 14 Aristotle, *On the Heavens*, trans. Stuart Leggatt (Warminster England: Aris & Phillips, 1995), 293. Burkert suggests that the Pythagorean cosmology, of which he views Philolaus as a practitioner, represents nothing more than mysticism. Walter Burket, *Lore and Science in ancient Pythagoreanism*, trans. E. L. Minar (Cambridge: Harvard University Press, 1972), 240, 67, 337-50. Many other scholars, however, point out that Philolaus and other later Pythagoreans approach cosmology in a scientific rather than religious manner. Guthrie, *A History of Greek Philosophy*, 326; Huffman, "Early Greek Philosophy," 83; Warren, *Presocratics*, 175-77.

97. Vlastos, *Plato's Universe*, 46 note 65.

98. Gregory points to several other empirical phenomena that Plato's cosmology could not account for and that would have been known to Plato. Andrew Gregory, "Eudoxus, Callippus and the Astronomy of the Timaeus," *Bulletin of the Institute of Classical Studies* 46, no. S78 (2003): 11-14.

99. Cornford argues that Plato's *Timaeus* does not make the claim that the celestial objects move in completely, orderly, circular, fashion. Instead he argues that all Plato meant was that the wandering objects have a set path, not that the set path was a perfect revolution. Cornford's position is a minority one and takes some license with the language of the text, which in my mind points rather clearly to the idea that the celestial objects moved in regular circular orbits Cornford, *Plato's Cosmology*, 92-93.

100. Simplicus, *On Aristotle's "On the Heavens 2.10-14"*, trans. Ian Mueller (Ithaca: Cornell University Press, 2005), 488.22-4. Wright argues that we should believe the historical accuracy of Simplicus' account. Wright, *Cosmology in Antiquity*, 146.

101. Cornford, *Plato's Cosmology*, 92-93.

102. Plato, "Sophist," in *Plato: Complete Works*, ed. John M. Cooper, trans., Nicholas P. White (Indianaopolis: Hackett, 1997), 242c-43a.

103. Cornford, *Plato's Cosmology*, viii-ix.

104. Simplicus, *On Aristotle's "On the Heavens 2.10-14"*, 488.22-4.

105. Plato, "Timaeus," 43d-44c, 47b-c; Plato, "Laws," 966e-68a.

106. Plato, *Republic*, 377c.

107. Plato, "Sophist," 242c-43a.

108. Plato, "Laws," 886a, 89b, 97a-98c, 967e.

109. Ibid., X 889 b,c

110. Vlastos, *Plato's Universe*, 23, 24.

111. Plato, "Laws," X 908e.

112. Glenn Raymond Morrow, *Plato's Cretan City: A Historical Interpretation of the Laws* (Princeton: Princeton University Press, 1993), 488.

113. Vlastos, *Plato's Universe*, 23.

114. Vlastos, "Ethics and Physics in Democritus," 382, 96-98; Taylor, "Early Greek Philosophy," 197; Wright, *Cosmology in Antiquity*, 175-77; Hussey, *The Presocratics*, 112-13; Warren, *Presocratics*, 11.

115. Diogenes Laertius, "The Lives of Eminent Philosophers," ed. Robert Drew Hicks (Cambridge, Mass.: Harvard University Press, 1925), Book IX.

116. Ava Chitwood, *Death by Philosophy: The Biographical Tradition in the Life and Death of the Archaic Philosophers Empedocles, Heraclitus, and Democritus* (Ann Arbor: University of Michigan Press, 2004), 100.

117. Morrow, *Plato's Cretan City*, 488.

118. For example, the ancient biographers claimed Aristotle wanted to burn Plato's dialogues and Protagoras wanted to burn Plato and Democritus' books. Chitwood, *Death by Philosophy*, 101.

119. Zeyl, *Timaeus*, xvii-xvii.
120. Johansen, *Plato's Natural Philosophy: A Study of the Timaeus-Critias*, 5.
121. Paul Scade, "Stoic Cosmological Limits and their Platonic Background," in *Aristotle and the Stoics Reading Plato*, ed. V. Harte, et al. (London: British Institute of Classical Studies, 2011), 161-70; Gretchen J. Reydams-Schils, *Demiurge and Providence: Stoic and Platonist Readings of Plato's Timaeus* (Brepols: Turnhout, 1999).
122. Blumenberg, *The Genesis of the Copernican World*, 20-24.
123. Whitehead, *Process and Reality: An Essay in Cosmology*, 39.

Chapter Two

The Dominance of the Unity Cosmology: From Plato to Galileo

"Do there exist many worlds, or is there but a single world? This is one of the most noble and exalted questions in the Study of Nature."[1] —St. Albertus Magnus, 1260 CE

Panhistoricism as a method necessitates the difficult undertaking of choosing particular case studies to express the continuity of ideas throughout history.[2] This chapter faces the formidable task of portraying the stability of an idea over the 2,000 years from Plato's Academy to the aftermath of the Copernican revolution. Under these circumstances the chapter could not possibly do justice to the complexity and differences of opinions within the period. Instead, it focuses primarily on two case studies. The first examines Christianity's melding of Platonic mythology onto the cosmology of Aristotle. The second examines the way the Copernican revolution reignited the plurality debate and thus challenged medieval values dependent on unity. Like the other case studies this book examines in-depth, both the reintroduction of the Platonic cosmology and the Copernican revolution represent areas of intellectual conflict, where powerful forces call unity into question. This helps reveal the often-unseen rhetorical work unity does in shaping society.

The Medieval period in between these events does not need as focused an examination precisely because the unity myth dominated academic and church culture for centuries. The success of the unity mythology in Medieval Europe makes it difficult to tease out precisely how it functioned. As E. M. W. Tillyard writes, "The conception of [cosmic] order is so taken for granted, so much part of the collective mind of the people, that it is hardly mentioned except in explicitly didactic writing."[3] The unity myth played a central role in anchoring this cosmic order despite often operating below the surface.

This chapter begins with an examination of Aristotle's reworking of Plato's cosmology. None of the other ancient cosmologies had nearly the impact on Medieval Europe that Aristotle's did. Aristotle, however, decoupled many of the rhetorical and mythic elements of the *Timaeus* from his discussion of unity. No longer did the celestial objects, geocentrism, or unity primarily reflect ethical and philosophical meaning for how to live. Instead, as Aristotle argued, these elements exist because physics demands the universe is structured as it is, effectively decoupling unity from the unity myth. This decoupling could have spelled the end of the unity myth, despite Aristotle's defense of unity as a fact.[4]

Before Aristotle's ideas could become a part of Medieval European thought, however, they had to conform to Christian dogma. Inconsistencies between Aristotle's cosmology and the Bible led to major conflicts between rival intellectual factions, the Averroists, philosophers who privileged Aristotelian thinking, and the Franciscans, a religious order who put the Bible ahead of science, in the twelfth and thirteenth centuries. Thomas Aquinas played a pivotal role in mediating between the warring groups by bringing Aristotelian science in line with Biblical teachings. He did this in part by rejoining the scientific defense of unity with the religious and philosophical implications for unity found in the *Timaeus*. Aquinas' reformulated cosmology merged Aristotelian logic with Platonic metaphysics and resulted in a system that proved remarkably stable for centuries.

This stable cosmology faded from the forefront, but continued to play an enormous role in shaping political, religious, and philosophical possibilities. Shakespeare, Dante, and many other authors could make casual allusions to the Aristotelian system, confident their audiences would get their references.[5] Monarchs justified their authority as grounded in Aristotelian structure of the cosmos.[6] In fact, individuals made use of the Aristotelian cosmology to maintain almost all the power structures at the time.[7] In effect, the Aristotelian cosmology became a scientific myth, with far-reaching implications.

The Copernican revolution challenged the Aristotelian cosmology and by extension the political, religious, and social structures that people had used it to justify. The close connection between Aristotle's cosmology and a variety of social structures explains why a discovery that overturned Aristotle's scientific system opened the door to sweeping changes across society. Most examinations of the effects of the cosmological upheaval focus predictably on the impact of heliocentrism. This chapter argues that heliocentrism represented such a threat to the established order, not only because of its physical relocation of the Earth and shattering of the celestial spheres, but also because it opened up the very real possibility of plurality. Unity always helped uphold the social order, but it became all the more important because it was the last element of the Aristotelian cosmology that Copernican science did not definitively disprove. Many individuals who wanted to preserve the existing social order grudgingly accepted heliocentrism, but refused to accept unity. The shift to heliocentrism and the threat it posed to unity reveals the rhetorical work that the unity cosmology had done in establishing the framework for debates about religion, philosophy and politics. The dominant medieval European culture grounded so many of its social constructs in its cosmology that the Copernican revolution and attendant shift to plurality ushered in a social as well as scientific revolution.

ARISTOTLE

In many respects Aristotle is the direct heir to the cosmology Plato outlines in the *Timaeus*. Aristole studied at Plato's academy for twenty years, until Plato's death, earning his place as a top student.[8] When Aristotle wrote his own cosmology, Plato's influence was obvious.[9] Like Plato, Aristotle argues for the ethereal nature of the celestial objects, geocentrism, orderly orbits, and unity.[10] Aristotle, however, constructs his cosmology in the context of his broader physics, which results in important changes from the *Timaeus*.

Although Aristotle defends unity, he does not justify it on the ground that unity represents perfection. This broke with the *Timaeus*, where Plato argues unity's connection to perfection provides proof against plurality, because a perfect god would create a universe as close as possible to perfection. Dick argues that Aristotle probably avoids this argument because it lacks scientific rigor.[11] Plato's argument depends on a perfect origin of the universe, the demiurge, and the intrinsic connection of unity to perfection. Presupposing all of these factors and then using them to prove unity represents precisely the type of unscientific argument that Aristotle generally eschewed.[12]

Aristotle defends unity on the basis of two primary arguments. The first relates to the way that Aristotle conceptualizes the nature of the basic units of matter. In the *Timaeus*, Plato identifies four components of matter: earth, water, fire, and air.[13] As mentioned in the previous chapter, these "forms" function similarly to the atoms of Democritus, in the sense that they provide the foundational building blocks of everything on the planet. Aristotle builds on Plato's theory and assigns the various forms different locations to which they naturally move. Fire, for example, moves away from our planet and the form of earth moves toward the center of our planet.[14] Aristotle argues that multiple worlds cannot exist, because the earth of the other world would be drawn to the center of our world, or vice versa.[15] The inevitable collapse of one world into another ensures the impossibility of the plurality of worlds.

The second argument Aristotle makes against plurality relates to the nature of action. He theorizes that since all action has a cause, the first action must have a cause. To explain the first cause, which puts into motion all other action, Aristotle posits a god-like prime mover. The prime mover can cause movement, but is not moved. He argues that each world would have to have its own prime mover. Aristotle treats the idea of multiple prime movers as philosophically impossible to justify, making it another argument against plurality.[16] The need for a unitary prime mover bears some resemblance to Plato's belief that unity connects with perfection. Aristotle's unitary prime mover, however, has less clear implications for humanity. The unity of the world in Plato's *Timaeus* provides a celestial model for ethical living; the prime mover of Aristotle's *de Caelo* exists in the background as a physical necessity and without a clear connection to values.

Aristotle's scientific approach to cosmology strips much of the mythic undercurrent from Plato's *Timaeus*. Aristotle removes the cosmological discussion from Plato's original context about living good lives. Many Neo-Platonists emphasize the mythic elements of the *Timaeus* and argue strenuously that one should not take it as a literal depiction of the universe.[17] Aristotle goes in the other direction and treats the mythic element as superfluous to the *Timaeus*' science.[18] Plato's rhetorical strategy depends on the intertwining of myth and science to accomplish what neither alone could. Both the Neo-Platonists and Aristotle threatened to undermine the strength of Plato's scientific myth by decoupling its science and myth. Christian thinkers, however, would rejoin the two by overlaying the mythic elements of the *Timaeus* onto Aristotle's new scientific justifications for unity.

It is difficult to overstate Aristotle's importance for the substantiation of unity myth in the Western world, but taken alone his writing was more or less agnostic on the unity myth. The persuasiveness of a scientific myth depends in large part on the acceptance of the science in which it is grounded. Aristotle's physics, which excluded the possibility of plurality, was the dominant scientific paradigm for centuries, creating the preconditions for the widespread adoption of the unity myth. As important as his work is for the propagation of the scientific idea of unity, however, Aristotle does not receive further analysis here because his work does not by itself advocate the unity myth. Ptolemy does not get extended treatment here for similar reasons. Ptolemy (90-168 CE) provided a geocentric unity model of the universe that was widely accepted as the most accurate account of the movements of the celestial bodies, until at least the time of Galileo. Like Aristotle, Ptolemy concerned himself primarily with the unity cosmology, rather than the unity myth. The evolving scientific debate behind unity and plurality positions is discussed extensively, elsewhere.[19] This chapter instead traces the solidification of the scientific myth of unity by examining how key players in the Western world harmonized unity with Christianity.

EARLY CHRISTIANITY AND COSMOLOGY

Philo (20 BCE-40CE), a Hellenized Jew who greatly influenced Christian thought, borrows heavily from the *Timaeus* in his writing.[20] He repeats Plato's argument for the connection of unity to perfection as manifest in God: "The creator is one, and he, making his creation to resemble himself in its singleness, employed all existing essence in the creation of the universe. For it would not have been complete if it had not been made and composed of all parts which were likewise whole and complete."[21] As if to make clear definitively from where he borrows the idea, Philo writes, "Some persons who believe that there are many worlds, and some who even fancy that they are boundless in extent, being themselves inexperienced and ignorant of the truth of those things of which it is desirable to have correct knowledge."[22] The language of the passage demonstrates that he references the *Timaeus* directly. Note the similarities between "ignorant of the truth of those things of which it is desirable to have correct knowledge" and Plato's remark about believers of plurality that they are "'unfinished' in things he ought to be 'finished' in."[23]

Despite Philo's embrace of the *Timaeus*, however, other Christians challenged the need to investigate the nature of the universe. Writing toward the end of the Roman Empire, St. Augustine gives a critique of the Greek cosmologists reminiscent of Plato's early dialogues.

> When, then, the question is asked what we are to believe in regard to religion, it is not necessary to probe into the nature of things, as was done by those whom the Greeks call *physici*, nor need we be in alarm lest the Christian should be ignorant of the force and number of the elements—the motion, and order, and eclipses of the heavenly bodies; the form of the heavens; the species and the natures of animals, plants, stones, fountains, rivers, mountains; about chronology and distances; the science of coming storms; and a thousand other things which those philosophers either have found out, or think they have found out.... It is enough for the Christian to believe that the only cause of all created things whether heavenly or earthly, whether visible or invisible, is the goodness of the Creator, the one true God; and that nothing exists by Himself that does not derive its existence from Him.[24]

Augustine's statement downplays the importance of science to Christians, whom he claims are better served by focusing on their faith in God. Augustine's criticism seems reminiscent of Plato's critiques of astronomy and cosmology in his early dialogues, in the sense that both men suggest that one is better spent learning about the Good through Christianity and philosophy respectively, than learning about the natural world. Augustine himself still read the ancient Greek texts, but many of his contemporaries and successors expand on his idea that a Christian need not take an empiricist approach to nature in order to advocate a hostile attitude to what we today would call scientific enterprises.[25]

Despite the disinterest in cosmology from some Christian voices, the *Timaeus* maintained a position of enormous respect between the fall of Rome and the twelfth century (the Early Middle Ages). The *Timaeus* was the only of Plato's dialogues continuously available in Latin throughout Western history and provided the foundation of much of Europe's ancient knowledge.[26] Few of Aristotle's cosmological works or any of the other Greek cosmologists made their way into the hands of Christian scholars, which meant the *Timaeus* had little competition.[27] Plato's cosmology became the leading cosmology at the time.

During Plato's lifetime there was debate about whether he meant readers to take the *Timaeus* literally, but during the Early Middle Ages the view of the *Timaeus* as myth faded. Two major translations of the *Timaeus* circulated, one by Calcidius and the other by Cicero. Both translations exclude much of the dialogue that did not examine the structure of the universe, like the myth of Atlantis. These translations presented the dialogue as a purely cosmological work, which in turn dampened the opinion that Plato meant his account of the universe as a myth.[28] Christian Platonists had another reason to take the cosmological account in the *Timaeus* as literal, rather than metaphorical: it harmonizes with the creation account in *Genesis*. Most non-Christian Platonists felt the need to read the *Timaeus* as a myth because they viewed the creation of the universe out of nonexistence as incompatible with science. The Christian Platonists, on the other hand, were looking for a scientific defense of *ex nihilo* creation and other elements of the Biblical account of *Genesis*.[29] Despite the reservations of Augustine's followers and others, the *Timaeus* continued to inform Christian cosmology.

The *Timaeus*' uncontested status ended at the close of the 12th century. In 1086, crusaders led by Alfonso VI captured the Muslim city of Toledo. Scholars from all over Christendom came to read the ancient Greek texts that the Islamic scholars preserved.[30] The introduction of previously absent Greek texts and commentaries on those texts by Arab, Jewish, and Persian scholars, like al-Kindi, al-Farabi, Ibn Gabirol, Ibn Sina (Avicenna), and Ibn Rushd (Averroes), in the end of the twelfth century and the beginning of the thirteenth century helped usher in a scientific revolution.[31] The new influx of texts included much of Aristotle's work, *de Caelo* being one of the most influential.[32] Aristotle's cosmology overtook Plato's *Timaeus* as the foundation for the scientific view of the universe held by the secular masters (the term for scientists at universities) and many of the Catholic orders.[33] Aristotle presents a more scientifically complete and integrated physics than that found in the *Timaeus*, which made the previous explanations of the natural world grounded in Platonic thinking seem haphazard.[34] Eventually, Aristotle's *de Caelo* would become the unquestioned cosmology until late in the seventeenth century, but first it faced a potentially devastating backlash from the Franciscan order and their allies.

Although the science of the Aristotelian cosmology better described the world, it did not always harmonize with Biblical teaching. Many secular masters, often referred to as Averroists after the famous Arab platonic scholar Averroes, argued that the physical laws of *de Caelo* and Aristotle's other works were absolute. This created two problems for widespread Christian adoption of Aristotelian ideas. First, there are specific elements of Aristotle's science that conflict with Christian teaching. Aristotle, for example, argues physics meant that the world existed eternally, which contradicts the description of creation found in *Genesis*.[35] The second, more general problem was that very idea of physical limitations place restrictions on God's power. Aristotle's claim that the natures of the elements prevent the possibility of a plurality of worlds, for example, denies the omnipotence central to the Christian God. If God wanted plurality, according to the Christian logic, God must be able to have it.

In an effort to harmonize Aristotle's science with Christian theology, Latin Averroists, like John of Jandun, Taddeo of Parma, and Angelo of Arrezo posited two truths: the philosophical truth that the universe has existed forever and the religious truth that God created the universe.[36] This angered the powerful Franciscan order of monks, led by general master Bonaventure. The Franciscans generally held Augustine's negative views on empiricism and viewed the double-truth belief as heresy.[37] The conflict between the two sides raged back and forth, resulting in major events like British Cardinal Robert of Courçon's ban on Aristotelian physics in 1215.[38] Thomas Aquinas, a rising star in the monastic orders, played a pivotal role in mediating the opposed religious factions and creating a hybrid system that merges Christianity with Aristotelian physics. This merger set the stage for the reemergence of Plato's mythic/scientific hybrid cosmology in the medieval period.

AQUINAS

Aquinas' background uniquely prepared him to offer a compromise between science and faith. From the ages of 5-14 he lived in a Benedictine environment where science, humanism, and religion coexisted harmoniously. He joined the Dominican order, notable for its commitment to teaching science, in 1243.[39] Albert Magnus, one of the first Europeans exposed to the full corpus of Aristotelian thought, became Thomas Aquinas' teacher. Aquinas rose through the ranks of the Church as unease began to grow among Franciscans and others about the compatibility of Aristotle's ideas with Christianity.

Aquinas made efforts to steer a middle road between the secular and religious forces. His works harmonize Aristotelian scientific thinking with Christianity and thus challenge the idea of a forced choice between the two.[40] Aquinas' middle path meant finding ways to merge the purportedly contradictory elements of Aristotle and Christianity. Sometimes this meant discarding specific elements of Aristotelian thought. In *On the Eternity of the World*, for example, Aquinas provides a justification for maintaining Aristotle's ideas broadly, but rejects his claim that the universe has existed eternally.[41]

The scientific necessity of unity provided a major area of conflict between the Averroists and the Franciscans, despite their general agreement that unity exists as a matter of fact. The Averroists argued that the nature of the four elements requires unity, because elements naturally gravitate to particular locations. They agreed with Aristotle that this makes unity impossible because if two planets existed they would converge on the same location. The Franciscans refused to accept any limitation on God's power and claimed that God could create a plurality of worlds and keep them from colliding if God chose to do so. At first glance it is strange that both sides fought so strongly over a question when they agreed on the fact of unity. The debate was not about the structure of the universe, however, but its nature. Did it follow immutable scientific laws or did it conform to the whim of God?

For Aquinas the controversy provided a major roadblock in his effort to synthesize the Bible with Aristotelian science. Aquinas had no problem jettisoning elements of Aristotle's thought to keep in line with Biblical teaching; for example, Aquinas' claim that the universe did not exist eternally. The nature of the elements was so much more central to Aristotle's scientific thought, however, which posed a dilemma. On the one hand, if Aquinas sided with the Franciscans in saying that unity was not a mandate of physics, it would render large parts of Aristotle's science incoherent. On the other hand, limits on the number of worlds God could create denied God's omnipotence, which was inconsistent with Aquinas' theology. The Averroists were unwilling to sacrifice the best scientific hypothesis at the time for the sake of comporting to the Bible. The Franciscans were unwilling to accept a limit on God's omnipotent power for the sake of a scientific theory.

Aquinas eventually draws on Plato's *Timaeus* to provide a solution acceptable to both sides. He defends unity on both the grounds of Aristotelian physics and because unity reflects the Good. God could create a plurality, but it would not be in keeping with the Good. This means that in reality God would never create a plurality, de facto preserving the immutable nature of Aristotle's scientific laws.

Aquinas outlines the argument in *de Potentia* (1265-1266), where he takes Plato to task for limiting God's power, much as the Franciscans accuse the Averroists of limiting God's power. Aquinas writes, "According to Plato

(*Tim.*) the best produces the best. Now the best can only be one. Since then God is best of all things, only one thing can be produced by him."[42] Aquinas puts forth Plato's position as a limit on God's power with the aim of refuting this proposition.[43] Aquinas' specific reference of the *Timaeus* indicates that the "one" produced by God references the world. Even though Aquinas supports unity as a fact of nature, he argues that God's power is not limited to creating unity. He says that arguments limiting God's ability to create plurality confuse the nature of cause and effect: "I answer that the impossibility of many things proceeding from one immediate and proper principle would seem to arise from the cause being determined to its effect, so that it would seem due and necessary that from such and such a cause such and such an effect should proceed."[44] In this sense, unity may relate to the Good but it does not place a limit on the power of God to create imperfect worlds.[45]

Aquinas, thus, dispatches what he suggests is Plato's argument for why a perfect god could only create unity. In actuality, Aquinas either misreads or misrepresents Plato's initial position. Plato never indicates in the *Timaeus* that God must create a unity, but rather, "being free of jealousy, he wanted everything to become as much like himself as was possible."[46] The unity of the universe models the demiurge's own Goodness, because of the desire of the demiurge, rather than a natural constraint. Aquinas misrepresents Plato's position in the *Timaeus* and ultimately goes on to reinvent Plato's original idea as a non-Aristotelian justification for unity.

Aquinas could not deny God the ability to create a plurality of worlds, because the Christian God has omnipotent power. The actual existence of a plurality of worlds, however, would create a serious problem for the Aristotelian physics at the heart of twelfth-century science. Aquinas needed an explanation for unity that did not rely on physics. In order to do this, Aquinas makes the case for the existence of unity on Platonic grounds.

> However, it should be realized that some prove the possibility of many worlds in other ways. In one way, as follows: The world was made by God; but the power of God, since it is infinite, is not limited to this world alone. Therefore it is not reasonable to say that He cannot make yet other worlds. To this it must be said that if God were to make other worlds, He would make them either like or unlike this world. If entirely alike, they would be in vain—and that conflicts with His wisdom. If unlike, none of them would comprehend in itself every nature of sensible body; consequently no one of them would be perfect, but one perfect world would result from all of them… it must be said that here it pertains to the goodness of the world to be one, because oneness possesses the aspect of goodness. For we see that through being divided some things lose their proper goodness.[47]

Aquinas indicates that God could create a plurality of worlds, but God did not, because unity best demonstrates Goodness.[48] Each world on its own would lack completeness and thus not reflect goodness. Even if the sum total of a plurality of worlds represents a perfect universe, each individual world would lack Goodness. Because Aquinas argues that if other worlds exist they would be isolated from one another, the Goodness of a universe of multiple worlds would be inaccessible to the individuals of each world. Aquinas jettisons what he suggests is Plato's mandatory connection of a perfect God and unity, but he maintains the belief that unity is the closest thing to perfection and thus the type of universe God did in fact create.

Aquinas reiterates unity's connection to perfection in his masterpiece the *Summa Theologica*, but goes beyond this claim to make the argument that unity provides the foundation for ethical behavior. In the work he explicitly cites the *Timaeus* as the genesis for the argument that unity of the world is modeled on God's unity.

> This reason proves that the world is one because all things must be arranged in one order, and to one end. Therefore from the unity of order in things Aristotle infers (*Metaph.* xii, text 52) the unity of God governing all; and Plato (*Tim.*), from the unity of the exemplar, proves the unity of the world, as the thing designed.[49]

The unity of God manifests itself in the unity of the world, even if God could theoretically create a universe not modeled on God's unity.[50]

Aquinas also raises the stakes for what unity meant for ethics, religion, and philosophy. He writes,

> The very order of things created by God shows the unity of the world. For this world is called one by the unity of order, whereby some things are ordered to others. But whatever things come from God, have relation of order to each other, and to God Himself, as shown above (11, 3; 21, 1). Hence it must be that all things should belong to one world. Therefore those only can assert that many worlds exist who do not acknowledge any ordaining wisdom, but rather believe in chance, as Democritus, who said that this world, besides an infinite number of other worlds, was made from a casual concourse of atoms.[51]

Aquinas much more explicitly channels Plato's rhetorical cosmology in this paragraph. In Aquinas' commentary on *de Caelo* he argues that a plurality of worlds would leave each individual world less than complete. Here he connects the unity of the world to God's inherent order, just as Plato's cosmos derives from its connection to being and Goodness. Plurality does not only mean that our world lacks Goodness, but denies the possibility of "any ordaining wisdom" and ushers in the materialism of Democritus. By pointing to divine sovereignty ("ordaining") and order ("wisdom"), Aquinas presents a universe completely at odds with the anarchic and chaotic cosmology of the atomists.

Raising the specter of atomist materialism would have been a devastating indictment at the time Aquinas wrote this passage. The only ideas of Democritus available at the time were filtered through his rival Aristotle's cosmological criticisms, which meant virtually none of his moral philosophy survived.[52] This positioned Democritus perfectly to fill the role of atheistic bogeyman in a time when religion reigned supreme.[53] In other words, the atomic system comes to stand in for moral relativism, despite Democritus personally rejecting moral relativism.[54]

The imperfection of plurality would implicate the universe as a model of Goodness. In the *Summa Theologica*, Aquinas writes, "the principal good in things themselves is the perfection of the universe."[55] Aquinas returns again to the idea of the universe as the model for ethical behavior in *Contra Gentiles* (1264):

> But things participate in the divine goodness to the extent that they are good, by way of likeness. Now, that which is the greatest good in caused things is the good of the order of the universe; for it is most perfect, as the Philosopher says. With this, divine Scripture is also in agreement, for it is said in Genesis (1:31): "God saw all the things He had made, and they were very good," while He simply said of the individual works, that "they were good."[56]

Aquinas claims both the Bible and philosophy (the "Philosopher" is likely a reference to Plato specifically) teach that the universe is the best example of absolute good. The idea that the universe represents the foundation for goodness in "things themselves," which includes humans, comes very close to the Platonic idea that humanity can look to the cosmos as a model for morality. Aquinas claims the universe is the best thing by virtue of its unified nature. Individual works (or humans) can get access to this goodness, by operating in conjunction with the *telos* of the universe. Plurality fragments the universe and destroys its unified purpose, undermining it as a model for how humans should live good lives.

By grounding unity in both Aristotelian physics and notions of the Good, Aquinas provides a third way between the Averroist and Franciscan factions. The Franciscans could take comfort in the idea that God created unity because it reflects the Good, rather than because the laws of physics demanded it. The Averroists could rest at ease knowing that because the Good demanded unity, the laws of physics effectively remain intact. This compromise provides the foundation of European cosmology for hundreds of years.

AQUINAS' LEGACY

In order to understand the impact of Aquinas' thinking on unity, one must first examine the success of his attempt to unify science and religion more generally. Etienne Gilson argues that clear-headed figures of the time realized Aristotle's system would inevitably win out because "the strictly physical and natural part of the doctrine presented a system so incomparably superior to the fragmentary and little coherent solutions proposed by the older schoolmen."[57] Little doubt exists that Aristotle's ideas represent the better science, but Gilson may engage in too Whiggish a history. The Franciscans proved a formidable foe of Averroist positions. In 1277, only three years after Aquinas' death, the Bishop of France condemned 219 Aristotelian ideas, many of which applied to Aquinas' thinking.[58] McInerny argues that if the "spirit of the condemnations had prevailed, the university would have turned its back on the enormous philosophical achievements of Aristotle and even more seriously called into question the assumption that faith and reason are complementary."[59] Absent interventions like Aquinas', one can imagine a world where conservative Christianity severely restricted medieval scientific advancement.

Ultimately, Aquinas' ideas did get the upper hand. In response to the 1277 ban, the order of preachers made the study of Aquinas mandatory. He posthumously received the rare title of Doctor of the Church, an honorific given for contribution to Catholic theology. In 1323 the Catholic Church canonized Aquinas and removed the condemnation of his work two years later.[60] Aquinas' successful harmonization of science and religion allowed Aristotle's ideas to flourish. Aristotle's works became the foundation of the university system and were studied second only to the Bible.[61] Just about everyone who learned to read and write also learned the basics of Aristotle's cosmology.[62] Aristotle's authority became near absolute, with the claim of *ipse dixit* (he [Aristotle] said it himself) serving as a phrase capable of ending arguments.[63] Amazingly, Aristotle's ideas gained the allegiance of humanists, scientists, and the Church leadership; even the Protestants maintained support for Aristotle's thinking after they split from the Catholic Church.[64]

Plato sought to motivate people to live good lives by putting forth a cosmology where Goodness is physically encoded into the universe. His cosmology merged science and myth into a rhetorical device, where unity demanded ethical behavior. The conflict between the Averroists and Franciscans split Plato's scientific myth into its component parts. The Averroists privileged the science embodied by Aristotelian physics, and the Franciscans privileged the myth now embodied by Biblical teachings. Aquinas rejoined myth and science into a revised scientific myth that both the Averroist and Franciscan factions could support. Scientific myth once more became a justification for ethical living.

With the ascension of Aquinas' compromise to religious dogma, Plato's vision of a society guided by scientific myth became a reality. The structure of knowledge production kept the Aristotelian paradigm relatively stable for centuries. Most scholars produced cosmological works in the style of commentaries. This technique for writing allowed individuals to disagree with Aristotle on particular points, but not create coherent theories to replace his. Nor did the authors of these commentaries cite one another. The large theory building Summas, like the kind Aquinas wrote, fell out of fashion and, when they did emerge, they mimicked the style of the commentaries rather than staying true to the format of the older Summas.[65] The fact that commentary writers rarely cited each other's work proved another major impediment to the development of a new cosmology. Had the commentaries cited one another, then the problems with Aristotelian physics would have been compiled in a cumulative way that better reflected the myriad of flaws within the system.[66] The writing style of the times resulted in lots of small debates about, for example, whether the celestial sphere exists as a fluid or series of solid masses or what caused an element to move from its natural place, but nothing that would challenge the core ideas of Aristotle.[67]

Unity was one of the core ideas of Aristotle's thought, which existed largely beyond reproach. Aquinas' return to Platonic justifications for unity, rather than reliance on Aristotle's physical laws, set the stage for this unquestioned acceptance. Debates did continue on how God could create a plurality of worlds, but these were motivated primarily by the decree of 1277 that made it an excommunicable offense to deny God's ability to create plurality.[68] The omnipotent nature of God represented one of the most important tenants of Catholic Christianity and the denial of the possibility of plurality challenged that power.[69] Many accounts that appeared positioned to defend the actuality of plurality ultimately refuse to do so, however.[70] A typical example, in the thirteenth century, Nicole Oresme wrote extensive rebuttals to Aristotle's scientific arguments against plurality, but concludes, "there has never been nor will there be more than one corporeal world."[71] Dick describes the strange phenomenon of arguing up to the point of plurality but refusing to endorse it as "a prime example of the uniquely medieval mixture of boldness and conservatism, of the appeal to reason still dominated by appeal to authority."[72] The appeal to Aristotle's authority certainly had much to do with the refusal to discard belief in unity.[73] Notwithstanding Aristotle's authority, Christian scholars had jettisoned elements of Aristotle's corpus that disagreed with Biblical teaching, like the belief that the universe existed eternally. This meant that unity must have, at the very least, been viewed as compatible with Christianity.

In fact, unity was not only viewed as compatible with Christianity, but many individuals also treated it as theologically crucial. It was likely the philosophical and theological implications of plurality prevented scholars from seriously advocating it. The 1277 edict forced a consideration of arguments for plurality, because to deny God the ability to create any universe he wished is heresy. The actuality of plurality, however, would call into question the Goodness of our universe and divine providence, for the reasons that Aquinas lays out so forcefully.

From Aquinas to the sixteenth century, only a handful of individuals argued that God *actually* had created a plurality. Cardinal Nicholas of Cusa (1401-1464), the most notable exception, not only asserted the existence of planets, but also their inhabitance.[74] Ingrid Rowland, however, argues Cusa wrote in a style that made his claims much more ambiguous than many scholars recognize. Following contemporary conventions, he used paradox to demonstrate his argument, which would make it difficult for hostile individuals to pin down his precise position.[75] Cusa also wrote at a time when the Catholic Church felt comparatively secure in its power, although it faced the Western schism, Waldensians in Italy, and Lollards in England, it did not face a challenge as great as later Protestantism. The Catholic Church's unchallenged status created conditions for more than intellectual freedom; this intellectual freedom would disappear during the Church's counterreformation response to the rise of Protestantism.[76]

Perhaps the most important and often overlooked reason that Cusa escaped censure is that he introduced a new conception of plurality. Previously plurality could best be understood as the existence of other universes, not the existence of inhabited planets within our own universe. Within the Aristotelian/Platonic cosmology, unity of the universe de facto disallowed the existence of alternate inhabited universes, because the celestial objects within our universe are by their very nature uninhabitable. Cusa posited that our universe was infinite and included other inhabited planets, but he only advanced the existence of one universe, a novel argument in the medieval period.[77] The opposition of the unity myth to the existence of other inhabited planets is clarified in the reaction to the cosmology of Bruno in the late 16th century.

Even if one views Cusa's work as a clear and open defense of the actual existence of plurality, he fits well under the adage, the exception that proves the rule. The fact that over the course of hundreds of years only one major figure bucked the unity consensus is remarkable, and it is a demonstration that plurality at the time existed as a minority discourse.

THE COPERNICAN SYSTEM

The sixteenth century began to show a few cracks in the unity argument's hegemony. John Major, a scholastic scholar, unambiguously argued for an infinite number of worlds (understood as universes) in the early 1500s.[78] In his texts he draws on the atomist Democritus, who he may have been introduced to by means of Lucretius's recently available *De rerum natura*. Overall, however, unity's dominance, both as a rejection of other worlds and other universes, held firm until the Copernican system's contestation of Aristotelian cosmology. The Copernican system and subsequent revolution in astronomy undermined two of the three elements of the Platonic cosmology: geocentrism and the ethereal, perfect nature of the planets. By suggesting that Earth orbited the sun, the Copernican system also opened up unity to question, because if the Earth orbited in the same manner as other planets, what was to say that other planets were not like Earth? Even as geocentrism and the orderly orbits of the planets fell by the wayside, many individuals rallied to defend unity as the last bastion of the stabilizing force of the Platonic scientific myth.

The Copernican model did not initially attract much attention from the Church, one of the groups with the greatest interest in preserving the Aristotelian cosmology. Copernicus' *On the Revolutions of the Heavenly Spheres* (Nuremberg 1543 and Basel 1566) laid out the case for his heliocentric universe, but he wrote the book in a style that made it inaccessible to all but a few astronomers.[79] The publisher, aware of the controversy the book would cause, added an introduction, "Mathematics is written for mathematicians; and among them, if I am not mistaken, my labours will be seen to contribute something to the ecclesiastical commonwealth."[80] The introduction, which reads as though Copernicus wrote it himself, suggests that the book is entirely theoretical and thus does not seriously challenge the prevailing cosmology.[81] Over the next few decades, scholars borrowed from Copernicus' diagrams and used his mathematical models, but did not endorse the heliocentric system. As the ideas of *On the Revolutions* slowly became more widely accepted, however, heliocentrism crept into scientific thinking.[82] Later Johannes Kepler and Galileo Galilei would complete the work of popularizing heliocentrism.

Some scholars (including many apologists for the Catholic Church) try to downplay the significance of the Copernican theory on the medieval cosmos and the religious and social structures it maintained.[83] These scholars claim that the Earth did not hold a privileged material position in the Platonic/Aristotelian cosmos. In these old systems the celestial objects represent perfection, whereas Earth represents the degraded "becoming." The move of the Earth from central to an orbit with the planets, thus, represents an increase in status if nothing else. This view is bolstered by the writings of Copernicus, Rheticus (Copernicus' lone student), and Galileo, who all view the heliocentric model as an elevation of Earth.

This reasoning holds certain logic, but one that views the move to a heliocentric as a self-contained measure, rather than as the beginning of a broad paradigm shift in astronomy. This paradigm shift did not elevate Earth to the realm of the celestial objects, but rather by making the celestial objects planets like Earth it degraded the whole universe.[84] The celestial objects were no longer ethereal models of perfection transmitting messages for how to live ethically. Instead they were merely planets, like Earth.

By making Earth a celestial object, it made it possible to study both the Earth and the other celestial objects in new ways. The heliocentric model led to William Gilbert's discovery that the Earth is not a perfect sphere and Giovanni Domenico Cassini's claim that Jupiter also was not spherical.[85] It also allowed Kepler to hypothesize elliptical orbits for the planets, rather than the perfect circular orbits that dated back to Plato's time. The Platonic myth encoded Goodness into the universe by making its structures fundamentally different from Earth. Heliocentrism undermined two aspects of the Platonic cosmology that had served as the foundation for the religious and social order: geocentrism and the ethereal celestial objects. The Copernican revolution also posed a serious threat to the idea of unity. If Earth did not have a central, special location, but instead orbited the sun just like Mars, Jupiter, Venus, etc., then one could reasonably infer that those bodies had similarities to Earth and perhaps even inhabitants.

Giordano Bruno, influenced both by Latin Averroesism and Platonism, helped demonstrate heliocentrism's connection to plurality and its attendant social consequences through his bold cosmological writings. Bruno, a mathematician, astronomer, and memory theorist, traveled across Europe and preached the existence of an infinite number of worlds. In his writings he flipped Plato's argument and said that a plurality of worlds, not unity, reflects the goodness of God:

> This argument is the more cogent since, if it is reasonable to postulate a finite goodness, a bounded perfection, all the more reasonable is the conception of an infinite goodness. For whereas finite goodness appeareth to us reasonable and convenient, the infinite is an imperative necessity.... We are then at one concerning the incorporeal infinite; but what preventeth the similar acceptability of the good, corporeal and infinite being? And why should not that infinite which is implicit in the utterly simple and individual Prime Origin rather become explicit in his own infinite and boundless image able to contain innumerable worlds, than become explicit within such narrow bounds? So that it appeareth indeed shameful to refuse to credit that this world which seemeth to us so vast may not in the divine regard appear a mere point, even a nullity?[86]

This infinite universe with its infinite worlds still technically met the original definition of Platonic unity found in the *Timaeus*. Bruno kept consistent with Cusa and the *Timaeus* by positing the existence of one universe. Bruno's unity, however, owed more to the atomist model that the *Timaeus* opposed than the Platonic model, as evidenced by Bruno's extensive citations of the atomist Lucretius throughout his cosmological works.[87] The infinity of worlds that the *Timaeus*, and to a lesser extent Aristotle's work, treat as universes become instead particular planets. Despite agreeing on the existence of a sole universe, "Bruno's cosmological works can only be understood as a reaction to the Aristotelian opposition to the plurality of worlds."[88]

Thus Bruno acts to conflate the ideas, such that unity as the opposition to plurality of universes becomes equivalent to opposition to a plurality of inhabited planets.

Bruno's created his own scientific myth around his version of plurality and the Copernican system that provided its foundation. He argued that his cosmological model set the stage for a revolutionary new order including the conciliation between Protestants and Catholics and the emergence of a Hermetic order.[89] These implications were anathema to the dominant power structures at the time, even if plurality itself was not.

While Bruno's other beliefs may have been sufficient to result in problems with the religious authorities, his vocal advocacy of his cosmology likely contributed to his demise; in 1600, the Papal inquisition sentenced Bruno to death by fire for heresy. Three years earlier, while still imprisoned by the Inquisition, he was told to abandon his belief in a plurality of worlds.[90] Why did Bruno's belief in plurality receive attention while Cusa's previous defense of plurality did not? Unlike Cusa, Bruno did not mask his work in paradox. For example, Bruno wrote, "Thus is the excellence of God magnified and the greatness of his kingdom made manifest; He is glorified not in one, but in countless suns; not in a single earth, a single world, but in a thousand thousand, I say in an infinity of worlds."[91] Bruno also based his cosmology on the Copernican system, a scientific framework, which lent legitimacy to his ideas, unlike Cusa's earlier work on plurality, which exists only as theological musings ungrounded in any scientific basis.[92] Most importantly, Bruno wrote at a time when the Catholic Church felt a serious threat from Protestantism, and thus the Inquisition was especially concerned about heretical doctrines.

Whether or not Bruno died for his belief in plurality, the Church opposed his cosmology. As Thomas Kuhn writes, "He had given [Copernicanism] a significance not to be found in *De Revolutionibus*."[93] The Copernican system now presented itself not only as a potential rival cosmology but also as a rival scientific myth. Bruno's infinite worlds not only called into question unity, he had put them in opposition to some of the prevailing social beliefs. The popular reaction to the news of Bruno's death also suggests that many people believed that he died for advocating plurality. Bruno's execution made Descartes hesitant to write about the plurality of worlds and made Bruno a martyr for plurality, famous among intellectuals like Edmund Spenser, Francis Godwin, and the School of Night, a group of loosely connected scientists and philosophers with atheist leanings.[94]

More importantly for our purposes, Bruno marks a shift in the plurality and unity discourse, where plurality and unity no longer primarily mean the existence of other universes, but instead the existence or absence of other planets within our universe. While the unity of the *Timaeus* had de facto ruled out other planets and the science of Aristotle had ruled out other planets explicitly, the religious and scientific debate up until this point had been on the existence of other universes. The Bruno affair helped redefine the unity and plurality debate as about the existence of other inhabited planets. It also served as a strong signal against the acceptability of a unity universe containing a plurality of inhabited planets.

GALILEO AND HIS TELESCOPE

Copernicus' theory gained ground because it was slightly more effective in explaining celestial movements than the heliocentric models were. Galileo's telescope provided much stronger proof for the Copernican system by furnishing difficult to refute evidence about the nature of the solar system. Through his telescope Galileo observed that the moon had topological features much like Earth and that Jupiter had a series of moons. The celestial bodies no longer appeared as perfectly formed objects, but more similar to Earth. Galileo's observations translated to the public much easier than Copernicus' mathematical charts; this popularity raised the notice of the Inquisition.

In 1616, the Catholic Church briefly banned Copernicus' book, because of the attention Galileo brought to the question of heliocentrism.[95] In 1632, Galileo published *Dialogue Concerning the Two Chief World Systems*, in which he depicts the geocentric view as foolish, resulting in his official condemnation by the Catholic Church. Kuhn remarks on the strange fact that the Church's reaction to heliocentrism occurred, not in the initial stages of the theory, but when the evidence had become all but overwhelming.[96] The specific charges brought against Galileo suggest this new concern with heliocentrism came in part from its increasing connection to the question of plurality.

Of the eleven charges brought against Galileo two dealt with the plurality of worlds. The eighth charge accuses Galileo of claiming the moon has Earth like qualities like water and mountains, which denies the ethereal nature of celestial bodies.[97] The belief that the moon has planet-like qualities sets the stage for the ninth charge, which indicates that Galileo's model opens up the possibility of "many worlds and earths and seas…and that there are human beings living there."[98] The Church argues that plurality contradicts scripture and raises questions about whether Jesus needed to be crucified multiple times in order to save each world's population.[99] The fact that two charges on plurality made their way into the trial suggests that plurality was one of the ultimate concerns of the Church.

Of course the issue of plurality was far from the only justification for the censure of Galileo and opposition to heliocentrism. There were after all nine unrelated charges. Various behind-the-scenes forces also acted to influence the Church's behavior. Santillana argues that many Aristotelians felt that heliocentrism would "subvert the vast and documented discourse of the schools, which allows us to account in orderly manner for nature and life and the soul itself" and helped to bring about the Church's hostile response.[100] Speller and Grant claim that Galileo offended the Pope by ignoring his advice for how to frame his book, resulting in something of a personal vendetta.[101] The literature on Galileo's treatment by the Catholic Church is vast. This chapter only endeavors to show that heliocentrism's plurality implications were both an official and unofficial concern of the Church and other important figures.

Galileo appeared to recognize the danger of supporting plurality, even as he defended a heretical heliocentric cosmology. In his early works he avoids any discussion of plurality. He did this despite the fact that his telescopic observations of Earth like qualities of the supposed "ethereal objects" had obvious implications for the question of plurality. Many of those who read his works made this connection and when they pressed Galileo to discuss it he gave noncommittal answers.[102] Guthke suggests the possibility that Galileo's evasions on the subject may reveal his true, pro-plurality position.[103] Whether or not Galileo was a pluralist, his refusal to discuss the idea despite its strong connection with heliocentrism tells us something important. Even the man bold enough not only to attack geocentrism but also depict its powerful defenders as "dull-witted, unimaginative... defenders of the indefensible," refused to get himself entangled in the plurality debate.[104]

Meanwhile, as Copernicanism gained ground, religious figures began to refer to plurality as the "new heresy."[105] Pierre de Cazre, rector of the College of Dijon, describes how the danger of heliocentrism is not the location of the Earth, but that the theory opens up the possibility of plurality.

> If the earth is doubtless one of the planets and also has inhabitants, then it is well to believe that inhabitants exist on the other planets and are not lacking in the fixed stars, that they are even of a superior nature and in proportion as the other stars surpass the earth in size and perfection. This will raise doubts about Genesis which says the earth was made before the stars and they were created on the fourth day to illuminate the earth and measure the seasons and years. Then in turn the entire economy of the Word incarnate and of scriptural truth will be rendered suspect.[106]

Pierre de Cazre writes that plurality would contradict the scripture, an argument also made against heliocentrism, but he also suggests that it would threaten humanity's special place in the cosmos. He worries that other planets and their inhabitants could be more perfect. Gabriel Naudé, physician to Louis XIII, declared plurality the greatest heresy:

> I am afraid that those old theological heresies are nothing compared to the new ones the astronomers are seeking to introduce with their worlds, or rather lunar and celestial earths. For the consequences of these heresies will be far more dangerous than those of the earlier ones and will bring about many stranger upheavals.[107]

To appreciate the boldness of Naudé's statement one must remember the ferocity with which the Church ferreted out and eliminated the "old" heretical views. Although he does not say why plurality poses a greater danger, one can imagine reasons related to the scientific nature of the controversy.

Plurality was potentially destabilizing for three primary reasons. First, it threw a number of Church doctrines into question. Did Jesus's death act to redeem all life in the universe, or would God need to become mortal to save the souls of each planet's inhabitants? Why did the Bible never mention life on these other planets? Eventually theologians found satisfactory answers to these questions, but initially they posed a threat to the Christian worldview.

Second, plurality was more deeply threatening to the idea of humanity's central place in the universe than geocentrism. Humanity held a sense of its own importance to the grand scheme of the universe. They existed at the center of the great chain of being at the level where all the action occurred; God and Satan did battle for the souls of humanity. In contrast, the central location in the universe is not a place of distinction in either a Christian or Aristotelian framework. In a Christian framework the center of the universe is Hell and the Aristotelian framework contrasts the imperfect materiality of Earth with the perfection of the celestial objects. So one could interpret heliocentrism as a promotion for humanity, whereas the possibility of multiple inhabited worlds deeply threatened humanity's sense of importance in the great chain of being.[108]

Third, Aquinas harmonized science and religion, just as Plato had done previously with philosophy and science. This harmonization placed a large emphasis on the role of unity as a demonstration of the goodness and perfection of God. Plurality suggested that each world lacked harmony and thus the universe could not serve as a message of God's perfection. These reasons prevented a broad acceptance of plurality even as heliocentrism became increasingly popular.

After Isaac Newton developed his new physics in *Philosophiæ Naturalis Principia Mathematica* (1687), the case for geocentrism became all but impossible to take seriously.[109] Combined with telescopic observations of the planets, two of Plato's three cosmic features collapsed. People no longer viewed Earth as the center of the universe or the celestial objects as perfect orbs moving in harmony. Unity, however, did not fade so easily, probably because the question of unity vs. plurality lacked much empirical evidence either way. Many individuals accepted heliocentrism but opposed plurality.

Figure 2.1. A wood carving symbolically depicting the transition from the Aristotelian to the Copernican cosmology. The caption translates as "A medieval missionary tells that he has found the point where heaven and Earth meet..."

Even though he accepted Copernicanism, Reverend Robert Jenkin provides a rather standard version of Plato's unity argument.

> Anaxagoras said, he was sent into the World to contemplate the heavenly Bodies: And the contemplation of them has contributed more than any one thing in Nature besides, to preserve a sense of Religion among Heathens... And these Parts of Nature which are so beneficial, and afford such satisfaction to Mankind in general, may fully answer all the Ends needful in their Creation, tho' they should not be designed for such farther uses, as some modern Philosophers have contrived for them.[110]

Jenkin argues that study of the universe instills proper values, even in those that had not been exposed to Christianity. For Jenkin, this represents the true purpose of the planets, rather than as a home to alien life. The use of the structure of the universe as a model for human goodness harkens back to Plato's *Timaeus*. Thomas Baker, a fellow of St. John's College in Cambridge, defends unity in his book *Reflections Upon Learning* (1699). He claims humanity serves as a worthy purpose of the universe.

> These World-Mongers are always objecting the improbability of GOD's framing so many vast and Glorious Bodies, only for the sake of the Earth, so inconsiderable a portion of the Whole... There is more Beauty and Contrivance in the Structure of a Human Body, than there is in the Glorious Body of the Sun; and more Perfection in one Rational Immaterial Soul, than in the whole Mass of Matter, be it never so bulky. There cannot then be any Absurdity in saying, That all Things were created for the sake of this inferior World, and the Inhabitants thereof.[111]

Contrary to pluralists that argue an empty universe would be a waste of space, Baker claims a single soul has greater value that the whole universe and thus it is not wrong to say that such a huge space was created for the inhabitants of a single planet.

Individuals like Baker and Jenkins try to preserve the socially stabilizing discourse of the Platonic/Aristotilean/Augustinian cosmology by salvaging the only aspect of it they could: unity. Plato provides a scientific justification of the Good that rested on ethereal celestial objects, orderly orbits, geocentrism, and unity, but Galileo and Newton undermine the case for ethereal celestial objects, orderly orbits, and geocentrism. Unity was all that was left to provide order and meaning to life.

THE SHIFT TO PLURALISM

The widespread acceptance of heliocentrism radically altered cosmological thinking in a way that opened the door to plurality.[112] Plurality initially began to gain a foothold among intellectuals like the Northumberland circle of natural philosophers in England.[113] But the popularization of plurality came from Bernard le Bovier de Fontenelle in his book *Conversations on the Plurality of Worlds* (1686). Fontenelle uses a fictional dialogue, the style of Plato and Galileo, between a scholar and a marquise that in casual, accessible prose explores the possibility of plurality.

Conversations on the Plurality of Worlds went on to sell out of numerous editions, and by 1800 it had been translated into Danish, Dutch, German, Greek, Italian, Polish, Russian, Spanish, and Swedish.[114] Fontenelle's work on alien life so fascinated the public that Samuel Pierpont Langley, the inventor of the bolometer, wrote over 200 years later that *Conversations on the Plurality of Worlds* is the first example of a popular science text.[115] It achieved this success despite the Vatican placing *Conversations* on its index of dangerous books, a status the book maintained until 1825, after which it was removed, only to be reinstated again in 1900.[116]

> *Conversations* begins by identifying the rhetorical stakes of the unity vs. plurality debate. The same desire which makes a courtier want to have the most honorable place in a ceremony makes a philosopher want to place himself in the center of a world system, if he can. He's sure that everything was made for him, and unconsciously accepts that principle which flatters him, and his heart will bend a matter of pure speculation to his interest.[117]

Fontenelle specifically highlights the strong perceived connection between human specialness and centrality and unity and later has the Marquise muse on the potential sense of pointlessness that may come with plurality:

> I'm beginning to see the Earth so frighteningly small that I believe hereafter I'll be impressed by another thing. Assuredly, if people have such a love of acquisition, if they make up plan after plan, if they go to so much trouble, it's because they don't know about vortices. I can claim that my new enlightenment justifies my laziness, and when anyone reproaches me for my indolence I'll answer: 'Ah, if you knew what the fixed stars are!'[118]

Fontenelle elsewhere, however, relishes in the overturning of the Aristotelian system and its replacement with a gigantic inhabited cosmos:

> When the sky was only this blue vault, with the stars nailed to it, the universe seemed small and narrow to me; I felt oppressed by it. Now that they've given infinitely greater breadth and depth to the vault by dividing it into thousands and thousands of vortices, it seems to me that I breathe more freely... The inhabitants of a planet in one of these infinite vortices see on all sides the lighted centers of the vortices surrounding them.[119]

Fontenelle not only made plurality a common concept, but also understood that the change in cosmology had implications for how people view their place in the universe. He relishes in the magnitude of space even as the plurality of worlds acts to decenter humanity. Fontenelle writes the first scientific myth associated with plurality to get a widespread hearing throughout the Western world since the time of the ancient Greeks.

While Fontenelle provides a publicly accessible introduction to plurality, Christiaan Huygens gave plurality scientific credibility.[120] Before he wrote about plurality, Huygens, a famous Dutch astronomer and mathematician, already had achieved fame for his scientific accomplishments like discovering the moon Titan. In his book *Cosmotheoros* (1698), he outlines arguments for plurality, much like Fontenelle, but with "a more scientifically developed presentation."[121] *Cosmotheoros* and *Conversations on the Plurality of Worlds* together spearheaded a tidal wave of pluralist thinking.[122]

Interestingly, belief in pluralism of worlds coincided with a pluralism of cultural, political, and religious power centers. Nation-states rose to replace the sprawling empires of the past. Vernacular language gained ground against the old Latin-based unification of learning. The Reformation had created a plurality of denominations to challenge Catholic ascendancy. The Ottoman Empire threatened Austria and Christendom more generally. Two whole continents of previously unheard of people had been discovered. The world moved ever more toward plurality of social structures.

By the early 1800s most scientists who studied the question of alien life believed the issue had been settled decisively in favor of plurality.[123] The idea that so much space could exist without inhabitants struck them as incredibly wasteful and inconsistent with the way that one finds life in even the remotest parts of the planet Earth. Most scientists advocated a principle of plentitude that said life would be found wherever it could possibly emerge.[124] The *Eclectic Review*, a highbrow British periodical, wrote, "the probability that the other orbs of our system are inhabited worlds must appear so great, that a direct revelation from heaven disclaiming the fact, would make but little difference in our assurance of it."[125] While the *Eclectic Review*'s defense of pluralism went beyond most others in terms of its bombast, most writers shared the general sentiment.[126] The offbeat theory of Fontenelle began to look more and more like scientific orthodoxy.

Not only did scientists accept pluralism, evidence suggests that the idea escaped from academic enclaves and seeped into mainstream thinking. In 1795, the internationally known propagandist and polemicist whose work was central to both the American and French revolutions, Thomas Paine, proclaimed,

> Though it is not a direct article of the Christian system that this world that we inhabit is the whole of the habitable creation, yet it is so worked up therewith, from what is called the Mosaic account of the Creation, the Story of Eve and the apple, and the counterpart of that story—the death of the Son of God—that to believe otherwise, that is, to believe God create[d] a plurality of worlds, at least as numerous as what we call the stars, renders the Christian system of faith at once little and ridiculous, and scatters it in the mind like feathers in the air. The two beliefs cannot be held together in the same mind; and he who thinks that he believes both has thought but little of either.[127]

Paine uses the common belief in plurality as a justification for an attack on Christianity. Paine, an avowed deist, opposed the belief that God actively interferes in the affairs of humans. He uses the prospect of a universe full of inhabited planets to suggest that it would be ridiculous to imagine God intervening in the affairs of Earth. He thus inverts Plato's argument: if unity makes humanity central, then plurality makes it mediocre.

Paine's *Age of Reason* went through eight editions in 1794, seven the following year and two in 1796.[128] Paine's work brought deist ideas to the common people in an enormously effective manner and became a best seller in France, England, and the United States.[129] Britain, fearful of radical ideas in the wake of the French Revolution, ultimately banned the book. And yet even banned, Paine's *Age of Reason* went on to reach a large audience in Britain where authorities prosecuted hundreds of booksellers over thirty years for selling the book.[130]

Rather than launch an attack on pluralism to combat its potential deist implications, like it had Copernicism, most elements of the Christian Church largely agreed with Paine's belief in plurality, but, of course, not his conclusion. A preacher of the Free Church of Scotland, Dr. Thomas Chalmers lectured extensively on the question in 1817 and claimed that no tension exists between Christianity and alien planets. In fact, argues Chalmers, pluralism demonstrates the awesome power of God.[131] Chalmers's lectures became a smashing success. Samuel Warren, an English barrister, wrote that as a boy he waited four hours to hear Chalmers lecture on pluralism and Christianity and even with the wait he did not come early enough to get a chair.[132] Chalmers collected his sermons on pluralism into a book, which sold 6,000 copies in ten weeks. By the end of a year it had sold 20,000 copies and gone through nine editions.[133]

Religious acceptance of pluralism paved the way for it to emerge in other contexts. Crowe argues that the astronomer Ormsby MacKnight Mitchel convinced the population of Cincinnati to fund the world's second largest refracting telescope in part by his appeal to pluralism.[134] Much as the Aristotelian cosmology became a commonplace of the medieval period, pluralism began to become a part of everyday life. A review of early American newspaper articles in the database *American Periodicals Series* (APS) showed fifty-seven results for the phrase "plurality of worlds," between 1774-1849.[135] Of the articles that took positions on the question, eight supported plurality and only one opposed it. This provides further proof of pluralist opinions in the general public.

The scientific discussion of plurality primed the American public to accept even the most bizarre tales of alien life. In 1835, the American journalist Richard Adams Locke, who claimed a Cambridge University pedigree, wrote a series of newspaper articles about discoveries of life on the moon, which he attributed to famed astronomer John Herschel. In reality, he completely fabricated the contents of the articles, which included absurd findings like the presence of blue unicorns and a race of bipedal tool-using beavers.[136] Locke meant the article to satirize the outlandish claims of the defenders of plurality, but the public largely accepted the story as true scientific discovery.[137] When an enterprising journalist uncovered the fact that the article contained no element of truth, newspapers denounced it as a hoax rather than a satire. Crowe explains the refusal to accept the article as satire: "It was not that Locke lacked the skills of a satirist; it was rather that pluralist preaching and pronouncements had so permeated the thought of his contemporaries that they first failed to see the articles as satire, and failed again as they branded them a 'hoax.'"[138] Fantastic pluralist ideas had taken such a hold on the public that they accepted Locke's absurd story as truth.

CONCLUSION

Prior to Copernicus, Plato's grand strategy for his rhetorical cosmology appears to have come true. He proposed a philosophical defense of the Good intertwined with scientific theory and a ban upon rival cosmologies. Aquinas' updated version of Aristotle's cosmology combined the latest scientific thinking with Plato's myths and philosophy from the *Timaeus*. The rejection of the dual truth approach of the Averroists joined Christianity with Aristotelian natural philosophy (the precursor to natural science). Rival cosmologies may not have faced an outright and explicit ban as is advocated in *Laws*, but they did not circulate widely and faced significant opposition when they emerged. Most of the Greek cosmologies opposed to Plato's did not make their way to Europe until the fifteenth century. The universities taught the Aquinian/Platonic/Aristotelian cosmology, while the threat of excommunication or death loomed over those that strayed too far from Church dogma. The actualization of the program Plato's outlines had exactly the kind of results

Figure 2.2. *The Sun* newspaper claimed this picture was based on observations of the moon via telescope.

his dialogues argue it would. The unity cosmology, as part of the broader Aristotelian cosmology, integrated itself so deeply into the popular imagination that it went without saying. This allowed it to offer a strong foundation for social stability, religion, and absolute ideals.[139]

The Copernican revolution represented the end for most elements of the Aristotelian cosmology. The scientific upheaval coincided with massive political, religious, and social change throughout Europe. Some individuals tried to maintain a version of Aristotle's cosmology and thus reinstate its rhetorical power to quell change by highlighting the one aspect that heliocentrism had yet to disprove: unity. Despite their efforts, pluralism gained enormous ground. Plurality went from something a few scholars theorized but did not actually defend, to a popular belief. This sets the stage for the 1800s, where two enormously important figures offer defenses of unity: William Whewell and Alfred Russel Wallace.

NOTES

1. Peter Swirski, *Between Literature and Science: Poe, Lem, and Explorations in Aesthetics, Cognitive Science, and Literary Knowledge* (Quebec City: McGill-Queen's University Press, 2000), 18.
2. Hawhee and Olson, "Pan-Historiography."
3. E. M. W. Tillyard, *The Elizabethan World Picture* (New York: Macmillan, 1944), 7.
4. Aristotle's *De Caelo* entered the Christian European world in a relatively complete form, through the Arab world in the twelfth and thirteenth centuries. I will engage a more detailed examination of its transmission later in the chapter.
5. Tillyard, *The Elizabethan World Picture*; C. S. Lewis, *The Discarded Image: An Introduction to Medieval and Renaissance Literature* (Cambridge: Cambridge University Press, 2002); Alison Cornish, "Dante's Moral Cosmology," in *Cosmology: Historical, Literacy, Philosophical, Religious, and Scientific Perspective*, ed. Norriss S. Hetherington (New York: Garland Publishing, 1993).
6. Arthur O. Lovejoy, *The Great Chain of Being: A Study of the History of an Idea* (Cambridge: Harvard University Press, 1942).
7. Edson and Savage-Smith, *Medieval Views of the Cosmos*, 9-20; Toulmin, *Cosmopolis*; Randles, *The Unmaking of the Medieval Christian Cosmos*, X-XI, 3; Hetherington, "Introduction: The Aristotelian World View," 177.
8. Mckeon, *The Basic Works of Aristotle*, xiv.
9. Johansen, *Plato's Natural Philosophy: A Study of the Timaeus-Critias*, 5.
10. Aristotle, *On the Heavens*, 270b1-25, 76a18-79b1.
11. Dick, *Plurality of Worlds*, 14.
12. Norriss S. Hetherington offers one interesting possible explanation of Aristotle's move away from the cosmological argument for the Good. He suggests that Plato had grown frustrated with the political process after his failure in Syracuse and thus turned to natural justifications of goodness. Aristotle had a completely opposite experience. He tutored Alexander the Great in the manner Plato recommended for a philosopher king and remained in his good graces throughout much of his life. Because of his success, he did not feel the need to make the move to cosmology as a justification for ethics that Plato made between the *Republic* and the *Timaeus*. Hetherington, "Aristotle's Cosmology," 97.
13. Plato, *Timaeus*, 54a-62e.
14. Aristotle, "Physics," in *The Basic Works of Aristotle*, ed. Richard McKeon, trans., R. P Hardie and R. K. Gaye (New York: Random House, 1941), 255b 13-30.

15. Aristotle, *On the Heavens*, 276a18-79b1.
16. Aristotle, "Metaphysics," xii 1074a.
17. John Dillon, "Timaeus in the Old Academy," in *Plato's Timaeus as Cultural Icon*, ed. Gretchen J. Reydams-Schils (Notre Dame: University of Notre Dame Press, 2003), 80-81.
18. This does not meant to imply that Aristotle had no interest in finding teleology in nature. In *Politics* for example he argues that plants existed for animals and "[nature] has made all animals for the sake of men." Aristotle, "Politics," 1256b 20-23. This passage and others like it have led scholars to argue Aristotle has a human-centered teleology, although most scholars go only so far as to say he has a natural teleology. David Sedley, "Is Aristotle's Teleology Anthropocentric," *Pronesis* 36, no. 2 (1991); John Cooper, "Aristotle's Natural Teleology," in *Language and Logos: Studies in Ancient Greek Philosophy*, ed. Malcolm Schofield and Martha Nussbaum (Cambridge: Cambridge Univeristy, 1982). A good source that defends the mainstream view of teleology without anthropocentrism, Christopher Shields, "Aristotle," *Stanford Encyclopedia of Philosophy*, http://plato.stanford.edu/entries/aristotle/. Whether or not Aristotle's teleology is anthropocentric it differs in important ways from Plato's. He disconnects teleology from cosmology preferring to find it in other examples, like the one above. Aristotle's teleology is also much less explicit and less directly connected to ethical living. Most importantly, as mentioned above, Aristotle does not connect Earth's unity with teleology, morality, philosophy, politics, or religion.
19. Crowe, "A History of the Extraterrestrial Life Debate."; Michael J. Crowe, *Theories of the World from Antiquity to the Copernican Revolution* (Mineola, NY: Dover, 2001); Dick, *Plurality of Worlds*; Dick, *The Biological Universe*.
20. Jaroslav Pelikan, *What Has Athens to Do with Jerusalem* (Ann Arbor: University of Michigan Press, 2000), 67-87.
21. Philo, "De Opificio Mundi," in *The Works of Philo: Complete and Unabridged*, ed. C. D. Yonge, trans., C. D. Yonge (Peabody, MA: Henderickson, 2000), 171.
22. Ibid.
23. Plato, "Timaeus," 55c-d.
24. Augustine and Thomas S. Hibbs, *Enchiridion on Faith, Hope, and Love*, trans. Thomas S. Hibbs (Washington, DC: Regnery, 2002), 9-10.
25. Thomas Kuhn, *The Copernican Revolution: Planetary Astronomy in the Development of Western Thought* (Cambridge, MA: Harvard University Press, 1970), 106.
26. Anna Somfai, "The Eleventh-Century Shift in the Reception of Plato's "Timaeus" and Calcidius's "Commentary'," *Journal of the Warburg and Courtauld Institutes* 65 (2002): 1.
27. This is not to say that other non-cosmological ancient did not circulate; for example Boethius translated Aristotle's *Logic*, and fragments came down through Plotinus and Proclus.
28. Somfai, "The Eleventh-Century Shift in the Reception of Plato's "Timaeus" and Calcidius's "Commentary'," 4. According to Somafi's account of the fragments available of the *Timaeus*, the sections on unity discussed in the previous chapter were present.
29. Andrew Gregory, *Ancient Greek Cosmogony* (London: Duckworth, 2007), 150.
30. Helen J. Nicholson, *The Crusades* (Westport, CT: Greenwood, 2004), 25-26.
31. Étienne Gilson, *The Philosophy of St. Thomas Aquinas*, trans. Edward Bullough (Freeport, NY: Books for Libraries Press, 1971), 13; Ralph McInerny, *Aquinas* (Cambridge: Polity Press, 2004), 10; Edward Grant, *Planets, Stars, and Orbs: Medieval Cosmos, 1200-1687* (Cambridge: Cambridge University, 1994), 12-13; Richard E. Rubenstein, *Aristotle's Children: How Christians, Muslims, and Jews Rediscovered Ancient Wisdom and Illuminated the Dark Ages* (Orlando: Harcourt, 2003), 15-16.
32. Grant, *Planets, Stars, and Orbs*, 14; Rubenstein, *Aristotle's Children*, 4-5.
33. Edward Grant, "A New Look at Medieval Cosmology, 1200-1697," *Proceedings of the American Philosophical Society* 129, no. 4 (1985): 428, footnote 12.
34. Gilson, *The Philosophy of St. Thomas Aquinas*, 13; Grant, "A New Look at Medieval Cosmology," 438 footnote 12.
35. Aristotle, *On the Heavens*, 279b4-83b17.
36. McInerny, *Aquinas*, 20-21; John D. North, "One Truth or More? Demarcation in the Universe of Discourse," in *Physics, Cosmology and Astronomy, 1300-1700: Tension and Accommodation*, ed. Sabetai Unguru (Boston: Kluwer, 1991).

37. Gilson, *The Philosophy of St. Thomas Aquinas*, 6; Dick, *Plurality of Worlds*, 26; McInerny, *Aquinas*, 20-21. The Franciscan disapproval of Aristotle was largely limited to his cosmology, they made use of his logic and ethics. Eugene Rathbone Fairweather, *A Scholastic Miscellany: Anselm to Ockham* (Philadephia: Westminster Press, 1956), 365.

38. Gilson, *The Philosophy of St. Thomas Aquinas*, 19. It is important to note that Cardinal Robert's edict still allows for Aristotle's ethics and logic.

39. Ibid., 2-3.

40. Ibid., IX-X, 6, 18; Kuhn, *The Copernican Revolution*, 110-11; J. D. North, *Stars, Minds and Fate: Essays in Ancient and Medieval Cosmology* (London: Hambledon, 1989), 403; Aidan Nichols, *Discovering Aquinas: An Introduction to His Life, Work, and Influence* (Grand Rapids, MI: William B. Eerdman, 2003), 14; Grant, *Planets, Stars, and Orbs*, 53.

41. Thomas Aquinas, "De Aeterniate Mundi," *Medieval Sourcebook*, http://www.fordham.edu/halsall/basis/aquinas-eternity.asp#f2.

42. Aquinas, *Quaestiones Disputatae De Potentia Dei*, q. 3, a. 16, o. 17.

43. Ibid.

44. Ibid., q. 2, a. 16, c.

45. Aquinas also writes, "The universe as created by God is the best possible in respect of the things that actually exist; but not in respect of the things that God is able to create." This passage lacks clarity; one can read it to suggest that God could create a perfect plurality. This interpretation, however, does not comport with the rest of Aquinas' thinking on unity vs. plurality as will be shown below. Instead, this passage likely means only that God could create a plurality of worlds, not that God in fact did so. If God is omnipotent, God can do anything God wills. Ibid., q. 2, a. 16, ad 17.

46. Plato, *Timaeus*, 29e.

47. Thomas Aquinas, *The Heavens*, trans. Fabian R.Larcher and Pierre H. Conway (Washington, DC: Priory of the Immaculate Conception), lec. 19, n. 197.

48. This represents a likely motivation for God's action rather than a restraint, in contrast to individuals like Michael Scot, William of Auvergne and Roger Bacon, who argued that it was impossible for God to create plurality. Dick, *The Biological Universe*, 14.

49. Aquinas, *The Summa Theologica of St. Thomas Aquinas*. Part, I, Q. 47, Art. 3, ad 1.

50. Interestingly, he also cites Aristotle making what Aquinas suggests is a similar claim. It is not clear what part of Book XII of *Metaphysics* Aquinas references, but it most likely is Aristotle's argument about how plurality would require multiple prime movers. As discussed before, Aristotle removed questions of the prime mover from the philosophical considerations Plato attributes to the demiurge. The unity of the prime mover occurred because of its logical necessity, not because of the "unity of order in things."

51. Aquinas, *The Summa Theologica of St. Thomas Aquinas. Part I QQ. L.-LXXIV*, q. 47, a. 3, c.

52. Furley, *The Greek Cosmologists*, 115.

53. The atomic theory fell by the wayside from the time of the ancients until the seventeenth century for precisely this reason. Dick, *Plurality of Worlds*, 21-23. The perceived connection between Democritus, atheism, and sophistry is explored more in the fourth chapter.

54. This represents a triumph for Plato's rhetorical strategy, which was contingent on the connection various cosmological models with a particular *telos* or lack thereof.

55. Thomas Aquinas, *Contra Gentiles Book Three: Providence (Q. 1-83)*, trans. Vernon J. Bourke (New York: Hanover House, 1956), III, c. 64, n. 9.

56. 1.0 Ibid., III, C. 64, N. 9

57. Gilson, *The Philosophy of St. Thomas Aquinas*, 16.

58. McInerny, *Aquinas*, 21.

59. Ibid.

60. Ibid., 141, 42. The belief that God *could* create other worlds continued to dominate Catholic thinking, despite removing the condemnation of Aquinas' works. Protestants, however, never accepted this reasoning Martin Luther and Philip Melanchthon (a major Protestant theologian) both claimed plurality was impossible. Dick, *Plurality of Worlds*, 88.

61. Edward Grant, "Aristotelianism and the Longevity of the Medieval World View," *History of Science* 16 (1978): 94; Dick, *Plurality of Worlds*, 24-25.

62. Grant, *Planets, Stars, and Orbs*, 59.

63. Francisco José Ayala and Theodosius Grigorievich Dobz, eds., *Studies in the Philosophy of Biology: Reduction and Related Problems* (Berkeley: University of California, 1974), 6.

64. Edward Grant, "In Defense of the Earth's Centrality and Immobility: Scholastic Reaction to Copernicanism in the Seventeenth Century," *Transactions of the American Philosophical Society* 74, no. 4 (1984): 4.

65. Grant, "Aristotelianism and the Longevity of the Medieval World View," 98-99.

66. Ibid.

67. Ibid., 96.

68. Grant, *Planets, Stars, and Orbs*, 155. The decree actually had the effect of spurring some scientific investigation, because scholastic needed to square the possibility of plurality with Aristotelian science. Debates also occurred on the method of justifying unity, like Giles of Rome attacks on Aquinas' argumentation, but not conclusion. Graham James McAleer, "Disputing the Unity of the World: The Importance of Res and the Influence of Averroes in Giles of Rome's Critique of Thomas Aquinas concerning the Unity of the World," *Journal of the History of Philosophy* 36, no. 1 (1998): 29-55.

69. Roland Hissette study of the condemnations suggests the theme of God's omnipotence in relation to science, runs throughout the articles. Roland Hissette, *Enquête sur les 219 articles condamnés à Paris le 7 mars, 1277* (Louvain: Publications Universitaire, 1977); Hans Thijssen, "Condemnation of 1277," *Stanford Encyclopedia of Philosophy*, http://plato.stanford.edu/entries/condemnation/.

70. Dick, *Plurality of Worlds*, 37; Pierre Duhem, *Medieval Cosmology: Theories of Infinity, Place, Time, Void, and the Plurality of Worlds*, trans. Roger Ariew (Chicago: University of Chicago Press, 1985), 449-72.

71. Nicole Oresme, *Le Livre du Ciel et du Monde*, trans. Albert D. Menut (Madison: University of Wisconsin Press, 1968), 179. William of Ockham, Jean Buridan, William Vorilong, and Guillaume de Vaurouillon represent other example of individuals defending the possibility, but not actuality of plurality. They opposed thinkers like Michael Scot, William of Auvergne and Roger Bacon, who suggested that plurality was simply impossible. Thomas F. O'Meara, "Christian Theology and Extraterrestrial Intelligent Life," *Theological Studies* 60 (1999): 13-15; Duhem, *Medieval Cosmology*, 442-49, 60-71. Grant argues the only scholastic to actually defend the existence (not just possibility of plurality) was John Major, who is discussed briefly later in the chapter. Grant, "A New Look at Medieval Cosmology," 155.

72. Dick, *Plurality of Worlds*, 37.

73. Rubenstein, *Aristotle's Children*, 67; Grant, "Aristotelianism and the Longevity of the Medieval World View," 94-96.

74. Nicolas Cusanus, *Of Learned Ignorance*, trans. Father Germain Heron (New Haven, 1954), 111-16.

75. Ingrid D. Rowland, "What Giordano Bruno Left Behind: Rome, 1600," *Common Knowledge* 14, no. 2 (2008): 425, 26 ; Louis Dupre and Nancy Hudson, "Nicholas of Cusa," in *A Companion to Philosophy in the Middle Ages*, ed. Jorge J. E. Gracia and Timothy B. Noone (Malden, MA: Blackwell, 2002), 468.

76. Toulmin, *Cosmopolis*, 77.

77. McColley, "The Seventeenth-Century Doctrine of a Plurality of Worlds," 402.

78. Grant, *Planets, Stars, and Orbs*, 166-68.

79. Kuhn, *The Copernican Revolution*, 185. Gingerich tracked down the original copies of *On the Revolutions* and argues that it was much more widely read that has previously been believed. Neil Harris disputes Gingerich's finding saying that he did not prove that the book had much impact outside a "tight-knit group of professional astronomers," who were more interested in its mathematical models of planetary movement then its argument for heliocentrism. Owen Gingerich, *The Book Nobody Read: Chasing the Revolutions of Nicolaus Copernicus* (New York: Walker, 2004); Neil Harris, "De Revolutionibus in Bibliography: Analysing the Copernican Census " *The Library* 7, no. 3 (2006): 324.

80. Nicolaus Copernicus and Stephen Hawking, *On The Revolutions of Heavenly Spheres* (Philadelphia: Running Press 2002), 5.

81. Gingerich, *The Book Nobody Read*, 141.

82. Kuhn, *The Copernican Revolution*, 185-91.

83. Guillermo Gonzalez and Jay W. Richards, *The Privileged Planet: How Our Place in the Cosmos Is Designed for Discovery* (Washington DC: Regnery Publishing 2004), 226-28; Lovejoy, *The Great Chain of Being*, 102; Guy Consolmagno, *Aquinas Lecture* (Atlanta: The Aquinas Center, Emory University, 2010), Audio Lecture; Alexandre Koyré, *From Closed World to the Infinite Universe* (Baltimore: John Hopkins University, 1982), 43.

84. Kuhn, *The Copernican Revolution*, 193.

85. Blumenberg, *The Genesis of the Copernican World*, 514.

86. Giordano Bruno, "On the Infinite Universe and Worlds," in *Giordano Bruno: His Life and Thought*, ed. Dorothea Waley Singer, trans., Dorothea Waley Singer (New York: Henry Schuman, 1950), 257.

87. Dick, *Plurality of Worlds*, 66.

88. Ibid.

89. The Hermetic tradition is a form of ancient mysticism that emphasized magical powers. At the time the Catholic Church did not look favorably on Hermeticism. Edwards A. Gosselin and Lawrence S. Lerner, "Introduction," in *The Ash Wednesday Supper* (Toronto: University of Toronto Press, 1995), 27-29, 41-53.

90. Blumenberg, *The Genesis of the Copernican World*, 374.

91. Giordano Bruno as quoted in Crowe, *Theories of the World from Antiquity to the Copernican Revolution.*

92. Karl S. Guthke, *The Last Frontier: Imagining Other Worlds, from the Copernican Revolution to Modern Science Fiction*, trans. Helen Atkins (Ithaca: Cornell University, 1990), 42.

93. Kuhn, *The Copernican Revolution*, 199.

94. Guthke, *The Last Frontier*, 76; Dick, *Plurality of Worlds*, 112.

95. Guthke, *The Last Frontier*, 95. The Catholic Church reapproved the book in 1620 with new language that made it clear heliocentrism was only a thought experiment, rather than a description of reality. An examination of extant texts, however, indicate that not many were actually censored outside of Italy. Owen Gingerich, "The Censorship of Copernicus' *De revolutionibus*," *JASA* 33 (1981): 58-60.

96. Kuhn, *The Copernican Revolution*, 199. One hypothesis for this is that Kuhn did not have the advantage of a couple of decades of the "history of the book" scholarship that shifts discussion from the ideas to the circulation of material texts and their reception practices.

97. Thomas Campanella, *A Defense of Galileo the Mathematician of Florence: Which is an Inquiry as to Whether the Philosophical View Advocated by Galileo Is in Agreement with, or Is Opposed to, the Sacred Scriptures*, trans. Richard J. Blackwell (London: University of Notre Dame Press, 1994), 44. The accusation refers to Galileo's *Sidereus nuncius* (1610), where he claims the moon has mountains, water, and an atmosphere. Not long after he wrote the book he abandoned his claims about water and atmosphere. Ibid., 135, footnote 25. The fact that Galileo never gives an indication that he believed the planet actually was inhabited suggests that the Inquisition had a problem with the implications of his theories for life and not Galileo's interpretation of his theory.

98. Ibid., 45.

99. Ibid.

100. Giorgio de Santillana, *The Crime of Galileo* (Chicago: University of Chicago Press, 1955), 34.

101. Jules Speller, *Galileo's Inquisition Trial Revisited* (Frankfurt: Peter Lang, 2008), 155-60, 375-96; Grant, "A New Look at Medieval Cosmology," 418.

102. Guthke, *The Last Frontier*, 96-97.

103. Ibid., 97.

104. Grant, "A New Look at Medieval Cosmology," 418.

105. Marjorie Nicolson, *Science and Imagination* (Ithaca: Cornell University Press, 1962), 55.

106. Camille Flammarion, *La Pluralité des Mondes Habités*, 2 ed. (Paris 1864), 341 Crowe, *Extraterrestrial Life Debate*, 16-17.

107. Gabriel Naudé to Ismael Boulliau, 15 August 1640, quoted in Guthke, *The Last Frontier*, 48.
108. Lovejoy, *The Great Chain of Being*, 102, 03.
109. Grant, "In Defense of the Earth's Centrality and Immobility," 67.
110. Jenkin, *The Resonableness and Certainty of the Christian Religion*, 205.
111. Thomas Baker, *Reflections Upon Learning: Wherein is Shewn the Insufficency Thereof, in its Several Particulars: In order to envince the Usefulness and Necessity of Revelation*, 4 ed. (London: Dyal and Bible, 1708), 119.
112. Toulmin, *Cosmopolis*, 82-87; Kuhn, *The Copernican Revolution*, 2.
113. Dick, *Plurality of Worlds*, 48.
114. Crowe, *Extraterrestrial Life Debate*, 19. It is worth noting that Fontenelle later wrote *Digression sur les Anciens et les Modernes* (1688). In the book he opposes absolute monarchy and makes a case that the new ideas of the present are better than classical ideals. Nina Rattner Gelbart, "Introduction," in *Conversations on the Plurality of Worlds* (Berkeley: University of California Press, 1990).
115. S. P. Langley, "The First 'Popular Science Treatise'," *Popular Scientific Monthly* 10 (1877).
116. Crowe, *Extraterrestrial Life Debate*, 19.
117. Fontenelle, *Conversations on the Plurality of Worlds* (Berkeley: University of California Press, 1990), 17.
118. Ibid., 63-64.
119. Ibid., 63.
120. Fontenelle also provided a clever response to the inquisitions claim that plurality would mandate multiple comings of Christ on other planets. He claimed that aliens would not be descendents of Adam and thus not effected by original sin. Ibid., 6.
121. Crowe, *Extraterrestrial Life Debate*, 20.
122. Ibid., 18-22; Guthke, *The Last Frontier*, 226-44.
123. Crowe, *Extraterrestrial Life Debate*, 263-64.
124. Isaac Todhunter, *William Whewell, D.D., Master of Trinity College, Cambridge; An Account of his Writings with Selections from his Literary and Scientific Correspondence*, vol. 1 (New York: Johnson Reprint Corporation, 1970), 185.
125. "Chalmer's Discourses on the Christian Revelation," *Eclectic Review*, no. 8 (1817).
126. Crowe, *Extraterrestrial Life Debate*, 190-91.
127. Thomas Paine, *The Age of Reason*, vol. 1 (London, R. Carlile, 1826), 38.
128. Gregory Claeys, *Thomas Paine: Social and Political Thought* (Florence, KY: Routledge, 2004).
129. Patricia Fara, "Heavenly Bodies: Newtonianism, Natural Theology, and the Plurality of Worlds Debate in the Eighteenth Century," *Journal for the History of Astronomy* 35, no. 2 (2004): 151.
130. Jacob Bronowski, *William Blake and the Age of Revolution* (London: Routledge, 1972), 81.
131. Chalmers, *A Series of Discourses*.
132. Samuel Warren, "Speculators among the Stars, Part II," *Blackwood's Edinburgh Magazine* 76 (1854).
133. William Hanna, "Memoirs of the Life and Writings of Thomas Chalmers, D. D., KK. D.," *The Living Age* 27, no. 333 (1850): 22.
134. Ormsby MacKnight Mitchel, *Orbs of Heaven* (London: Bradbury and Evans, 1851), 221-23; Philip S. Shoemaker, " Stellar Impact: Ormsby MacKnight Mitchel and Astronomy in Antebellum America" (Diss., University of Wisconsin-Madison, 1991), 186-231; Crowe, *Extraterrestrial Life Debate*, 234.
135. The fifty-seven results are after removing advertisements and book lists. Another three of those fifty-seven results do not refer to plurality in the sense of alien worlds.
136. "Great Astronomical Discoveries Lately Made by Sir John Herschel, L.L.D, F.R.S, &c. At The Cape of Good Hope," *New York Sun*, August 25, 1835. The story ran for several days and gave a major sales boost to Benjamin Day's *New York Sun* (notable as the first penny

paper). Brian Thornton, "The Moon Hoax: Debates About Ethics in 1835 New York Newspapers," *Journal of Mass Media Ethics* 15, no. 2 (2000): 89-100.

137. Crowe, *Extraterrestrial Life Debate*, 210-15; Steven W. Ruskin, "A Newly-discovered Letter of J.F.W. Herschel Concerning the 'Great Moon Hoax'," *Journal for the History of Astronomy* 33 (2002): 71-72.

138. Crowe, *Extraterrestrial Life Debate*, 215.

139. Edson and Savage-Smith, *Medieval Views of the Cosmos*, 9-20; Toulmin, *Cosmopolis*; Tillyard, *The Elizabethan World Picture*; Randles, *The Unmaking of the Medieval Christian Cosmos*, X-XI, 3; Hetherington, "Introduction: The Aristotelian World View," 177; Cornish, "Dante's Moral Cosmology."; Lewis, *The Discarded Image*.

Chapter Three

William Whewell and Alfred Russel Wallace

Unity Cosmology in the Modern Era

"They say in Cambridge that Dr. Whewell's book, 'Plurality of Worlds,' reasons to this end: The planets were created for this world; this world for man; man for England; England for Cambridge; and Cambridge for Dr. Whewell!"[1]
—Maria Mitchell, American astronomer, November 1857

"Who is man, and what his place?
Anxious asks the heart, perplex
In this recklessness of space,
Worlds with worlds thus intermixt;
What has he, this atom creature,
In the infinitude of Nature?"
—F. T. Palgrave, British poet[2]

The collapse of the geocentric model of the universe helped to usher in the modern era.[3] Without a celestial justification for the status quo, new ideologies, politics, and philosophies emerged more easily. Modernity, at its foundation, represented an opening up of all the previous ideologies to challenge.[4] Utilitarianism and classical liberalism challenged the dominant philosophical and political orders. Deism questioned the need for God's active presence in the universe. The French Revolution undermined belief in the stability of monarchies across Europe. For many individuals, the totality of these (and many more) radical changes represented a dangerous move away from traditional ideas.[5]

Some critics of the post-Copernican transformation employ a rhetorical strategy centered on a return to the stabilizing function of Plato's unity cosmology. This strategy, however, could not simply restore the Aristotelian cosmology of the medieval ages. Galileo's models, combined with the explanatory power of Newton's physics and the increased observational power of new telescopes, meant that no serious scholar could defend many discredited elements of Plato and Aristotle's cosmology.[6] The new science definitively disproved the physical centrality of the Earth and the ethereal, perfect nature of the celestial objects and their orbits. The question of a plurality of worlds, however, had yet to be definitively resolved by science. This chapter examines two case studies of individuals using the unity cosmology in order to argue for absolute values against modernist instantiations of relativism.[7]

The first case study examines the famous Anglican philosopher William Whewell (1794-1886). His book *Plurality of Worlds* (1853) ruptured the illusion of plurality consensus by providing a well-argued case for the unity cosmology from a respected intellectual figure. Early in his career, Whewell accepted and even defended the pluralist position. Like Plato, whose dialogues he translated and published (1861), Whewell grew concerned with the ways that pluralism provides an argument against a teleological and perfectible universe. Both forwarded unity as an argument against their intellectual opponents. While Plato battled the sophists, Whewell engaged in an ongoing academic debate with the utilitarians, e.g., Bentham, Mill, and Locke.[8] As discussed below, Whewell explicitly viewed the utilitarians as modern-day sophists, tied together by their shared advocacy of a relativistic and materialist worldview. Whewell, a scientist, did not deny the importance of empirical observation of the material world, but advocated a hybrid method that mixed observation with the idealism of Plato and Kant.[9] In the *Plurality* Whewell defends a reworked version of Plato's unity cosmology, which he uses as an argument for absolute values. While he made few converts in the scientific world, the dialogue and debate that emerged after his book revealed the cracks in the plurality consensus and helped erect a bulwark against utilitarian ideas.

The second case study investigates Alfred Russel Wallace (1823-1913), the co-discoverer of the theory of evolution via natural selection. Wallace, like Whewell, also acted to revive the unity cosmology as a method to advocate for his political and social beliefs. Wallace deeply opposed capitalism and colonialism, but over time he became more and more convinced that the Darwinian theory he helped discover justified these very political ideologies.[10] He sought a scientific counterweight to temper the social implications of orthodox Darwinism and return humanity to a privileged position. In his books *Man's Place in the Universe* (1903) and *The World of Life* (1910) he uses unity cosmology to achieve this end.

The 1700s saw the collapse of much of the cosmology of Plato, Aristotle, and Aquinas and ushered in a host of radical social changes culminating in the French Revolution and Napoleonic Era.[11] Wallace and Whewell act to revive unity, the one aspect of medieval cosmology not definitively disproven by heliocentrism and Newtonian physics. Both argued that scientific paradigms provided an intellectual justification for ethics and religion, and maintained that only unity could justify absolute values. Whewell and Wallace argued that plurality gave credence to the positions of their opponents, whether or not their opponents made this connection.

WILLIAM WHEWELL

Whewell's defense of the unity cosmology came as a shock to his peers. By the early 1800s most scientists who studied the question of alien life believed the issue had been settled decisively in favor of plurality.[12] The *Eclectic Review*, a highbrow British periodical, wrote, "The probability that the other orbs of our system are inhabited worlds must appear so great, that a direct revelation from heaven disclaiming the fact, would make but little difference in our assurance of it."[13] While the *Eclectic Review*'s defense of pluralism went beyond most others in terms of its bombast, most writers shared the general sentiment.[14] Whewell's first biographer suggests that only Whewell's mental anguish at the untimely death of his wife could explain such a strange break from the commonly accepted science.[15] Whewell's drastic change of opinion appeared to come out of nowhere, which must have proved a major surprise, not only because of the heterodox nature of his opinion, but also because of the stature of the opinion holder.[16]

By the time Whewell wrote his defense of the unity cosmology in 1853 he had achieved considerable success in life. He held the position of Master at Trinity College in Cambridge on conservative Prime Minister Robert Pell's recommendation to the Queen, was a founding member of the British Association for the Advancement of Science, a fellow of the Royal Society, and a president of the Geological Society.[17] He acquired the nickname the "harmonious blacksmith" from Erasmus Darwin after he saw Whewell respectively listen, "all ear (not all assent)" at a lecture of Thomas Carlyle.[18] He would eventually become one of the most respected intellectual figures in England, but his rise to prominence did nothing to foreshadow his future public support for unity.

Whewell straddled two major competing intellectual traditions. The empiricist school epitomized by John Locke and David Hume advocated an observational approach to knowledge that deemphasized epistemology not grounded in sensation. The idealist school, epitomized by Immanuel Kant and Samuel Taylor Coleridge, believed that true knowledge could be intuited by humanity and that structural limitations severely handicapped an empiricist method. Like Thomas Reid, who greatly influenced his work, Whewell criticized both schools and sought a middle path that took ideas from both idealism and empiricism.[19] This interest in a hybrid of idealism and empiricism drew Whewell to scientific debates like plurality that blended scientific/empiricist and philosophical/idealist implications and methodologies.

In 1833, Whewell published *Astronomy and General Physics*, a volume for the prestigious *Bridgewater Treatises*, in which he defended the possibility of plurality. Reverend Francis Henry, Earl of Bridgewater, commissioned the compendium in his 1825 last will and testament for the purpose of collecting the scientific evidence for the existence of God. Each of the *Bridgewater Treatises* covered a different scientific field and was written by an expert in that field. The resulting documents represented a "largely nontechnical, politically conservative, and religiously safe compendium of contemporary science."[20] Given the popularity of plurality and the unadventurous nature of the *Bridgewater Treatises*, one should not be surprised that Whewell's volume on astronomy and general physics remained open to the idea of plurality

In his treatise, *Astronomy and General Physics*, Whewell defends the standard science of the time, which included the high probability of plurality. He writes, "We may thus have in the universe worlds, no one knows how many, no one can guess how varied."[21] Whewell showed no signs of his later concern that plurality would pose a concern for teleology and ultimately Christianity. In fact, he explicitly denounces this possibility when he says, "*Numbers* [author's italics] are nothing in themselves; and when we reject the known, but unessential limits of our own faculties, it is quite as allowable to suppose a million millions of earths as one, to be under the moral government of God."[22] When Whewell later attacked the plurality position, at least one commentator mentioned the irony that one could easily mistake Whewell's *Astronomy and General Physics* for a work of any number of other pluralist writers of the time.[23]

Although his peers recognized Whewell as an expert on astronomy, he also engaged in a variety of other academic pursuits. Before publishing his attack on plurality, Whewell wrote many important texts on philosophy, religion, and in scientific fields other than astronomy.[24] In fact, between his *Bridgewater Treatise* and *Plurality* Whewell shifted his research focus to questions of practical morality and even thought about leaving Cambridge to take up a parish.[25] One theme, however, united his works across various disciplines: Whewell's prolonged conflict with utilitarian ideas.[26]

WHEWELL AND UTILITIARIANISM

During the time Whewell wrote, utilitarianism had garnered enough power to present a serious challenge to traditional ethics with its focus on sensationalism and materialism. In the fields of ethics, language, and philosophy of science, utilitarian philosophers made great strides. William Paley's *Principles of Moral and Political Philosophy* laid the foundation for philosophers to attack predetermined morality.[27] John Horne Tooke offered a linguistic theory that used etymology to highlight the social constructs behind words like "right," in order to show that morality exists only as a linguistic construct.[28] J.S. Mill championed an empiricist science that claimed experience determined reality.[29] As isolated developments, these proved problematic for Whewell; taken together they represented a frontal attack on Whewell's philosophical beliefs and a serious implicit threat to the foundations of Christian belief by supplanting the traditional spiritual order with a deist or material one.

Like the rise of sophism, the growing materialism posed a threat to the existent social order, a social order that Whewell was very invested in maintaining. Utilitarians pushed for measures that would overhaul higher education, shifting the focus from classics and mathematics to politics, economics, and modern philosophy, while decentralizing institutional power.[30] Perry Williams suggests that Whewell connected utilitarian beliefs to the French Revolution and unrest in England, which jeopardized the place of the Church and the established social order.[31] Whewell's writing suggests that he believed the impetus for radical social change began in the academy with the work of the founder of utilitarianism, Jeremy Bentham, and that winning the battle of ideas in academia could help reverse what he saw as negative political developments.[32]

Evidence suggests that Whewell viewed his dispute with utilitarianism as parallel to Plato's own conflict with the sophists. The prevailing historical view held the sophists as dangerous radicals interested in commercialism, skepticism, and populism.[33] Liberal thinkers, however, began to offer positive portrayals of the sophists, which they connected with ancient Greek democracy.[34] In the 1850s the liberal historian and utilitarian sympathizer George Grote would defend Protagoras' belief of "man as the measure of all things," as foundational to good governance and utilitarianism.[35] One would expect that Whewell would oppose the rehabilitation of the sophists, but in fact he praises Grote "for having rejected an established system for vilifying and misrepresenting Plato's opponents, the Sophists, and ascribing to them in every thing that they say, Sophistry in its modern English usage."[36]

In order to understand why Whewell welcomed this new take on the sophists, it is important to remember that scholars had previously treated them as hucksters with no underlying philosophy beyond their own self-interest.[37] As Whewell writes, the very term *sophist* had become synonymous with deceit. It is unthinkable that Whewell sympathized with the positions of the sophists. It is more likely that the caricature made the debates between Socrates and the sophists in Plato's dialogues uninstructive for modern times, because it amounts to Socrates triumphing over a collection of straw persons. Whewell treated the debate between Plato and the sophists as a serious intellectual battle between two radically opposed schools of thought; a battle that Whewell suggests was replaying itself in his own time. With radicals increasingly taking on the mantle of sophists, Whewell was free to cast himself as a modern-day Plato.[38]

As early as 1839, Whewell began contextualizing the current rise of utilitarianism with the debates of the ancient world. He delivered a lecture that connected his beliefs in absolute morality with Plato's.

> We maintain, with Plato, that reason has a natural and rightful authority over desire and affection; with [Joseph] Butler, that there is a difference of kind in our principles of action; with the general voice of mankind [*sic*], that we must do what is right, at whatever cost of pain and loss. We deny the doctrine of the ancient Epicureans, that pleasure is the supreme good; of Hobbes, that moral rules are only the work of men's mutual fear; of Paley, that what is expedient is right, and that there is no difference among pleasures except their intensity and duration; and of Bentham, that the rules of human action are to be obtained by casting up the pleasures which actions produce.[39]

Whewell does not mention the sophists by name in this passage, but doing so is unnecessary. Plato wrote his philosophy in dialogue form; Plato's defense of "reason," "over desire and affection," occurred in debates with the sophists. By aligning himself with Plato, Whewell links his philosophical opponents with the sophists. Rather than treating history as a series of unique and historically contextualized debates, Whewell links defenders of absolute values and relative values across time. In the narrative, Whewell, Plato, Butler, and "the general voice of mankind" stand in opposition to the sophists, Epicureans (ancient philosophers that emphasized physical pain and pleasure and defended the atomist cosmology) and in more recent times Paley, Hobbes, and Bentham.[40]

Whewell analogizes the materialists of his day and the sophists of ancient Greece in the 1861 introduction to his translation of *Protagoras*, in the *Platonic Dialogues*, in a gratuitous peroration.

> One sentence of [Protagoras'] is attacked by Plato, and is understood by the commentators, as if it meant that we have no knowledge except by sensation. He said that "Man is the measure of all things." It is plain that this may mean that man's faculties are the measure of human knowledge above all things; a very blameless doctrine, as seems to me. And even if the expression means that all our knowledge is derived from sensation, it conveys a doctrine which though, as I conceive, false, is extremely prevalent among many of the most moral, clear-headed, and right-minded persons among ourselves.[41]

The reference to sensationalists likely refers to the growing epistemology of materialism in England. Whewell treats Protagoras' dictum not as a defense of ethical relativism, but either as "a blameless doctrine" or a myopic attitude toward knowledge.

Whewell's charitable claim that the sensationalists make up "many of the most moral, clear-headed, and right-minded persons among ourselves" suggests he references a group big enough to include Unitarians and religious scientists who deployed a materialistic epistemology.[42] Whewell saw many of these individuals as allies on the question of ethics even if they held problematic scientific views, just as Plato agreed with the atomists on many ethical points but not their cosmology. Whewell would not grant the title of "moral, clear-headed, and right-minded persons" to utilitarians. He viewed Locke's followers as providing the intellectual grounding for the political and metaphysical revolutions that Whewell so despised.[43]

In contrast to the misguided sensationalists that favored a materialist epistemology, the dangerous sensationalists of the time were the followers of proto-utilitarian philosopher John Locke.[44] Whewell accuses them of giving "exclusive authority of the senses in [an] extreme unmitigated manner" and refers to them as the "sensationalist school."[45] Again he cast the debate between his more idealist views and the sensationalists in term of the ancients.

> [The opponents of pure sensationalism] knew, too, that many of the plausible tenets of the new philosophy were revivals of fallacies which had been discussed and refuted in ancient times. But the advocates of mere experience came on with a vast store of weighty truth among their artillery and with the energy which the advance usually bestows. The ideal system of philosophy could, for the present, make no effectual resistance; Locke, by putting himself at the head of the assault, became the hero of his day.[46]

Whewell argues that "weighty truth" (likely a reference to recent advances in science) gave renewed credence to the ancient, once discredited ideas of the sensationalists. Whewell suggests that the sensationalists, buoyed by the new scientific paradigm, rushed to revive the materialistic philosophy of the ancient sophists, in effect advancing scientific myths. Whewell argues that these scientific myths, however, have failed to properly contend with the arguments of Plato and others.

Whewell in effect identifies two problematic intellectual movements. First are the utilitarians like Locke, who use a materialistic epistemology to advocate for radical political reform and a new system of ethics. Second are groups like the Unitarians and religious scientists who deployed a materialistic epistemology, but did not necessarily agree with the implications the utilitarians find in their epistemology. For Whewell, the epistemological materialists may be "right-minded" in their ethical and religious views, but their embrace of materialism acted as a justification for Lockean politics and ethics. Whewell makes it clear that he wrote on the question of epistemology for the express purpose of attacking the "ultra-Lockian" position, which shows that Whewell treats morality and politics as inextricably linked with scientific and epistemological questions.[47]

The theme of Plato-versus-the-Sophists as a historical analogue for Whewell's intellectual conflict with the utilitarians emerges again in "Influence of the History of Science on Intellectual Education," a lecture he delivered in 1854, a year after the publication of *Plurality*. Whewell positions himself as a modern-day Plato battling the utilitiarians' sophistic educational program.

> The first great attempt made for the improvement of intellectual education, so far as history tells us, was that undertaken and prosecuted with preserving vigor by Socrates and Plato… the teachers whom Socrates and Plato perseveringly opposed—have been habitually called *the Sophists* … the education which these teachers professed to give, and frequently gave, was precisely what we commonly mean by a *good education.* It was an education enabling a young man to write well, speak well and act efficiently, on all ordinary occasions, public and private. The moral doctrines which they taught, even according to the most unfavorable representation of them, were no worse than the moral doctrines which are most commonly taught among ourselves at the present day,—the morality founded upon *utility* [his emphasis]; but many of them repudiated this doctrine as sordid and narrow, and professed higher principles.[48]

Whewell directly connects the sophists and utilitarians like Locke. Whewell in this passage even goes so far as to suggest that the utilitarianism represents a force more "sordid" then that of the sophists, because at least some of the sophists believed in some form of moral code. Given the then-prevalent historiography, Whewell's placement of the utilitarians below the sophists represents a serious insult.

Whewell makes the connection between his intellectual battle with the utilitarians and Plato's struggle with the sophists, and, like Plato, Whewell engaged his opponents across the spectrum of his work. Whewell followed Plato, too, in emphasizing a strong mathematics education as a bulwark against relativism. The utilitarians and other radical groups sought to shift focus away from math, where scholars could produce little new knowledge, to philosophy where new developments constantly occur. Whewell fiercely opposed this plan, because he felt the eternal truths of math engrained a sense of absolute moral values, whereas the deluge of new developments in philosophy encouraged relativism by demonstrating that what people believe is true today can change tomorrow.[49] Whewell did not passively declare his opposition to the shift away from a mathematics curriculum but became active in the public debate by writing position papers opposed to the proposed changes. He made the case that the "poll men," students who took a less rigorous set of courses at the university, should have to take math classes.[50]

Whewell felt that academia should make math education available beyond the elite, because it could steel the general population against the assault of utilitarianism. He also wrote works on ethics, as replacements for William Paley's utilitarian textbook that was required reading for undergrads.[51] Finally, he publically advocated an inductive scientific method that he believed delegitimized materialism.[52] The span of Whewell's efforts against materialism, relativism, and utilitarianism reflect Plato's lifelong struggle against these forces in his own time.

THE VESTIGES OF CREATION

Despite Whewell's best efforts, utilitarian thinking continued to gain ground. In 1825, J.S. Mill wryly observed that his writings had become "the textbooks of the young men… at Cambridge."[53] The publication of *The Vestiges of Creation* (1844), anonymously authored by successful publisher Robert Chambers, however, proved more problematic to Whewell than the decline in math education or the popularity of Mill's essays.[54] Whewell invested heavily in the idea that science and math revealed eternal immutable truths, which would both prove God's design and highlight appropriate universal ethics. In contrast, *The Vestiges of Creation* outlined a universe where life, planets and stars all emerged from a process of blind, impersonal evolution. In effect, it stole the appeal to the eternal cosmos that undergirded Whewell's entire rhetorical attack on the utilitarians by making the fixed and divine appear mutable and mechanical.

Vestige's evolutionary model harkens back to the cosmology of the ancient Greek atomists, who argued that the universe emerged from chaos by means of random chance. Chambers offers a modified account of 18th century Swedish scientist Emanuel Swedenborg's nebular hypothesis, which argued nebulae (cloud-like astronomical bodies) could not be reduced to stars by further magnification of telescopes.[55] Instead, these clouds of astronomical dust are engaged in a process of development whereby natural forces turn them into solar systems like our own. Once these solar systems form, then life would develop in an evolutionary-like fashion (what Chambers called transmutation of species). This cosmogony went beyond even the deistic interpretations of the creation of the universe in removing the role of God.

The popularity of the *Vestiges* outstripped any other scientific work at the time. The book captured the attention of the educated public, selling 40,000 copies in Britain alone. It became a fashionable subject at parties, a popular item charged out at libraries and a major source of ire for conservative theologians and academics who wrote pamphlets and preached sermons about the dangers of the book.[56] Ultimately *Vestiges* would play a significant role in setting the groundwork for Darwin's evolutionary theory.[57]

Anticipating the controversy and inevitable accusations of atheism, Chambers attempted an explanation of why the impersonal cosmology he advocated did not oppose Christian theology.[58] He also made a series of rhetorical moves to hide the radical nature of his thesis. He used family metaphors to explain complicated scientific ideas in ways that made them appear less threatening and quoted respectable sources, including a response to Whewell's *Bridgewater Treatise* by respected scientist Charles Babbage.[59] Chamber's private correspondence suggests that he carefully calculated *Vestiges*' deferral to religion in order to avoid the censure that had fallen upon other radical books.[60]

Despite these moves to harmonize his scientific beliefs with prevailing morality, Chambers likely welcomed the possibility that his cosmology would call into question the current social order. He engaged in liberal politics and supported secular education, free trade, and financial reform. He also may have abandoned his earlier Christian religious views.[61] He surrendered much of his own share of *Vestiges*' sales in order give away free books and pushed for a cheap edition to reach a mass audience, suggesting that he wanted to maximize the political and social impact of his book.[62] The anonymous authorship of the text, however, hid the source of these political motives from the wider public.[63]

Even without knowledge about the author, Whewell understood the danger of *Vestiges*, because he had long connected science with morality. Whewell initially refused offers to respond in review form, because he did not want to raise the profile of the book.[64] He eventually acquiesced and wrote *Indications of the Creator* (1845), a collection of parts of his previous writings that he believed refuted the *Vestiges*.[65] The text makes it clear that Whewell did not take the author of the *Vestiges*' word that his cosmology did not seek to supplant religion.

> Innumerable questions of the same kind might be asked, and the conclusion to be drawn is, that every new physical theory which we include in our view of the universe, involves us in new difficulties and perplexities, if we try to erect it into an ultimate and final account of the existence and arrangement of the world in which we live. With the evidence of such theories, considered as scientific generalizations of ascertained facts, with their claims to a place in our natural philosophy, we have here nothing to do. But if they are put forwards as a disclosure of the ultimate cause of that which occurs, and as superseding the necessity of looking further or higher; if they claim a place in our Natural Theology, as well as our Natural Philosophy; we conceive that their pretensions will not bear a moment's examination.[66]

Whewell makes an important distinction between natural philosophy and natural theology that he uses as an indictment of the *Vestiges*. Whewell claims that new understandings of the nature of the universe that emerge from "generalizations of ascertained facts" represent the harmless advance of natural philosophy. Here natural philosophy is understood as understanding the world through observation and logic. Whewell argues that *Vestiges* is dangerous because it goes beyond natural philosophy and seeks to supplant natural theology. Natural theology is, in simple terms, the study of how the nature of the universe reflects the existence of God, the kind of project Whewell participated in with his Bridgewater treatise. The *Vestiges* threatens natural theology because it offers a mechanistic hypothesis for the creation of the universe in tension with the idea that God created the universe. The mechanistic "ultimate cause" of the *Vestiges* "supersed[es] the necessity of looking further or higher." In other words it makes natural philosophy unnecessary.

Going back to his volume of the *Bridgewater Treatises*, Whewell had defended and participated in the practice of natural theology. He wrote, "We may thus, with the greatest propriety, deny to the mechanical philosophers and mathematicians of recent times any authority with regard to their views of the administration of the universe."[67] A mechanized worldview makes its holder, "more than common men, liable to miss the road to truths of extreme consequence."[68] By "truths of extreme consequence," Whewell refers to religion, which means that the mechanized scientific framework threatens religious belief.[69] Although Whewell elsewhere says that the nebular hypothesis does not necessarily conflict with Christianity, his suggestion that it could undermine natural theology indicates that he held deep concerns about the *Vestiges*.[70]

Whewell closes the book with a discussion of the eternal, god-derived nature of good and evil. This appeal to universal and transcendental truth follows in the mold of previous Whewellian attacks on utilitarian social thought.[71] Whewell's inclusion of this passage at the end of a scientific response to the nebular hypothesis suggests he understands the potential political implications of Chambers' cosmology. Whewell, like Chambers, recognized that cosmology could play a pivotal role in shaping the political environment. He treated the *Vestiges* as a scientific defense of the political and philosophical views of John Locke and his materialist allies and used his response to attack Lockean ideas.[72] The way Whewell's discussion of the *Vestiges* spills over to deal with philosophical concerns foreshadows how he uses unity as a foundation to discuss metaphysical issues in *Plurality*.

In the contest for the dominant cosmology, however, Chambers continued to take a commanding lead. Whewell's work, *Indications of the Creator*, did little to stem the popularity of *Vestiges*. The third edition of the *Vestiges* sold out almost immediately and writers in the popular press continued to write positive reviews, eclipsing Whewell's response.[73] As the work continued to gain popularity, Whewell marveled at how the *Vestiges* achieved a success that "no really philosophical book could have had."[74] The success of Chamber's hybrid work of cosmology, politics, and religion, despite vicious attacks by the scientific community, foreshadows Whewell's mimicry of Chamber's popular style in *Plurality*.[75]

As Whewell spent more time and effort on the philosophical difficulties posed by *Vestiges*' success, he began to connect the problematic messages in *Vestiges* to the broader plurality thesis.[76] He came to the conclusion that both gave credence to materialism and decentralized humanity. Given this view, the popularity of plurality made things look bleak for the success of Whewell's philosophical and religious views. He claims to have written *Plurality* with the hope that it "might have some value as a strong case exactly opposed to [the *Vestiges*]."[77] Since *Vestiges* did not deal primarily with the question of plurality, this line suggests a more fundamental attack on the philosophical underpinning of the book.

Before publishing the *Plurality*, Whewell sent drafts to intellectuals whose opinions he respected. When his friend Stephen James wrote him that he should focus more on the scientific arguments and less on the religious elements, Whewell replied,

> With regard to the relation between the Christian scheme and the Plurality of Worlds, I must remark that it is by no means introduced for the first time in pages 246 and 247. On the contrary, *the difficulties* belonging to that relation are the starting point of my whole essay... the topic cannot be excluded for it is in fact *the* topic of my essay.[78]

There is no reason not to take Whewell at his word. The idea that Whewell's book is primarily a defense of Christianity through means of a scientific myth harmonizes well with the events preceding the book's publication. Whewell was engaged in a scientific debate over the cosmology of the *Vestiges,* a battle that he viewed as having philosophical and religious implications. It is reasonable to imagine his defense of unity as an extension of this debate. Whewell's thinking at the time was heavily indebted to Plato, who had presented a cosmology in support for his philosophical program. Finally, it would offer an explanation for his radical shift from his previous pro-plurality position from his *Bridgewater Treatise*.[79] Whewell's revival of the unity myth acted as an argument against both the atheism Whewell saw in the *Vestiges* and the philosophical and political program of the utilitarians.

UNITY AS AN ANSWER TO UTILITY

In 1853, Whewell published the *Plurality*, in which he argues that Earth serves as the only habitation of intelligent, spiritual life in the universe. The book is a strong rebuke not only to the *Vestiges*, but the plurality thesis in general. A significant part of the book reads in a dense scientific prose, which camouflages its function as a philosophical argument. A close reading of *Plurality*, however, reveals that Whewell rather explicitly targets utilitarianism in a way meant to be accessible to a wide audience. Whewell's book followed *Vestiges* in becoming a bestseller, going through numerous editions.[80] Amazingly, direct, published responses to *Plurality* included over eighty articles and twenty books.[81]

If one only read the middle chapters of *Plurality*, he or she would confront a rather dry scientific manuscript, in which Whewell exposes many of the weaknesses of the plurality position. Pluralists following the Copernican revolution rely heavily on arguments based in analogy, for example:

A) Earth is a planet.
B) Earth has life.
A) Mars is a planet.
B) Thus Mars should have life.

The argument suggests that because other planets have qualities similar to Earth that one should assume they are inhabited. Whewell highlights numerous dissimilarities between the other planets and Earth that call into question such an analogy. The pluralists also defended plurality by claiming that the presence of so many planets would represent a waste of space. Whewell makes the rather clever case that wide swaths of time existed before humans came into existence, but that time does not constitute a waste.[82] He also argues that nature has lots of waste, like embryos that fail to emerge.[83] All of these arguments do work to undermine the arguments at the heart of the pluralist case.

Writing hundreds of years after the controversy, William C. Heffernan claims that from a scientific perspective Whewell's arguments demolish the pluralist position.[84] Heffernan makes a compelling argument, as many of the pluralist arguments lacked strong empirical evidence and failed to conform to any kind of clear methodology. And, of course, history has largely vindicated Whewell's position with humanity's space exploration having yet to confirm any life (much less intelligent life) on the planets in our solar system. Given the lack of methods to verify the claim of plurality empirically, however, it is unsurprising that Whewell could poke so many holes in the pluralist case. Crowe makes note of the "confidence with which many accepted pluralism" despite the "paucity of evidence available to support it."[85] What is interesting for this study is not how strong Whewell's arguments were *against* the pluralist position, but how strong they were *for* the unity position.

Whewell went beyond attacking the case for pluralism to argue strenuously for a unity cosmology. Heffernan treats Whewell primarily as a "methodological dissenter," rather than a "proponent… of a negative answer to the age-old question of life elsewhere."[86] This chapter disagrees, arguing that Whewell clearly makes the case for unity as likely fact, rather than mere possibility. This pro-unity thesis frequently put Whewell on scientific ground as shaky as the pluralists; the passion for unity in the absence of clear evidence suggests that Whewell had extra-scientific reasons for his defense of unity. Heffernan acknowledges, but ultimately downplays Whewell's philosophical and religious motivations, "Whewell himself was not without important convictions about man's status or about the significance of the cosmos, but he understood full well that these considerations were not relevant to the empirical question of life elsewhere."[87] Whewell's letter to Stephen James, however, provides a compelling argument that religion was central to his case for unity. *Plurality* also contains numerous connections between religion and philosophy and the unity cosmology, which make treatment of the book absent of its mythic undertones myopic. The fact that religion and philosophy do not hold an even more central role in *Plurality* may owe itself to the constraints placed on Whewell, by his audience.

In constructing the argument that plurality undermined religion and absolute ethics, Whewell faced a difficult rhetorical task. The problem occurs, because, as mentioned at the end of the last chapter, the only other individuals making this argument at the time were deists like Thomas Paine. By contrast, prominent Christian figures, like Thomas Chalmers, had defended plurality as consistent, if not required, by Christianity.[88] Whewell wanted to defend the premise that plurality renders Christianity suspect and humanity insignificant (and the reverse that unity made Christianity true and humanity important), but combined it with a scientific case against plurality and for unity. Whewell shared Chalmers' religious beliefs, but he agreed with the deists that plurality undermined Christianity.

The beginning of *Plurality* navigates this tension by slowly building the connection between unity and religion through a dialogue with Chalmers' work. Chalmers' first lecture on the connection between plurality and Christianity begins with a quote from Psalms,

> When I consider thy heavens, the work of thy fingers, the moon and the stars, which thou hast ordained; What is man, that thou art mindful of him? And the son of man, that thou visitest him? — Psalm viii. 3, 4.[89]

Chalmers argues that the size of the universe does render humanity small in the grand scheme of things, but that God ultimately redeems humanity's value because "I am as much known to him as if I were the single object of his attention."[90] He laments that "infidels" have turned the Psalmist's wonder at the size of the universe (taken in the modern scientific context of plurality) as an argument against Christianity.[91] Although he does not mention any infidels by name he almost certainly refers to Thomas Paine, given the popularity of his work.

In a nod to Chalmers, Whewell also begins his book by focusing on the question posed by the Psalmist.

> The earth is not at rest, with the celestial luminaries circulating above it, as the ancients believed, but itself moves in a circle around the sun, in the course of every year; and the other planets also move around the sun in like manner, in circles, some within and some without that which the earth describes. This collection of planets, thus circulating about the sun, is the Solar System: of which the earth thus forms a very small part. Jupiter and Saturn are much larger than the earth. Mars and Venus are nearly as large. If these be inhabited, as the Earth is, which the analogy for their form, movements and conditions, seems to suggest, the population of the earth is a very small portion of the population of the solar system. And if the mere number of the subjects of God's government could produce any difficulty in the application of his providence to them, a person to who this view of the world which we inhabit had been disclosed, might well, and with far more reason than the Psalmist, exclaim, 'Lord, what is man, that thou art mindful of him? the inhabitant of this Earth, that thou regardest him?'[92]

At this point in the text Whewell does not provide his own answer to the Psalmist's rhetorical question, although this passage does suggest it raises problems for religion. Whewell quotes the Psalmist several more times in the first chapter without assenting to what Chalmers called the "infidel" position. By the end of the book, however, it becomes obvious that Whewell fears plurality renders humanity insignificant and Christianity absurd.[93]

Early in the text Whewell seems leery of scaring off readers by defending the idea that plurality is proof against Christianity. Whewell must have imagined that doing so would cause readers to believe that he was himself opposed to Christianity. In order to prevent guilt by association, he requests that the audience read his objection to plurality (spoken through the words of the Psalmist) as "difficulties of religious men, [rather] than as objections of irreligious men."[94] This suggests he wants the audience to read his book as a dialogue with men like Chalmers, rather than a defense of deists like Paine. Whewell hides behind the Psalmist's rhetorical question, but it becomes clear that he agrees with the "infidels" second premise, that plurality undermines humanity's place in the universe. Whewell attempts to revive the intuitive connection between plurality and human insignificance in the face of objections from individuals like Chalmers. Whewell's decision to begin his book with a defense of the deist objection to Christianity signals to readers that they should not interpret *Plurality* as simply a scientific text. Instead, it indicates that the question of cosmology has enormous bearing on the place of humanity in the universe.

WHEWELL'S UNITY AND THE *TIMAEUS*

Whewell's connection of unity as a necessary condition for Christianity harkens back to Plato and Aquinas. The evidence suggests this connection does not occur as a coincidence and that Whewell drew heavily from Plato's cosmological rhetorical strategy as a model. Early in the final chapter he denotes the two sides in the debate, "The two doctrines which we have here to weigh against each other are the Plurality, and the Unity of the World."[95] What the term "Plurality of Worlds" meant would have been clear enough to general readers in the time it was published given the popularity of the works of Fontenelle and others. In fact, Whewell choose to name his work the "Plurality of Worlds," even though he argued against the proposition, precisely because of the term's familiarity.[96] In order to understand the term "Unity of the World," however, readers would have needed to understand where Whewell borrowed the term.

Although Whewell never explicitly makes the connection, it is likely that he took the term from the *Timaeus*. Here is a passage from Whewell's translation of the *Timaeus* where he connects the term unity with Plato.

> Of one: since it is made after the model. For that [model] which includes all Intelligible animals, cannot exist along with a second. For then, [to have a model really including all,] we must have one including those two, of which they would be parts; and then the universe would be rightly made after the likeness of this Including [model], not of those Included. And thus that the world, by its unity, should resemble the supremely perfect animal [or living thing], the Creator did not make either two or an infinite number of worlds; but, on the contrary, this world is and ever shall be the one created world.[97]

Whewell's clear familiarity with the *Timaeus*, as well as Whewell's use of the term "unity," leaves little doubt that the idea that Earth's singularity as the foundation for teleology emerges from it. In a letter to Herschel, he describes the Greek classics as "essential… [to] the liberal education of the present age" and claims

> To suppose that man will ever become, in these respects, independent of the history and antiquities of literature, seems to me to be equivalent to supposing that he will cease to keep his footing in the path of intellectual progression, of advancing civilization, of the mind and feelings, which began with the philosophical and poetical age of Greece.[98]

In the years prior to writing *Plurality*, Whewell made a transition from a Kantian to a Neo-Platonist, which gives further credence to his deep familiarity with Plato's major dialogues, like the *Timaeus*.[99] In the year he published *Plurality*, Whewell responded to a friend's request to give a lecture by saying, "The only subject on which I could lecture with any satisfaction is

Socrates and Plato, of whom my thoughts are full, and likely to be so for the next year."[100] This letter shows the influence of Platonic thought on Whewell at the time he wrote *Plurality*. In later editions of *Plurality*, Whewell quotes Plato's *Timaeus* before the new sections of text.[101] The clearest sign of the influence of the *Timaeus*, however, comes from the way *Plurality* echoes Plato's arguments for why unity provides a model for ethics.

Whewell adopts Plato and Aquinas' idea that completeness represents a critical component of perfection. Thus, only a singular inhabited world could be perfect, whereas plurality made each planet and its inhabitants imperfect. Whewell writes in *Plurality*, "instead of manufacturing a multitude of worlds on patterns more or less similar, He has been employed in one great work, which we cannot call imperfect, since it includes and suggests all that we can conceive of perfection."[102] Just as in Plato's *Timaeus* and the Thomistic cosmology of the Middle Ages, humanity's singularity is crucial for a universe encoded with perfection.[103] This perfection is a counterargument to utilitarian relativism, because it points to the possibility of transcendent values beyond the human senses. In later editions of the book, Whewell writes,

> The accumulation of innumerable worlds of moral and intellectual creatures, without connexion or subordination, may rather be termed a chaos; 'A mighty maze, and *all* without a plan,' so far as we can see, or even guess. When the term *Cosmos* was invented and applied, those who used it never dreamt of other inhabited worlds besides the earth, and other rational beings besides man.[104]

The quote "A mighty maze, and *all* without a plan" comes from the first edition of Alexander Pope's *Essay on Man*. Pope changed the line to "A mighty maze! But not without a plan," in later editions in response to the suggestion from a friend that the original line was impious.[105] Whewell's use of Pope's original line points to the impiety of plurality.[106] The chaos of plurality could never be the beautiful order (cosmos) to which philosophers like Plato refer.

Whewell makes it clear that the perfection of unity has implications for ongoing philosophical debates. He devotes a section of his book to Plato's notion of ideal knowledge, as opposed to the relativistic thinking of what he calls the "Democritic" philosophers. Whewell says that the Democritic philosophers believe that truth does not exist independently from humanity, but instead that all knowledge derives from human minds.[107] If one views *Plurality* as a purely scientific text on the structure of the cosmos, the discussion of epistemology appears out of place. If one reads Whewell's book as a defense of absolute values, however, the discussion of epistemology fits. Like Plato's *Timaeus*, Whewell's *Plurality* uses cosmology as a bulwark against relativism. The unity of the world provides justification for Whewell's belief in absolute ideals over the relativism and materialism of his opponents.

The origin of the term "Democritic" philosophers helps support this view. In his later supplement to the original text, Whewell explains that he borrowed the term from Richard Owen, English biologist and outspoken critic of evolution. Whewell then acknowledges that Owen borrowed the term from Ralph Cudworth, English philosopher and leader of the Cambridge Platonists.[108] Cudworth places Democritus, Leucippus, and Protagoras under the banner of Democritic philosophers. The term "philosopher" is important, because contrary to most historians today, Cudworth argues that the "physiology" of atomism predates Democritus and Leucippus.[109] Cudworth has a healthy respect for atomism and refuses to believe an atheist could have devised the system, writing "we are of Opinion that neither Democritus, nor Protagoras, nor Leucippus was the first Inventour of this Philosophy and our reason is, because they were all three of them Atheists... and we cannot think that any Atheists could be the Inventours of it, much less that it was the Genuine Spawn and Brood of Atheism."[110] Instead he claims that the Democritic philosophers seized upon atomism to spread skepticism and atheism, despite "rightly understood [atomism being the] most effectual Engin against Atheism that can be."[111] Whewell by contrast would have no trouble viewing Democritus and Leucippus as the founders of atomism, because he viewed atomism as an atheistic system.[112] Whewell's use of Democritic "philosophers" treats the atomists as more than mere scientists; he approaches them as cosmologists in the mold of Plato, with a linked philosophy and cosmic worldview.[113]

Of course it makes little sense for Whewell to devote so much time to the ancient past without it having some relevance to his own time. Cudworth wrote his book in response to the ethical theories of Thomas Hobbes, who he viewed as a spiritual successor to the philosophy of Democritic materialism.[114] Whewell was engaged in an ongoing battle with utilitarians, who he likely regarded as the Democritic philosophers of his own time. Whewell suggests the Democritic philosopher views truth as emerging from human minds, which parallels his earlier critique of utilitarians and materialists for treating ethics and science as emerging solely from human reason. Unity stands as an argument against this proposition in the spirit of natural theology. A single world serves as a connection to a preexisting universal perfection and as a demonstration of the existence of a prefect creator.

While the middle of *Plurality* reads as a rather dry scientific treatise, the ethical implications Whewell outlines in the introduction and conclusion make clear his philosophical program. And while Plato left no clear primary documentation of his motivations for writing the *Timaeus*, Whewell's letters suggest he consciously deployed the strategy of linking scientific cosmology and ethics. Whewell's arguments effectively attack the claims for plurality, but do not provide strong argumentation for unity. Rather than make the case that not enough information existed to make the case for unity or plurality, however, Whewell sides definitively with unity. Whewell entered the cosmological debate on plurality as someone more interested in the practical application of philosophy than the scientific process.[115] Previously he wrote that cosmological debates did not confine their influence to the heavens, but instead had the potential for enormous implications for how people thought about politics, religion and morality. Without diminishing the quality of Whewell's scientific arguments, it is fair to say that *Plurality* primarily serves as a reinvention the unity myth for a post-Copernican world.

WHEWELL'S UNITY

One major change Whewell makes from Plato's *Timaeus* and the reigning cosmology of the Middle Ages is allowing for the possibility that planets may have animal life. Plato, Aquinas, and Philo had disallowed the possibility of other planets, much less inhabited ones. The terms of the scientific debate had shifted following the Copernican revolution, however. With telescopes, astronomers could ascertain that objects like Mars and Venus were likely planets, rather than ethereal spheres. This led to rampant speculation about the possibilities of their inhabitants as detailed at the end of the previous chapter.

Whewell's *Plurality* argues that the question of other planets was never the heart of the unity myth. What matters instead was the idea that humanity holds a central location by merit of its uniqueness. In this spirit, Whewell's cosmology allows for the possibility of animal life in space as long as these animals do not live under a "moral law."[116] Whewell can allow for the possibility of animal life, because he does not see animals as posing a challenge to a human-centered teleology:

> As we have said, we have no insuperable difficulty in conceiving other parts of the Universe to be tenanted by animals. Animal life implies no progress in the species.... Progress implies, or at least suggests, a beginning and an end. If the mere existence of a race imply [*sic*] a sustaining and preserving power in the Creator, the progress of a race implies a guiding and impelling power; a Governor and Director as well as a Creator and Preserver. And progress, not merely in the exercise of bodily faculties, but in the exercise of mental faculties, in the intellectual condition of a portion of the species, still more implies a special position and character of the race; which cannot, without great license of hypothesis, be extended to other races; and which, if so extended becomes unmeaning, from the impossibility of our knowing what is progress in any other species;—from what and towards what it tends.[117]

For Whewell, animal life may be the product of God's creation, but he denies animals receive God's special guidance.[118] Given the distinction he makes between human-like creatures and animals, Whewell argues that Earth's status as the home of a "World of Mind" makes it more valuable "than thousands and millions of stars and planets, even if they were occupied by a myriad times as many species of brute animals as have lived upon the earth since its vivication."[119]

Whewell's dismissal of the value of animal life as threatening to humanity's special status in the universe was consistent with his reaction to utilitarianism.[120] The belief that animals deserved moral consideration was one of the tenets of utilitarianism, which stemmed from Jeremy Bentham's belief that pain and pleasure operate as the proper metric of calculating just action.[121] Whewell's rejection of utilitarianism included a specific refutation of this proposition. For Whewell, the human capacity for reason and moral judgment represents the crucial distinction between humans and animals. He claimed anything that blurred the distinction would justify utilitarianism.[122]

Whewell shifts the terms of the unity/plurality debate in a way that has implications for those that follow him. Plato's *Timaeus* connected plurality with a lack of perfection, because each world would be incomplete. Whewell echos this argument, but why does he allow an exception for animals? Whewell's divisions of human/animal, intelligent/nonintelligent, spiritual/nonspiritual suggest that the completeness of Earth is a function of its status as "World of Mind." Earth gains the status of "World of Mind" by means of human behavior. Only another planet with "moral and intellectual creatures" threatens to render Earth incomplete.[123]

What constitutes "moral and intellectual creatures?" In some respects it is a literal take on Protagoras' maxim, "man is the measure." For Whewell extraterrestrial life is assessed by the presence of humanity's characteristics of intelligence, progression, and spirituality that set a benchmark for what constitutes meaningful life. Later chapters examine other unity thinkers that deploy a similar strategy and argue that life forms without some set of human characteristics do not meaningfully threaten humanity's central place in the universe. This is not to say that all future unity thinkers allow for the possibility of extraterrestrial animal life, as the second part of this chapter demonstrates.

RECEPTION AND RESPONSES

The sheer number of responses to Whewell's *Plurality* suggest that the book had an enormous impact. Todhunter, Whewell's first major biographer, found nine books and twenty-two articles written in response to *Plurality*. Todhunter remarked at the time that "rarely in recent times has a book received so much attention from reviewers as [Plurality]."[124] Crowe expands that number to twenty books and fifty-four articles. Searching on the *American Periodicals Series* (APS) database reveals another twenty-four articles.[125] The discovery of new articles in the APS database further reflects the importance of the debate, in that it spread from England into the United States at a volume much greater than previously revealed. The fervor surrounding Whewell's book reached such a pitch that one journal lamented the attention spent on the plurality question, "to [*sic*] many men of scientific ability and reputation...devote more time to controversy and speculation on subjects of no practical benefit whatever... as it can neither be settled by argument nor science, in its present state.[126] Most reviewers, however, treated the question as enormously important.

David Brewster, Scottish astronomer and President of the British Association for the Advancement of Science, wrote a fiery review condemning Whewell's book, and a year later published his own book-length response, *More Worlds Than One: The Creed of the Philosopher and the Hope of the Christian* (1854). Brewster defends a robust pluralism that included the existence of life on the sun and moon. He brutally attacks Whewell's intelligence and character. Brewster refers to Whewell's arguments as "the most ingenious, though shallow, piece of sophistry which we have ever encountered in modern dialectics."[127] Other reviewers latched on to Brewster's accusation of "sophistry."[128] Brewster also accuses Whewell of being in league with the atheistic philosophy of the *Vestiges*, "[Whewell] tak[es] for granted the truth of the nebular theory, adopted by the author of the Vestiges of Creation, and maintained only by persons who have very erroneous ideas of creation.... Sir Isaac Newton considered the nebular theory as tending to Atheism."[129] Again other reviewers echoed this charge.[130]

The accusations of sophistry and support for the *Vestiges* must have stung Whewell. He had set out precisely to undermine sophistry and the cosmology of the *Vestiges*. Whewell found the tone of Brewster's reply incredibly disagreeable, and he wrote to a friend, "Why'd [Brewster] have to be so savage?"[131] John Hedley Brooke suggests that Brewster's personal grievances with Whewell motivated the attack, as the two had clashed on numerous occasions about education policy and Whewell had written Brewster a poor review.[132] Although this feud almost certainly added incentive for Brewster's caustic word choice, Brewster's daughter wrote that her father hated the book before he knew Whewell authored it.[133] Furthermore, Brooke's thesis does not explain the hostility shown by other reviewers.

A more likely explanation is that pluralism had firmly engrained itself into the scientific community. Todhunter indicates that Whewell's private correspondences shows "no eminent scientific name" wrote him in support of his thesis.[134] Of the scientists Crowe examines who responded to Whewell's book, 83 percent favored pluralism.[135] The enmity of the reviews could easily have emerged from scientists attacking a heterodox position.

Even though most of these scientists shared Whewell's opposition to sophistry, *The Vestiges*, and in many cases utilitarianism, they still viewed the unity proposition as deeply problematic. This may be because these scientists failed to perceive that advances in scientific knowledge that decentralized humanity's place in the universe would become weapons against absolute values and religion. Whewell wrote his book at the end of the period of natural theology, when some scientists wrongly assumed the relations between science and religion would always be harmonious.[136] Religion served as a potential mechanism to escape any sympathy individuals had with the philosophical and theological arguments Whewell raised against the plurality thesis. Two reviews seem at first to concede Whewell's point that a plurality of worlds undermines humanity's value. *The Christian Observer* writes,

> Earth was esteemed but a spark in Jehovah's realm, which might be blotted out of existence and scarcely missed in the universe of matter. So few, also, are its intelligent inhabitants, compared with the mighty whole, that they might be annihilated and their loss in the universe of intelligences would scarcely be discerned—perhaps no more than a drop of water would be missed from the ocean, or a grain of sand from the globe.[137]

One finds a very similar statement in *The Baptist Quarterly*: "For, as we have shown, the earth is but one of many earths. If it were annihilated, its loss would scarce be observed more than the fall of one leaf in a forest."[138] Ultimately, however, both articles claim that Christianity allows the simultaneous existence of intelligent aliens and humanity's privileged position. Brewster makes a similar move, by arguing that of all the planets in the universe Jesus visited Earth and only Earth, thereby redeeming all life on all planets.[139]

Brewster, Chalmers, and the writers for the *Baptist Quarterly* and the *Christian Observer* did not appear troubled by the prospect of human marginalization in a crowded universe, because the Bible clearly outlined God's personal care of humanity. Whewell, however, did not write his book only to preach to natural philosophers. Like his forays into math education and inductive philosophy, Whewell sought a wider audience. He defended a unity cosmology in order to provide a scientific refutation of atheistic utilitarianism. Brewster felt confident that

> A mere inference or a theory in science, however probable, must ever give way to a truth revealed; but a scientific truth must be maintained, however contradictory it may appear to the most cherished doctrines of religion. In freely discussing the subject of a plurality of worlds, there can be no collision between Reason and revelations.[140]

Whewell, however, still reeling from the success of the *Vestiges* and his place on the losing side of the battle to reform education, saw that scientific theories are not neutral. He understood that while natural theologians may be able to harmonize any scientific discovery with religion in their own minds, scientific theories that people viewed as marginalizing humanity's special nature would invite atheism and utilitarianism in the larger public.

Despite his failure to attract much support from the scientific community, Whewell did have his share of supporters.[141] The book achieved major popular success, going through numerous editions. Crowe finds religious journals tended to support Whewell's position about fifty percent of the time.[142] Ten years after Whewell's book, a journalist declared a "state of deadlock" between plurality and unity thought.[143] More importantly from Whewell's perspective, he succeeded in expanding the debate from academia into public discourse. A mere three years after Whewell's book, Anthony Trollope could refer to the Whewell-Brewster debate in his book *Barchester Towers*, confident that the audience would get the reference.[144] In a list describing important new books, a writer for a literary magazine says *Plurality* "need no comment," implying the book's fame would make any remark redundant.[145] The last review of Whewell's book came out in 1867, a startling fourteen years after the original publication of *Plurality*.[146]

Whewell's book reintroduced the elements of the Platonic unity myth into the public debate. The Copernican revolution had smashed the crystalline heavens and opened up the physical space to populate the cosmos. Popular figures like Fontenelle and Huygens soon filled the universe with all manner of intelligent life, which became the scientific and religious orthodoxy. The success of Whewell's work, however, suggests the intuitive and ancient connection between unity and human specialness held strong in the public imagination. The power of the nonscientific aspects of unity ensured that despite widespread scientific opposition, Whewell's work would flourish. The religious and moral stakes of the question of alien life also set the stage for a future revival of the public debate over unity.

A. R. WALLACE: BRIDGE TO THE 20TH CENTURY

By the turn of the century, the combination of the rise of Darwinism, belief in the nebular hypothesis, and the prevalence of a empirical methodology made the idea that "life is not only a possible implication but also a basic property of the universe" the predominant "worldview."[147] But only three years into the twentieth century, Alfred Russel Wallace wrote a book that could easily be mistaken as a sequel to *Plurality*.[148] Wallace's book *Man's Place in the Universe* (1903) revived the unity cosmology in the 20th century. In it, he not only argued Earth was the only planet inhabited by intelligent life, but as

Whewell's *Plurality* did, he connected this proposition with universal values and human purpose.

Man's Place in the Universe proved to be a smash success. It went through seven editions in five years, including cheap ones that the publishers made for a wider non-specialist audience. Despite the disinterest of the academic community, the unity cosmology continued to strike a chord among readers. Wallace's impeccable scientific credentials meant that even people opposed to his beliefs took notice of his unorthodox cosmology.

In 1870, the Entomological society elected Wallace President, he had articles regularly in *Nature*, the most prestigious science journal of the time, and in 1872 he was elected a fellow at the Linnaean society, the world's most prominent society that studied natural history. Alfred Russel Wallace's fame emerged primarily for his contributions to evolutionary theory. He independently developed a theory of evolution remarkably similar to Charles Darwin's and their ideas were presented together at a meeting of the Linnaean Society in 1858.[149] Wallace became a figure at the forefront of defending Darwinism and hewed so strongly to the hypothesis of survival of the fittest that he quipped, "Some of my critics declare that I am more Darwinian than Darwin himself, and in this, I admit, they are not far wrong."[150] Wallace's work on evolution put him in personal contact with many of the scientific luminaries of the time like Thomas Huxley.

Wallace's connection to Darwinism makes his convergence with Whewell's ideas a surprising one. Whewell so deeply despised Darwinism that he tried to ban *On the Origin of Species* from the library at Trinity University.[151] In fact, the differences between the two thinkers extend across a range of issues. Wallace spent his youth absorbed in books like Thomas Paine's *The Age of Reason* and attended lectures on the teachings of Robert Owen, a secularist and socialist.[152] Robert Chamber's *Vestiges* had an enormous impact on Wallace's early thinking about evolution, and he actually became friends with Chambers later in life.[153] Wallace developed an interest in spiritualism that would last throughout his life.[154] As a spiritualist, socialist, secularist, and evolutionist Wallace appears the perfect opposite of the religious, politically conservative William Whewell.

The two did unite on one very important issue aside from cosmology, however. Both figures engaged heavily in political activism and came to the conclusion that science and politics were deeply intertwined. While scholars primarily remember Wallace for his scientific contributions, social activism played a large role in his life. Of his 747 articles, essays, reviews, commentaries, interviews, and letters 25 percent focused on social issues, only a little less than the 27 percent about evolution.[155] Wallace pushed publically for socialism, nationalization of land, rights for women, and against mandatory vaccination and colonization.[156] He held residency in the Land Nationalization Society and corresponded with leading economists and political figures on issues of social justice.[157] He even helped create plans to set up a utopian commune in an uninhabited part of Africa.[158] These political goals represent, in most cases, the polar opposite of Whewell's, but both felt the prevailing scientific and philosophical beliefs of the time threatened their political program. Whewell feared that materialist epistemology would justify utilitarianism, socialism, and secularism. Wallace believed orthodox Darwinism could justify exploitation and inequality at the expense of socialism. Both wanted a scientific defense of their ethical beliefs.

SOCIAL DARWINISM

The very Darwinian scientific revolution that Wallace had helped usher in became a serious threat to his political ideals. Many of his important contemporaries believed that Darwinism gave scientific credence to anti-egalitarianism, scientific myth in action.[159] Wallace himself came to this conclusion, while researching in the Malay islands.[160] He found much to appreciate in the local Malay cultures:

> It is a state of individual freedom and self-government, rendered possible by the equal development and just balance of the intellectual, moral, and physical parts of our nature—a state in which we shall each be so perfectly fitted for a social existence, by knowing what is right and at the same time feeling an irresistible impulse to do what we know to be right that all laws and all punishments shall be unnecessary.... Now it is very remarkable, that among people in a very low stage of civilization, we find some approach to such a perfect social state.[161]

This group of people, which many Europeans viewed as savage, in Wallace's mind exemplified a "perfect" society.[162] Whereas Europeans "have progressed vastly beyond the savage state in intellectual achievements, we have not advanced equally in morals... the mass of our populations have not at all advanced beyond the savage code of morals, and have in many cases sunk below it."[163] Despite the moral inferiority of Europeans, however, it was clear to Wallace that they would triumph over the local cultures either by assimilating the population or exterminating them.[164]

This violent colonization, while an abomination to Wallace, seemed to him inevitable through the lens of Darwinian logic. If nature preserved the "fittest" at the expense of the unfit, then the morally unscrupulous, but technologically superior Europeans, appear destined to destroy the morally just, but technologically inferior, Malaysians. Wallace's fear of the political implications of Darwinism went well beyond the Malaysians. In his autobiography, Wallace attributed many of the negative aspects of America, its "landlordism" and "capitalism" that destroyed the environment and impoverished the lower classes, to a Darwinian process run amuck.[165] Wallace did not need to draw the connection between Darwinism and politics on his own. Darwinism became a buzzword used to justify many of the hegemonic practices of the time, like racism, militarism, and imperialism.[166] Many individuals, both for and against institutions like capitalism, argued that not only were *laissez faire* capitalism and many other political structures inevitable on the grounds of Darwinism, but also ethically were justified.[167]

Francis Galton, the cousin of Charles Darwin, first popularized the potential political implications of Darwinism grounded in the "differential birth rate."[168] He felt that productive individuals (those with good genetics) had few children, while the poor and unproductive (those with bad genetics) had many children.[169] Society thus had artificially changed the natural order from the survival of the fittest to a system where the least fit would have disproportionally high rates of reproduction. The belief in the harm of the differential birth rates caused Galton and others to oppose socialism, because by increasing equality it would further remove checks on the unfit propagating.[170]

Galton's ideas had a major impact on Wallace's thinking about the political ramifications of Darwinism. Galton ranks eight in an analysis of the biggest influences on Wallace's writing.[171] Despite his disdain for coercive eugenic programs and his support for socialism, Wallace always treated Galton's ideas with respect.[172] Wallace felt some level of appreciation for the argument at the foundation of social Darwinism even though he opposed it and sought to find alternative interpretations of the science more consistent with his political beliefs.

Wallace's response to the more extreme proponents of Social Darwinism helps elucidate his thinking on the manner. *Might Is Right* (1890) is one of the most radical works ever written linking Darwinism to ethics and politics. The book attacked socialism, morality, religion, and empathy and sought to instantiate behavior grounded in the survival of the fittest:

> All ethics, politics and philosophies are pure assumptions, built upon assumptions. They rest on no sure basis. They are but shadowy castles-in-the-air erected by day-dreamers, or by rogues, upon nursery fables. It is time they were firmly planted upon an enduring foundation. This can never be accomplished until the racial mind has first been thoroughly cleansed and drastically disinfected of its depraved, alien, and demoralizing concepts of right and wrong...We must be, like nature, hard, cruel, relentless.[173]

Wallace viewed such a message as anathema to his own beliefs and wrote a letter of response in the journal *The Eagle and the Serpent*. Despite his disagreement, he begins with a note of sympathy, writing, "I can understand Dr. Redbeard's position, though I cannot accept it."[174] Here Wallace acknowledges the intuitive connection between Darwinism and the *Might Is*

Right philosophy, but he disagrees with Redbeard, because he views humans as different from other animals:

> If men were only "herds of animals" his view might be the true one. But the mere fact that men, everywhere, and throughout all history, have had words and ideas corresponding to truth, justice, virtue, right, and that there have always been men who would sacrifice even their lives for these ideas, proves that mankind is more than an aggregation of "herds of animals."[175]

Wallace used the distinctions between humanity and other animals as an argument against the "survival of the fittest" justifying behavior he viewed as horrific. In crucial areas, these human/animal distinctions began to affect his beliefs on the mechanics of evolution.

WALLACE'S SCIENTIFIC EVOLUTION

Wallace's concerns about the political implications of Darwinism began to alter his scientific beliefs. At first Wallace maintained that all successful evolutions better equip a creature to survive: "the assertion of 'inutility' in the case of any organ or peculiarity which is not a rudiment or a correlation, is not, and can never be, the statement of a fact, but merely an expression of our ignorance of its purpose or origin."[176] In other words, animals do not have any characteristics that are evolutionarily disadvantageous or even any characteristics that are not evolutionarily advantageous. Wallace began to carve out an exception for human traits, however, making the case that one could not explain traits like consciousness and hairlessness through competition alone. He argued that while these mutations would benefit humans in the long term, in the short term they would have no effect or even a negative effect on their survival.[177] Given the presence of mutations in humans that would not initially have had utility, Wallace argued that a force beyond survival of the fittest must have guided human evolution. The non-utilitarian evolutions served as proof for a teleological undercurrent to evolution, which Wallace associated with his political and spiritual beliefs.[178] The presence of a guiding force beyond strict competition offered an alternative model of nature, which could justify politics of equality and cooperation.

Wallace's change of mind came as a shock to Darwin who wrote, "But I groan over Man—you write like a metamorphosed (in a retrograde direction) naturalist, and you the author of the best paper that ever appeared in the Anthropological Review! Eheu! Eheu! Eheu! —Your miserable Friend, C. Darwin."[179] Darwin reacted so hostilely in part because he, Thomas Huxley, and other prominent scientists sought to create a scientific discipline totally removed from politics and religion.[180] This vision of science had no place for the metaphysical guiding hand that Wallace saw shaping human evolution.

When Wallace wrote his cosmology in *Man's Place in the Universe* it is likely that he did not accept the bifurcation between science and the other disciplines. Many scholars claim that Wallace's change of heart on evolutionary theory reflects a move toward a more traditional natural philosophy approach, which integrated science, religion, and politics.[181] As Fichman eloquently articulates, "Wallace's evolutionary cosmology, with its mix of sociopolitical reformism, theism, spiritualism, and ethical philosophy, abandoned any pretext of ideological neutrality."[182] Of course, Darwin and Huxley had their own ideological agenda, but they pushed it while maintaining the opinion that science was neutral and separate from other modes of inquiry.[183]

Given the number of scholars who suggest that Wallace integrated science, politics, and spiritualism, it should come as no surprise to find political and spiritual implications in his cosmology. However, while many scholars recognize the political and spiritual undertones of *Man's Place in the Universe*, they do not offer much of an explanation of how the book connects with Wallace's broader agenda.[184] Remember that religious opinion split on the question of plurality and prominent Marxist figures like Friedrich Engels had no problem reconciling plurality with a call for social justice. This returns us to William Whewell, with whom Wallace begins his book.[185]

Wallace proclaims his intellectual debt to Whewell's *Plurality of Worlds* early in *Man's Place in the Universe*. In the introduction this occurs mostly in terms of referencing specific scientific arguments that Whewell had made. As one reads the book, however, it is clear Wallace also views the implications of the plurality debate in similar terms. When discussing the Copernican revolution Wallace says that it "seemed to upset the whole accepted order of nature, and to degrade man by removing his dwelling-place, the earth, from the commanding central position it had always before occupied."[186] Wallace must have been acutely aware that the Darwinian revolution he helped engineer had a similar "degrading" effect on humanity. His cosmology represents an attempt to blunt the materialism that many saw as an inevitable product of the new scientific developments.[187] Wallace ends his book by making this claim explicit:

> All this life upon our earth has led up to and culminated in that of man. It has been, I believe a common and not unpopular idea that during the whole process of the rise and growth and extinction of past forms, the earth has been preparing for the ultimate—Man. Much of the wealth and luxuriance of living things, the infinite variety of form and structure, the exquisite grace and beauty in bird and insect, in foliage and flower, may have been the mere by-products of the grand mechanism we call nature—the one and only method of developing humanity. And is it not in perfect harmony with this grandeur of design (if it be design), this vastness of scale, this marvelous process of development through all the ages, that the material universe needed to produce this cradle of

> organic life, and of a being destined to a higher and a permanent existence, should be on a corresponding scale of vastness, of complexity, of beauty? Even if there were no such evidence as I have here adduced for the unique position and exceptional characteristics which distinguish the earth, the old idea that all the planets were inhabited, and that all the stars existed for the sake of other planets, which planets existed to develop life, would, in light of our present knowledge, seem utterly improbable and incredible. It would introduce monotony into a universe whose grand character and teaching is endless diversity. It would imply that to produce the living soul in the marvelous and glorious body of man—man with his faculties, his aspirations, his powers for good and evil—that this was an easy matter which could be brought about anywhere, in any world. *It would imply that man is an animal and nothing more* [italics mine], is of no importance in the universe, needed no great preparations for his advent, only, perhaps, a second-rate demon, and a third or fourth-rate earth.[188]

In the passage, Wallace addresses the concerns of the "degrading" implications of the heliocentric model. Humanity may not inhabit the physical center of the universe, but the universe exists "to produce this cradle of organic life." Humanity holds the spot in the spiritual, metaphysical, or metaphorical center, depending on which specific frame one chooses to deploy.

Beyond heliocentrism, the passage tackles Darwinism by portraying the process of evolution as centered upon humanity. As the *telos* of the universe, humanity becomes elevated from another mere animal produced by the Darwinian process to something special. Wallace addresses this point in a reverse fashion. Absent unity, humanity would be "an animal and nothing more," which implies the opposite that in unity humanity is not just another animal species. Wallace argues Darwinism on a universal scale represents the true threat to spirituality and social progress, by justifying social Darwinism. Darwinism on a universal scale introduces "monotony" into a "diverse" system and makes the process of Darwinism, rather than its result, humanity, the central story of existence. Darwinism on a planetary scale avoids these problems because it merely serves as the mechanism for bringing about humanity, not an end in and of itself.[189]

Survival of the fittest on many planets also opens up the possibility for evolution that results in animals radically different from humanity, just as a non-teleological evolution allows for human futures that do not result in the type of socialist, ecologically minded, egalitarian society that Wallace desired. In fact, a non-teleological world would only bring destruction, as Wallace writes, "If man is a product of blind forces and unconscious laws acting upon non-living matter, then, as he has been produced by physical law, so he will die out by the continued operation of the same laws, against which there is no appeal."[190] For Wallace, humanity's very future depends on its distance from evolution, and humanity gains this distance from its place in the cosmos.

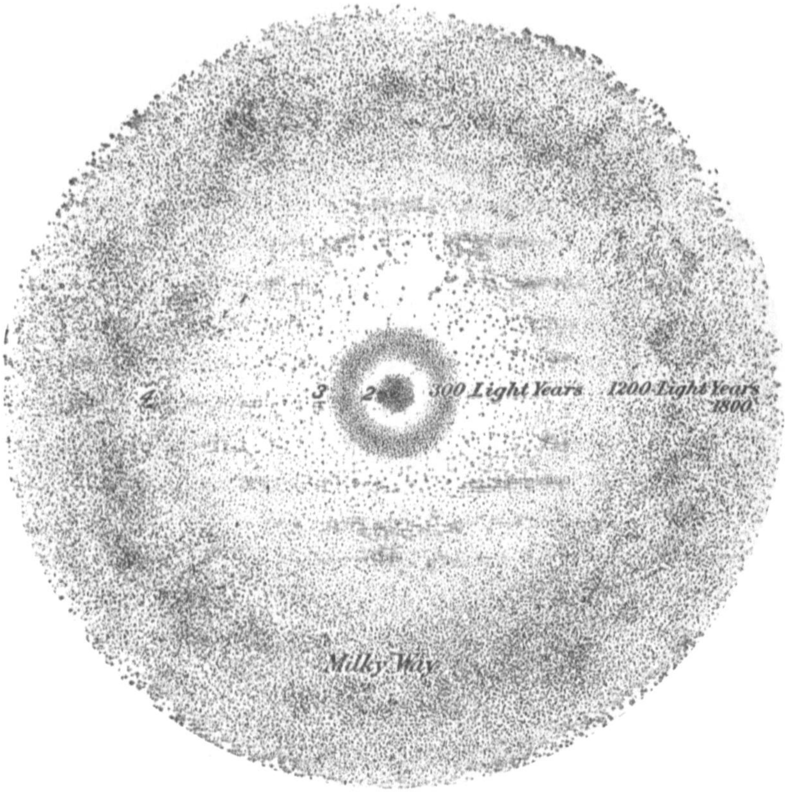

Figure 3.1. Alfred Russell Wallace's depiction of the universe from his book *Man's Place in the Universe*. While Earth does not have a perfectly central location in Wallace's cosmology it still maintains a relatively central location.

In *The World of Life* (1910), which Wallace "considered supplementary" to *Man's Place in the Universe*, he expands on this argument.[191] He shows that the combination of the plurality cosmology and Darwinism provides the justification for some scientists to devalue humanity. Wallace quotes the well-known German biologist Ernst Haeckel to emphasize his point: "Our own human nature sinks to the level of a placental mammal, which has no more value for the universe at large than the ant, the fly of a summer's day, the microscopic infusorium, or the smallest bacillus."[192] While Wallace claims he finds Haeckel's argument "beyond my comprehension," Wallace's earlier grappling with the implications of Darwinism suggest that he took it seriously. Wallace uses unity as physical proof that Haeckel's view of "man's place in the universe" is wrong. In a Darwinian plurality it may be that humanity has no more value than bacteria. However, in the unity cosmology, humanity exists as the central focus of the universe.

Wallace gives another reason that unity provides the necessary foundation for spiritualism and ethics: Darwinism without teleology not only lacks purpose, but also represents a system of profound purposeless suffering. He explains,

> The idea, therefore, that the whole system of nature from the remotest eons of the past—from the very first appearance of life upon earth—has been founded upon destruction of life, on the daily and hourly slaughter of myriads of innocent and often beautiful living things, in order to support the lives of other creatures, which others are specifically adapted to destroy them, and are endowed with all kinds of weapons in order that they may the more certainly capture and devour their victims.[193]

Wallace describes evolution as a mechanism for perfecting violence and the capacity to inflict pain and suffering. All life, including humanity, cannot escape being "founded" on this suffering. Wallace follows this quote with similar ones from Thomas Huxley and J. Arthur Thomson in order to make the case that many prominent scientists view evolution as a cruel product of nature, rather than a system consistent with divine benevolence.[194]

The view of human life as a random product of cruelty and suffering, rather than divine guidance proves deeply problematic for Wallace, because he cannot reconcile it with a benevolent religion or a socialist future.[195] He writes, "All this is so utterly abhorrent to us that we cannot reconcile it with an author of the universe who is at once all-wise, all-powerful and all-good."[196] The belief that massive suffering of evolution disproves a loving God had high-profile adherents, like Huxley, who Wallace quotes, "were our ears sharp enough, we should hear sighs and groans of pain like those heard by Dante at the gate of Hell, the world cannot be governed by what we call benevolence."[197] As previously discussed, Wallace claims that the suffering intrinsic to Darwinism can also call into question progressive politics.[198] If all life exists on a foundation of pain and suffering then one could easily dismiss Wallace's socialist hopes as counter to biological reality, as many did.[199]

The unity cosmology helps ameliorate the Darwinian threat to Wallace's beliefs on religion and politics. In a unity cosmology, the process of evolution to create humanity only needs to occur once. Wallace writes that as terrible as the evolutionary process may be, having it take place once would be justified because humanity has the power to reduce suffering,

> that the universe had no designer or creator, but has always existed; and that the life-pageant, with all its pain and horror, has been repeated cycle after cycle from eternity in the past, and will be repeated in similar cycles for ever.

> We have here presented to us one of the strangest phenomena of the human mind—that numbers of intelligent men are more attracted by a belief which makes the amount of pain which they think does exist on the earth last for all eternity in successive worlds without any permanent and good result whatever, than by another belief, which admits the same amount of pain into one earth only, and for a limited period, while whatever pain there is only exists for the grand purpose of developing a race of spiritual beings, who may thereafter live without physical pain—also for all eternity![200]

Evolution works to create humanity, who through the political process will bring about an end to pain on earth and through spiritual piety end it for all eternity by achieving access to some form of afterlife. Although evolution had its problems, Wallace writes that it must "be *the best*, and almost certainly the *only* method, that could have subsisted through the immeasurable ages and could have then produced a being capable, in some degree, of comprehending and appreciating it. For *that* is surely the glory and distinction of man."[201] Evolution, thus, represents the least necessary evil to create humanity, which could ultimately end the world's suffering through spiritualism and progressive politics.

Plurality called into question the idea of evolution as the least necessary evil, because it required repeating the evolutionary process on different planets. Wallace, like Whewell, viewed the existence of more races of intelligent beings as superfluous. For Wallace, the existence of multiple species "comprehending and appreciating" creation did not offer any significant advantage over only a single species doing it. The idea that the evolutionary process happens on a plurality of worlds, however, would greatly magnify the concerns raised by Huxley. In plurality the horrors of "survival of the fittest" would occur endlessly across the universe, rather than once on earth.

Wallace's view that the unique creation of humanity acts as a justification for the suffering caused by evolution, and explains why he did not join Whewell and some later unity thinkers in allowing for animal life on other planets. If other planets had animals that existed without the presence of spiritual and ethical beings like humanity, then the suffering of those animals would have no purpose. If those animals were fated to evolve into human-like creatures by the teleological process that Wallace claims created humanity, then Wallace would no longer be defending unity.

PUBLIC BATTLE

Wallace saw the unity cosmology as central to his spiritual and political mission. He wrote in his autobiography that the cosmological turn in *Man's Place in the Universe* represents "the 'third chapter of my book,' that is to say of my literary work."[202] He argued that scientists had overstepped their bounds in pushing a materialistic and ethically blind worldview.[203] The public, meanwhile, accepted these pronouncements according to Wallace, because "when a man becomes widely known as a great authority in any department of science, [many people] accept him as a safe guide in any other departments on which he expresses his opinions."[204] Despite his criticism of the mixture of science with ethical and political pronouncements, Wallace took up a similar strategy of using a scientific cosmology to provide justification for his social beliefs. Given Wallace's change of view on the nature of evolution to harmonize his scientific beliefs with his political and spiritual beliefs, this should come as little surprise.

The desire to continue persuading the public ensured that Wallace would persist beyond *Man's Place in the Universe* and *The World of Life.* He vigorously defended his scientific position in the press, writing two articles in response to his critics.[205] In 1906, the famous scientist Percival Lowell wrote a follow up book to his argument that telescopes revealed that Mars had a series of intricate canals.[206] Lowell argued that only an advanced civilization of intelligent creatures could have created these canals. Wallace responded a year later with a book-length review that refuted the science behind Lowell's claims, cementing himself as one of Lowell's most prominent critics.[207] For Wallace, the existence of canals on Mars implied the existence of intelligent life, which threatened spiritualism and socialism.

By integrating science, religion and ethics into his scientific understanding of the world, Wallace created a foundation for social or spiritual change. Social Darwinist thought, with which Wallace was intimately familiar, provided a powerful example of how an ideology attached to science could quickly gain power.[208] Wallace's desire to lend scientific authority to his religious and political arguments help explain why he rejected the label "Wallaceism." He preferred "Darwinism," despite the many differences he and Darwin had over evolution. He wanted to maintain the persuasive capital of Darwinism in order to make the case for his political and religious views, rather than have to publically legitimize a new scientific term, before he could use it rhetorically.[209]

Wallace's political, philosophical and spiritual views impacted his approach to science. He began as an empirically minded scientist, who bragged about being more Darwinian than Darwin. The connections that he and others drew between Darwinism and social order changed the way Wallace engaged in science. He began by trying to make exceptions to the theory of evolution for human intelligence in order to place humanity outside the survival of the fittest framework, which justified exploitation. The unity cosmology served as another scientific justification for humanity's privileged position. Wallace believed unity could provide a foundation for ethics grounded in transcendental values.

Figure 3.2. Percival Lowell observing the planets from a telescope at the Lowell Observatory he established in Arizona.

Whewell and Wallace failed to convert many others to the unity cosmology within the scientific community.[210] Despite the scientific opposition to unity, however, Wallace's cosmological writings achieved major commercial success.[211] Something about unity resonated with the public, despite its rejection among the intelligentsia. Wallace's book delighted Calvinist Christians who continued to reject the peace the Catholic Church had made with plurality.[212] Many presumed the new science of the modern era had definitively enshrined plurality, but nothing could be further from the truth.

CONCLUSION

The scientific myth found in the *Timaeus* reached its zenith in medieval Europe, but the advent of modernity did not spell its end. While scientists relegated orderly orbits, geocentrism, and ethereal celestial objects to the dustbin of history, unity remained an open question. Whewell and Wallace recognized unity's value as a human-privileging science in a time when biology, geology, and physics all increasingly marginalized the species. While fellow scientists may have shunned their work, Whewell and Wallace kept the plurality debate alive in the public imagination.

Whewell and Wallace shifted the unity myth in important ways. Unity in the pre-Copernican era primarily meant opposition to other universes. Following the Copernican revolution, plurality proponents focused almost exclusively on the question of other inhabited planets. Despite this shift, Whewell's *Plurality* was the first major work defending the unity myth in terms of planets, a move followed by Wallace as well as the other case studies examined in this book. Whewell also raises the nature of life on other planets as an important consideration for the question of unity, by denying cosmic importance to extraterrestrial animal life.

The defense of unity by figures with political and religious views as divergent as Whewell and Wallace demonstrates the rhetorical value of unity. If they had lived during the same time period, Whewell and Wallace would likely have fiercely opposed one another politically. Whewell favored conservativism, whereas Wallace supported radical social change. They both believed that their ideologies depended on an inherent dignity of humanity and the existence of transcendent values, however. And both men viewed unity as a scientific proof for these two foundational beliefs.

The way the unity/plurality debate intertwined with broader philosophical and religious issues in the 19th and early 20th century meant that despite the paucity of evidence for either side, strong opinions abounded. A writer from the *National Magazine* cautioned that since the Bible did not explicitly take a position on plurality it was best to keep an open mind, but such an even-handed approach proved exceedingly rare.[213] Most of those who mentioned plurality had strong opinions on the manner. One astute writer for *Littell's Living Age* suggested, "[the question of plurality] will never be settled, it may form a point of comparison for the minds, the methods, and the states of opinion in different ages."[214] The next chapter examines unity in the late 20th century and finds that it is deployed in ways that do indeed reflect the "minds, the methods, and the states of opinion" of its time.

NOTES

1. Maria Mitchell, *Maria Mitchell: Life, Letters, and Journals* (Boston: Lee and Shepard, 1896), 121.
2. Wallace, Alfred Russel *Man's Place in the Universe*, 1903
3. Kuhn, *The Copernican Revolution*, 2; Toulmin, *Cosmopolis*, 82-87; Guthke, *The Last Frontier*, 47-48.
4. Jonathan I. Israel, *Enlightenment Contested: Philosophy, Modernity, and the Emancipation of Man, 1670-1752* (Oxford: Oxford University Press, 2006), 1-15.

5. Arthur Versluis, "Antimodernism," *Telos* 137, no. 96 (2006); Israel, *Enlightenment Contested*, IX-X, 10.

6. Of course, geocentrist hold-outs remained in the 1800s, just as there remain a few during our own time. In 2007, Robert Sungenis and Robert Bennett released two extensive tomes in defense of geocentrism. These authors organized a conference, to discuss geocentrism, held on November 6, 2010. Robert Sungenis and Robert Bennett, *Galileo Was Wrong: The Church Was Right: The Scientific Case for Geocentrism*, vol. 1 (State Line PA: Catholic Apologetics, 2008); Robert Sungenis and Robert Bennett, *Galileo Was Wrong: The Church Was Right: The Historical Case for Geocentrism*, vol. 2 (State Line, PA: Catholic Apologetics, 2008); "Galileo Was Wrong," http://galileowaswrong.com/galileowaswrong/. If the paucity of geocentrists did not adequately reflect the disrepute of these theories their questionable associations certainly do. I visited the website of Catholic Apologetics, the group founded and currently run by Robert Sungenis, who published the books. The organization's book of the month was notable holocaust "revisionist" Thomas Dalton's *Debating the Holocaust: A New Look at Both Sides*. "Bellarmine Theological Forum," *Catholic Apologetics International Publishing*, http://www.catholicintl.com/.

7. William C. Heffernan, "The Singularity of Our Inhabited World: William Whewell and A.R. Wallace in Dissent," *Journal of the History of Ideas* 39, no. 1 (1978); Zachary Sayre Schiffman, *On the Threshold of Modernity* (Baltimore: John Hopkins University Press, 1991).

8. Laura J. Snyder, *Reforming Philosophy: A Victorian Debate on Science and Society* (Chicago: University of Chicago, 2006), 204-321; Cymbre Quincy Raub, "Robert Chambers and William Whewell: A Nineteenth-Century Debate Over the Origin of Language," *Journal of the History of Ideas* 49, no. 2: 294-95.

9. Richard R. Yeo, "Whewell's Philosophy of Knowledge," in *William Whewell: A Composite Portrait*, ed. Menachem Fisch and Simon Schaffer (Oxford: Clarendon Press, 1991), 191; Menachem Fisch, "A Philosopher's Coming of Age: A Study in Erotetic Intellectual History," in *William Whewell: A Composite Portrait*, ed. Menachem Fisch and Simon Schaffer (Oxford: Clarendon, 1991), 294; Gerd Buchdahl, "Deductivist versus Inductivist Approaches in the Philosophy of Science as Illustrated by some Controversies between Whewell and Mill," in *William Whewell: A Composite Portrait*, ed. Menachem Fisch and Simon Schaffer (Oxford: Clarendon, 1991), 321-22.

10. Francis Galton, Charles Darwin's cousin, is credited with founding the social darwinism, understood as concern of differential birthrates between the fit and unfit members of society. Diana B. Paul, "Darwin, Social Darwinism and Eugenics," in *The Cambridge Companion to Darwin*, ed. Johnathan Hodge and Gregory Radick (Cambridge: Cambridge University Press, 2009), 221. This led many individuals to advocate against programs to help the poor as will be discussed later in the chapter.

11. Peter Gay, *The Enlightenment, An Interpretation: The Rise of Modern Paganism*, vol. 1 (New York: Norton, 1977); Peter Gay, *The Enlightenment, An Interpretation: The Science of Freedom*, vol. 2 (New York: Norton, 1977).

12. Crowe, *Extraterrestrial Life Debate*, 263-64.

13. "Chalmer's Discourses on the Christian Revelation."

14. Crowe, *Extraterrestrial Life Debate*, 190-91.

15. William Whewell, Janet Mary Douglas, and Stair Douglas, *The Life and Selections from the Correspondence of William Whewell* (London: C. Kegan Paul, 1881), 221-22. Crowe finds no evidence of this theory, however. Crowe, *Extraterrestrial Life Debate*, 305.

16. "Speculators Among the Stars: Part II," *Littell's Living Age* 8, no. 555 (1855): 73; "Art. II," *British Quarterly Review* 20 (1855); David Brewster, *More Worlds than One: The Creed of the Philosopher and the Hope of the Christian* (New York: Carter & Brothers, 1854).

17. Laura J. Snyder, "William Whewell," *Stanford Encyclopedia of Philosophy*, http://plato.stanford.edu/entries/whewell/.

18. Thomas Carlyle and James Anthony Froude, *Reminiscences* (New York: Charles Scribner's Sons, 1881), 434.

19. Yeo, "Whewell's Philosophy of Knowledge," 191; Snyder, *Reforming Philosophy*, 36-50; Fisch, "A Philosopher's Coming of Age," 294; Buchdahl, "Deductivist versus Inductivist Approaches," 321-22.

20. Jonathan R. Topham, "Beyond the 'Common Context': The Production and Reading of the Bridgewater Treatises " *Isis* 89, no. 2 (1998): 241.
21. William Whewell, *Astronomy and General Physics: Considered with Reference to Natural Theology.*, vol. 3, *The Bridge Water Treatises* (London: William Pickering, 1833), 270.
22. Ibid., 281-82.
23. "Speculators Among the Stars.—Part I," *Littell's Living Age* 44, no. 554 (1855): 5.
24. William Whewell, *The Elements of Morality, Including Polity*, vol. 1 (London: John W. Parker, 1845); William Whewell, *The Philosophy of the Inductive Sciences: Founded Upon Their History*, vol. 1 (London: John W. Parker, West Strand, 1840).
25. Michael Ruse, "Introduction," in *Of the Plurality of Worlds. 1853 A Facsimile of the First Edition with an introduction by Michael Ruse*, ed. Michael Ruse (Chicago: University of Chicago Press, 2001), 11-12.
26. John Hedley Brooke, "Indications of a Creator: Whewell as Apologist and Priest," in *William Whewell: A Composite Portrait*, ed. Menachem Fisch and Simon Schaffer (Oxford: Clarendon, 1991), 155.
27. William Paley, *Principles of Moral and Political Philosophy*, seventh ed. (Philidelphia: Thomas Dobson, 1788); Yeo, "Whewell's Philosophy of Knowledge," 499-500.
28. John Horne Tooke, *Epea Pteroenta or the Diversions of Purley* (London: J. Johnson, 1805); Yeo, "Whewell's Philosophy of Knowledge," 499-500; Andrew R. Cooper, "'Monumental Inscriptions': Language, Rights, the Nation in Coleridge and Horne Tooke," *ELH* 66, no. 1 (1999).
29. Richard Yeo, "William Whewell, Natural Theology and the Philosophy of Science in Mid Nineteenth Century Britain," *Annals of Science* 36 (1979): 499-500; Elizabeth S. Anderson, "John Stuart Mill and Experiments in Living," *Ethics* 102, no. 1 (1991).
30. Perry William, "Passing on the Torch: Whewell's Philosophy and the Principles of English University Education," in *William Whewell: A Composite Portrait*, ed. Menachem Fisch and Simon Schaffer (Oxford: Clarendon, 1991), 132-40.
31. Ibid., 135; Jonathan Hodge, "History of the Earth, Life, and Man," in *William Whewell: A Composite Portrait*, ed. Menachem Fisch and Simon Schaffer (Oxford: Clarendon Press, 1991), 258-60.
32. William, "Passing on the Torch," 132-36. The utilitarians in Britain did advocate numerous political reforms, like economic liberalization and church reform. Leslie Stephens, *The English Utilitarians*, vol. 2 (London: Continuum International Publishing Group, 2005), 41-73. Although they aligned with the radicals, they still maintained a more centrist political position than the Chartists with whom they often feuded over strategy. Leslie Stephens, *The English Utilitarians*, vol. 3 (London: Duckworth, 1900), 31-32, 36.
33. Karen E. Whedbee, "Making the Worse Case Appear the Better: British Reception of the Greek Sophists prior to 1850," *Rhetoric & Public Affairs* 11, no. 4 (2008): 608-13; Giovanni Giorgini, "Radical Plato: John Stuart Mill, George Grote and the Revival of Plato in the Nineteenth-Century England," *History of Political Thought* 30, no. 4 (2009): 621.
34. Whedbee, "Making the Worse Case Appear the Better," 614-22.
35. Giorgini, "Radical Plato," 643-56; Jarratt, *Rereading the Sophists: Classical Rhetoric Refigured*, xxi, 5.Some utilitarians were quick to distance themselves from the sophists. John Stuart Mill attacks the sophists, and claims Plato as a utilitarian, on the very first page of his book *Utilitarianism*, "Socrates listened to the old Protagoras, and asserted (if Plato's dialogue be grounded on real conversation) the theory of utilitarianism against the popular morality of the so-called sophist." John Stuart Mill, *Utilitarianism* (Chicago: University of Chicago Press, 1906), 1.
36. Todhunter, *William Whewell*, 241.
37. Whedbee, "Making the Worse Case Appear the Better," 608-14.
38. Even if one accepts that Whewell did hold a positive view of the sophists this only applied to them as an aggregate class. He held Protagoras's ideas in poor regard as I will discuss when I examine his discussion of the "Democritic" philosophers later in the chapter.
39. William Whewell, *Lectures on the History of Moral Philosophy in England* (Cambridge: Cambridge University Press, 1852), ix-x.

40. Bentham, Hume, Smith, and Paley all drew from Epicurean ideas to bolster their utilitarian views. J. S. Mill in particular made heavy reference to Epicureanism in his works on utility. Frederick Rosen, *Classical Utilitarianism from Hume to Mill* (London: Routledge, 2003), 172-74.

41. William Whewell, *The Platonic Dialogues*, vol. 2 (Cambridge: Macmillan, 1860), 8-9.

42. Whewell at one time held a empiricist epistemology in the mold of Francis Bacon at the same time as he maintained his fervent religious beliefs, which would suggest some sympathy to this point of view. Fisch, "A Philosopher's Coming of Age," 45-47.

43. Frederick Denison Maurice, *Moral and Metaphysical Philosophy: Fourteenth century to the French with a Glimpse into the Nineteenth Century*, vol. 2 (London: Macmillan, 1873), 434.

44. 1.0 A. P. Brogan, "John Locke and Utilitarianism," *Ethics* 69, no. 2 (1959).

45. William Whewell, *The Philosophy of the Inductive Sciences: Founded Upon Their History*, vol. 2 (London: John W. Parker, 1840), 457-69.

46. Ibid., 458.

47. William Whewell, *The Philosophy of the Inductive Sciences: Founded Upon Their History*, 2 ed., vol. 1 (London: John W. Parker, 1847), IV.

48. William Whewell, *Influence of the History of Science upon Intellectual Education: A Lecture Before the Royal Institution of Great Britain* (Boston: Gould and Lincoln, 1854).

49. Michael Ruse, "William Whewell: Omniscientists," in *William Whewell: A Composite Portrait*, ed. Menachem and Fisch and Simon Schaffer (Oxford: Clarendon, 1991), 113; William, "Passing on the Torch," 143. Whewell would approvingly tell the story of Plato refusing students who had not studied geometry. Todhunter, *William Whewell*, 213. In *Elements of Morality,* Whewell uses geometry as a foundational metaphor for his ethical system. William Whewell, *The Elements of Morality including Polity*, vol. 1 (London: John W. Parker, 1845), v-x. He hopes physics and mathematics will provide the foundation of a philosophy based on unchanging truths, rather than experiments Whewell, *The Philosophy of the Inductive Sciences*, 427-28.

50. Fisch, "A Philosopher's Coming of Age," 48; William, "Passing on the Torch," 127-29.

51. Snyder, *Reforming Philosophy*, 205.

52. Buchdahl, "Deductivist versus Inductivist Approaches," 312-22, 35; William, "Passing on the Torch," 117-47.

53. John Stuart Mill to John Ramsey McCulloch, 18 Aug. 1825, quoted in Alexander Bain, *James Mill: A Biography* (London: Longmans, Green and, Co., 1882), 292.

54. Robert Chambers, *Vestiges of the Natural History of Creation* (London: John Churchill 1845).

55. James A. Secord, *Victorian Sensation: The Extraordinary Publication, Reception, and Secret Authorship of Vestiges of the Natural History of Creation* (Chicago: University of Chicago, 2000), 50.

56. Ibid., 2,3, 130-34, 38-40

57. Ibid., 39.

58. Chambers, *Vestiges*, 25-26, 155-59, 384-86.

59. Secord, *Victorian Sensation*, 101-03, 07, 09.

60. Ibid., 103.

61. Ibid., 50, 85, 89-90, 399-400. John M. Lynch argues that Chamber's never abandoned his Christian faith.John M. Lynch, *'Vestiges' and the debate before Darwin*, vol. 6 (Bristol, England: Thoemmes, 200), xxi.

62. Secord, *Victorian Sensation*, 125, 51. Chambers' effort to push for mass readership over financial profit reflects a "social sense" model of authorship that was not uncommon during the period. Ronald J. Zboray and Mary Saracino Zboray, *Literary Dollars and Social Sense: A People's History of the Mass Market Book* (New York: Routledge, 2005), xii-xxi.

63. Chamber's anonymity breaks with the prevalent practice of anonymous texts because of the depth of secrecy. It was rare that a text was really anonymous, in that the author was typically known by at least some circles. Chamber's achieved a "deep" anonymity. Secord, *Victorian Sensation*, 19-20.

64. Richard Yeo, "Science and Intellectual Authority in Mid-Nineteenth-Century Britain: Robert Chambers and 'Vestiges of the Natural History of Creation'," *Victorian Studies* 28, no. 1 (1984): 13.
65. Snyder, *Reforming Philosophy*, 186.
66. William Whewell, *Indications of the Creator: Extracts, Bearing upon Theology, from The History and Philosophy of the Inductive Sciences* (London: John W. Parker, 1845), 18, 19.
67. Whewell, *Astronomy and General Physics*, 334.
68. Ibid., 338.
69. This fits within Whewell's broader resistance of mechanized views of nature. Michael Ruse, "The Relationship between Science and Religion in Britain, 1830-1870," *Church History* 44, no. 4 (1975).
70. Whewell, *Indications of the Creator*, 17-19.
71. Ibid., 167-68.
72. Raub, "Robert Chambers and William Whewell," 292-95.
73. Secord, *Victorian Sensation*, 231.
74. William Whewell to Richard Jones, 18 July 1845, in Isaac Todhunter, *William Whewell, D.D., Master of Trinity College, Cambridge; An Account of His Writings with Selections from His Literary and Scientific Correspondence*, vol. 2 (New York: Johnson Reprint Corporation, 1970), 327.
75. In *Plurality,* Whewell moves away from the overly scientific prose of *Indications* and begins by interacting with the work of Thomas Chalmers. Given the popular success of Chalmers' lectures this move was likely designed in part to draw in lay readers. We know from his lectures to commoners and his development of popular science textbooks that Whewell had both the interest and capability to reach out to nonacademic audiences.
76. Laura J. Snyder, "'Lord Only of the Ruffians and Fiends'? William Whewell and the Plurality of Worlds Debate," *Studies in History and Philosophy of Science* 38 (2007): 586-87; Ruse, "Introduction," 12-13; John Hedley Brooke, "Natural Theology and the Plurality of Worlds: Observations on the Brewster-Whewell Debate," *Annals of Science* 34, no. 5 (1977); Ruse, "The Relationship between Science and Religion in Britain," 517. Michael Crowe attributes the change to Whewell's growing belief that Christianity conflicted with plurality. He tries to distinguish this from the thesis that the *Vestiges* was responsible for Whewell's defense of unity, but I believe that these are complementary arguments. Crowe, *Extraterrestrial Life Debate*, 265-67.
77. William Whewell to J. D. Forbes, February 1854, in Todhunter, *William Whewell*, 400.
78. William Whewell to Stephen James, 25 October 1853 in ibid., 392.
79. Snyder suggests that Whewell had never really committed to the plurality thesis in his *Bridgewater Treatises*. Under this interpretation of history, Whewell never radically changed his mind, but instead moved toward unity over time. This distinction does not make much of a difference for my argument, in either case, Whewell eventually came to the conclusion that unity helped counter materialism and thus embraced it. Snyder, "Lord Only of the Ruffians and Fiends," 584-85.
80. A WorldCat review of books the firm published reveals that they had published Whewell's work *English Moral Philosophy* the previous year, as well as other scientific treatises. They also published in 1851 John Stuart Mill's *A System of Logic, Ratiocinative and Inductive, Being a Connected View of the Principles of Evidence, and the Methods of Scientific Investigation* In the future they would publish important works like Coventry Patmore's megahit *The Angel in the House.* On Parker, see Ruth Richardson, *The Making of Mr. Gray's Anatomy* (New York: Oxford University Press, 2008), 60-92—Whewell's relationship to the publisher is discussed on 65-67.
81. Crowe, *Extraterrestrial Life Debate*, 300. My research has revealed more articles as detailed later in the chapter.
82. Whewell, "Of the Plurality of Worlds," 52-105.
83. Ibid., 224.
84. Heffernan, "The Singularity of our Inhabited World."
85. Crowe, *Extraterrestrial Life Debate*, 264.
86. Heffernan, "The Singularity of Our Inhabited World," 83.

87. Ibid., 86.
88. Whewell's letters reveal that he had Chalmers in mind when he wrote the first chapters of his book. William Whewell to James Stephen 13 September 1853 in Todhunter, *William Whewell*, 380-81.
89. Chalmers, *A Series of Discourses*, 17.
90. Ibid., 53.
91. Ibid., 53-54.
92. Whewell, "Of the Plurality of Worlds," 5.
93. Ibid., 1, 5, 8, 15.
94. Ibid., 22.
95. Ibid., 248.
96. William Whewell, *Of the Plurality of Worlds: An Essay Also, A Dialogue on the Same Subject* (London: Longmans, Green, Reader, and Dyer, 1867), 339.
97. William Whewell, *The Platonic Dialogues: For English Readers*, vol. 3 (Cambridge: Macmillan, 1861), 357, 58.
98. William Whewell to William Herschel, 9 April 1836, quoted in Todhunter, *William Whewell*, 233-36.
99. Ruse, "William Whewell: Omniscientists," 113-14.
100. William Whewell to James Garth Marshal, 19 August 1854 quoted in William Whewell and Janet Mary Douglas, *The Life and Selections from the Correspondence of William Whewell*, 2 ed. (London: Kegan Paul and Trench, 1882), 436.
101. Whewell, *Of the Plurality of Worlds*, 320.
102. Whewell, "Of the Plurality of Worlds," 243.
103. Whewell continues the post Copernican practice of associating perfection with our habitable Earth, rather than the ancient tradition of viewing the plurality of worlds in terms of separate universes. A few philosophers, like Immanuel Kant, maintained the idea that the unity (of the universe) connected with perfection while believing in a plurality of inhabited planets. Andrew Norris Carpenter, "Kant's Earliest Solution to the Mind/Body Problem" (PhD diss., University of California at Berkeley, 1998), 197-200.
104. Whewell, *Of the Plurality of Worlds*, 384-85.
105. William Empson, *Seven Types of Ambiguity* (London: Chatto and Windus, 1949), 203-04.
106. A few years before Whewell's *Plurality of Worlds* was published, reverend George Townsend uses Pope's original line to reference a Godless and chaotic world. George Townsend, "Preliminary Dissertation," in *The Acts and Monuments of John Foxe: A New and Complete Edition*, ed. Stephen Reed Cattley (London: R. B. Seeley and W. Burnside, 1841), 5.
107. Whewell, "Of the Plurality of Worlds," 249-50.
108. Whewell, "A Dialogue on the Plurality of Worlds," 35, 36. All page numbers from this book are from the original first edition, not Ruse's updated page numbers.
109. Ralph Cudworth, *The True Intellectual System of the Universe: The First Part Wherein All the Reason and Philosophy of Atheism is Confuted, and its Impossibility Demonstrated*, 2 ed. (London: J. Walthoe, 1743), xxxi.
110. Ibid., 12.
111. Ibid., 11-12, 17.
112. Whewell seemed to share the prevalent view that Democritus did in fact come up with the idea of atomism. Whewell, *The Philosophy of the Inductive Sciences*, 409-14.
113. Whewell uses the terms "atomic doctrine" and "atomic theory" when referring simply to the scientific, rather than the philosophical and ethical, ideas of Democritus. Ibid.
114. W. M. Kyle, "British Ethical Theories: The Intuitionist Reaction Against Hobbes," in *Thomas Hobbes Critical Assessments*, ed. Preston King (New York: Routledge, 2000), 14-15.
115. Whewell's scientific arguments are hardly distinguishable from his opponents. This reflects the blurriness of the induction/deduction distinction that he tries to draw between his approach and that of his opponents. William Whewell, *Of Induction: With Especial Reference to Mr. J. Stuart Mill's System of Logic* (London: John W. Parker, 1849).
116. Whewell, "Of the Plurality of Worlds," 30.
117. Ibid., 35-36.

118. Whewell presents no real guideline for what level of intelligence or civilization an alien species would have to hold to merit equality to humanity. A charitable reading, however, suggests that Whewell would count any human level of social organization as signs of an intelligent alien species. This seems true in light of his conference of equal value to all humans, regardless of whether others at the time would have considered said people uncivilized or in/ subhuman.

119. Whewell, "Of the Plurality of Worlds," 37, 244.

120. Whewell quotes Bentham's argument for respecting the pleasure and pain of animals as a reductio ad absurdum against the entire philosophy of utilitarianism. J. S. Mill takes him to task for this move.John Stuart Mill, "Whewell on Moral Philosophy 1852," in *The Collected Works of John Stuart Mill*, ed. John M. Robson (Toronto: University of Toronto Press, 1985).

121. The utilitarian concern for animals can be seen as the fringe element of a broader movement towards more liberal treatment of classes previously believed unworthy of consideration. Thomas L. Haskell, "Capitalism and the Origins of the Humanitarian Sensibility: Part 1," *The American Historical Review* 90, no. 2 (1985): 339-61; Thomas L. Haskell, "Capitalism and the Origins of the Humanitarian Sensibility: Part 2," *The American Historical Review* 90, no. 3 (1985): 547-66.

122. Brooke, "Indications of a Creator," 155; Snyder, *Reforming Philosophy*, 250-51.

123. Whewell, *Of the Plurality of Worlds*, 384.

124. Todhunter, *William Whewell*, 190.

125. "Plurality of Worlds," *Circular* 3, no. 150 (1854); "Editors' Book Table: The Plurality of Worlds," *The Independent* 6, no. 290 (1854); "Editors' Table: The History of the Planets. The Elm-Tree," *Godey's Lady's Book and Magazine* 49, no. 173 (1854); W, "The Plurality of Worlds," *Christian Watchman and Reflector* 35, no. 31 (1854); "Book Notices," *The National Magazine* 5 (1854); "The Plurality of Worlds," *Christian Inquirer* 8, no. 47 (1854); "The Plurality of Worlds," *Christian Inquirer* 13, no. 39 (1859); "Literary and Scientific," *Christian Advocate and Journal* 29, no. 40 (1854); "The Plurality of Inhabited Worlds," *The Theological and Literary Journal* 7, no. 2 (1854); , *Christian Examiner and Religious Miscellany* 57, no. 3 (1854); "The Plurality of Worlds," *New Englander* 12, no. 48 (1854); "Plurality of Worlds."; "Publications," *The Monthly Religious Magazine* 12, no. 6 (1854); "The Plurality of Worlds," *Christian Watchman and Reflector* 35, no. 49; "Speculators Among the Stars: Part II."; "Speculators Among the Stars: Part I," *Littell's Living Age* 8, no. 554 (1855); "Worlds in the Sky," *Littell's Living Age* 12, no. 611 (1856); "The Master of Trinity's Plurality of Worlds," *Littell's Living Age* 12, no. 612 (1856); "Review: Plurality of Worlds," *Mercersburg Quarterly Review* (1855); "The Habitability of Worlds," *Christian Review* 80 (1855); "The Unity of the World, Versus the Plurality of Worlds," *The Albion* 14, no. 20 (1855); "The Plurality of Worlds," *The Eclectic Magazine of Foreign Literature* 37, no. 1 (1856); "Literary Notices," *Zion's Herald and Wesleyan Journal* 30, no. 17 (1859); "New Books," *The Albion* 13, no. 27 (1854); "Who shall be Leader?," *Circular* 3, no. 127 (1854).

126. "Are the Stars Inhabited," *Scientific American* 9, no. 44 (1854).

127. Brewster, *More Worlds Than One*, 208.

128. Brewster's charge of sophism was repeated favorably in *Little's Living Age*, "Speculators Among the Stars: Part II," 73. Another reviewer independently leveled the charge of sophistry. "Art. II," 46.

129. Brewster, *More Worlds Than One*, 232.

130. "Review 0," *The Theological and Literary Journal* 7, no. 1 (1854).

131. Whewell to Murchison, 30 May 1854 in Crowe, *Extraterrestrial Life Debate*, 300.

132. Brooke, "Natural Theology and the Plurality of Worlds."

133. Margaret Maria Gordon, *The Home Life of Sir David Brewster* (Edinburgh: Edmonston and Douglas, 1870), 246-47.

134. Todhunter, *William Whewell*, 203.

135. Crowe, *Extraterrestrial Life Debate*, 352.

136. Susan Faye Cannon, *Science in Culture: The Early Victorian Period* (New York: Dawson and Science History Publications, 1978), 1-3.

137. "Hashshamayun; or the Heavens; The Plurality of Worlds," *Christian Observer* 37, no. 23 (1858).

138. Winkler, "Religion and Astronomy," 72.
139. Brewster, *More Worlds Than One*, 145-46.
140. Ibid., 133.
141. Despite the failure to attract much scientific support Whewell never surrendered the scientific case for unity in order to focus solely on religious and moral arguments. In his supplement to *Plurality* he carefully responded to the objections of all his major critics. Todhunter, *William Whewell*, 186.
142. Crowe, *Extraterrestrial Life Debate*, 352.
143. "Les Pluralite Des Mondes Habites," *Little's Living Age* 83 (1864): 363.
144. Anthony Trollope, *Barchester Towers*, vol. 1 (Leipzig: Bernhard Tauchnitz, 1859), 226-27; Crowe, *Extraterrestrial Life Debate*, 351.
145. T. W. M., "One Letter From Boston," *American Publishers' Circular and Literary Gazetter* 2, no. 7 (1856): 94.
146. Crowe, *Extraterrestrial Life Debate*, 359.
147. Dick, *The Biological Universe*, 27-35.
148. The publisher of Wallace's book Chapman and Hall is notable for their publication of other major works, like H. G. Well's *Anticipations of the Reaction of Mechanical and Scientific Progress Upon Human Life and Thought* (1902), *Mankind in the Making* (1903), and *A Modern Utopia*. (1905).
149. Neither Wallace nor Darwin made the event, Wallace because he was in the Malay Archipelago and Darwin because he was sick.
150. James Marchant, *Alfred Russel Wallace: Letters and Reminisences* (New York: Arno Press, 1975), 274, 75.
151. Michael Shermer, *In Darwin's Shadow: The Life and Science of Alfred Russel Wallace* (Oxford: Oxford University Press, 2002), 204.
152. Peter Raby, *Alfred Russel Wallace: A Life* (Princeton: Princeton University Press, 2001), 14, 27, 191; David A. Stack, "The First Darwinian Left: Radical and Socalist Responses to Darwin, 1859-1914," *History of Political Thought* 21, no. 4 (2000): 690.
153. Shermer, *In Darwin's Shadow*, 183; John R. Durant, "Scientific Naturalism and Social Reform in the Thought of Alfred Russel Wallace," *British Journal for the History of Science* 12, no. 1 (1979): 37, 38.
154. Raby, *Alfred Russel Wallace*, 22-23; Martin Fichman, *An Elusive Victorian: The Evolution of Alfred Russel Wallace* (Chicago: University of Chicago Press, 2004), 72-73.
155. Shermer, *In Darwin's Shadow*, 17. Shermer characterizes Wallace's other works as dealing with biogeography and natural history (29%), spirituality and phrenology (7%), and anthropology (12%).
156. Stack, "The First Darwinian Left," 690. Wallace believed in a socialism modeled on Edward Bellamy's book *Looking Backward* (1888), which he believed struck a balance between individual rights and economic justice. Fichman, *An Elusive Victorian*, 250-51.
157. Fichman, *An Elusive Victorian*, 216, 30, 33; Makolm Jay Kottler, "Alfred Russel Wallace: The Origin of Man, and Spiritualism," *Isis* 65, no. 2 (1974): 162.
158. Shermer, *In Darwin's Shadow*, 247.
159. Gregory Claeys, "The "Survival of the Fittest" and the Origins of Social Darwinism," *Journal of the History of Ideas* 61, no. 2 (2000): 226. Lessl provides the most comprehensive examination of the origins of the scientific myth of Darwinism, which he calls "evolutionism." Thomas M. Lessl, *Rhetorical Darwinism: Religion, Evolution, and the Scientific Identity* (Waco, TX: Baylor University Press, 2012).
160. Wallace exemplified the "field scientist" approach mixing observation and high order science. Jeremy Vetter, "Wallace's Other Line: Human Biogeography and Field Practice in the Eastern Colonial Tropics," *Journal of the History of Biology* 39, no. 1 (2006).
161. Alfred Russel Wallace, *The Malay Archipelago: The Land of the Orang-utan, and the Bird of Paradise* (New York: Harper & Brothers, 1869), 596, 97.
162. Differences of opinion between Wallace and the mainstream scientific organizations of the time, stemming in large part from his observations in Malay, contributed to his embrace of radical leftist politics. Jeremy Vetter, "The Unmaking of an Anthropologist: Wallace Returns from the Field, 1862–70," *Notes and Records of the Royal Society* 64 (2010): 34-35.

163. Wallace, *The Malay Archipelago*, 597.
164. Ibid., 595, 96.
165. Alfred Russel Wallace, *My Life: A Record of Events and Opinions*, vol. 2 (New York: New York, Dodd, Mead, 1905), 192-98. His negative views of structures like "landlordism" came in large part from his reading of Henry George, who he later met with on his trip to the United States. Fichman, *An Elusive Victorian*, 220-24; Shermer, *In Darwin's Shadow*, 274.
166. Claeys, "The "Survival of the Fittest" and the Origins of Social Darwinism," 226. There did exist Darwinian socialists, who downplayed the Malthusian and natural selection elements of Darwinism to emphasize other potential evolutionary factors. Paul, "Darwin, Social Darwinism and Eugenics," 237.
167. Bernard Lightman, "Ideology, Evolution and Late-Victorian Agnostic Popularizers," in *History, Humanity and Evolution: Essays for John C. Greene*, ed. James Richard Moore (Cambridge: Cambridge University Press, 1989), 296-97.
168. Paul, "Darwin, Social Darwinism and Eugenics," 221.
169. Socialism and liberalism also latched onto Darwinian ideas in order to justify their beliefs. But for the most part these efforts amounted to little more than grafting the Darwinian buzzwords onto a preexisting ideology. Stack, "The First Darwinian Left."; Paul, "Darwin, Social Darwinism and Eugenics," 233. With a few exceptions (like Wallace) most of those really applying the principles of Darwin to social practices advocated forms of eugenics and class warfare against the poor. Ibid., 234; R. J. Halliday, "Social Darwinism: A Definition," *Victorian Studies* 14, no. 4 (1971): 403-04. This is demonstrated by the entrants for a contest on the meaning of Darwinism for social practices where despite a few liberal and socialist outliers, the vast majority of entrants advocated state controlled eugenics. Paul, "Darwin, Social Darwinism and Eugenics," 238-39.
170. Halliday, "Social Darwinism: A Definition."
171. Smith counts the number of times Wallace's references an individual and gives each reference further weight if the individuals name is bolded or underlined. Charles Smith, "The Most Important People in Wallace's Intellectual Life," *The Alfred Russel Wallace*, http://people.wku.edu/charles.smith/index1.htm.
172. Diana B. Paul, "Wallace, Women, and Eugenics," in *Natural Selection and Beyond*, ed. Charles H. Smith and George Beccaloni (Oxford: Oxford University Press, 2008), 264-68.
173. Ragnar Redbeard and Darrell W. Conder, *Might is Right or Survival of the Fittest*, New ed. (Springfield, MO: Dil Pickle Press, 2005), 3.
174. Alfred Russel Wallace, *The Eagle and the Serpent* (1900), http://people.wku.edu/charles.smith/wallace/S580.htm.
175. Ibid.
176. Alfred Russel Wallace, *Darwinism: An Expositon of the Theory of Natural Selection with Some of its Applications* (London: Macmillan, 1889), 137.
177. Kottler, "Alfred Russel Wallace," 154-56.
178. A similar politically inspired about face occurred over the question of the role sex-selection in evolution. Wallace initially believed that only adaptations that increased survival would propagate themselves and he downplayed the idea that some traits would exist only because they appealed aesthetically to members of the opposite sex. Shermer, *In Darwin's Shadow*, 212-13.
179. Marchant, *Alfred Russel Wallace*, 206.
180. Martin Fichman, "Biology and Politics: Defining the Boundaries," in *Victorian Science in Context*, ed. Bernard V. Lightman (Chicago: University of Chicago Press, 1997), 103.
181. Joel S. Schwartz, "Darwin, Wallace, and the "Descent of Man"," *Journal of the History of Biology* 17, no. 2 (1984): 283-86; Durant, "Scientific Naturalism and Social Reform in the Thought of Alfred Russel Wallace," 40-53; Fichman, *An Elusive Victorian*, 268-77.
182. Fichman, *An Elusive Victorian*, 6.
183. Fichman, "Biology and Politics: Defining the Boundaries," 103-05.
184. Shermer, *In Darwin's Shadow*, 22, 231; Raby, *Alfred Russel Wallace*, 272 ; Fichman, *An Elusive Victorian*, 288, 301-13.
185. Wallace, *Man's Place in the Universe*, 12-24.
186. Ibid., 8.

187. It should come as no surprise that Wallace initially titled the book *Universe for Man*, a name that foreshadows the heavily anthropocentric nature of the work. Steven J. Dick, "The Universe and Alfred Russel Wallace," in *Natural Selection and Beyond*, ed. Charles H. Smith and George Beccaloni (Oxford: Oxford University Press, 2008), 325.

188. Wallace, *Man's Place in the Universe*, 316-17.

189. It is important to stress that not all socialists held this view. In fact, many felt that Darwinism justified the Marxist belief in progress. Stack, "The First Darwinian Left."

190. Alfred Russel Wallace, "Man's Place in the Universe," *Fortnightly Review* 74, no. 435-411 (1903): 397.

191. Wallace, *The World of Life*, VIII.

192. Ibid., 403.

193. Ibid., 398.

194. Ibid., 399-400.

195. This utilitarian view of the pain inflicted by nature was bolstered by Darwinism. Many individuals at the time felt that Darwin's theory provided scientific proof for orthodox utilitarian beliefs. Darwinism did alter utilitarian thinking in that it caused many proponents to view individuals as a product of their genetic makeup (and thus the choices of their ancestors), rather than fully autonomous individuals. Halliday, "Social Darwinism: A Definition," 395-96. This gives further credence to the idea that Wallace wanted a view of evolution where a break occurred between humanity and its animal past, in order to separate humanity from the baser instincts of its ancestors.

196. Wallace, *The World of Life*, 398.

197. Ibid., 400.

198. The reason the prospect of a cruel natural order represented such a deep concern for Wallace and did not bother many other liberal Darwinist probably stems from his belief in natural philosophy. Wallace, like Whewell, Plato, and Aquinas, believed the natural order shaped the moral order. Other liberal Darwinists could maintain socialist beliefs in conjunction with their belief in evolution with no difficulty, because they viewed human social order as disconnected from the natural order (although some viewed the evolutionary aspect of Darwinism as a natural indication of progression that would culminate in socialism). Stack, "The First Darwinian Left."

199. Lightman, "History, Humanity and Evolution," 296-97; James Allen Rogers, "Darwinism and Social Darwinism," *Journal of the History of Ideas* 33, no. 2 (1972).

200. Wallace, *The World of Life*, 400, 01.

201. Ibid., 403.

202. Wallace, *My Life*, 416.

203. Wallace, *The World of Life*, 7.

204. Ibid.

205. Alfred Russel Wallace, "Does Man Exist in Other Worlds "A Reply to my Critics"," *The Daily Mail*, Nov. 12, 1903; Alfred Russel Wallace, "Man's Place in the Universe: A Reply to Criticisms," *The Independent* Aug. 27, 1903.

206. Percival Lowell, *Mars and Its Canals* (London: Macmillan, 1906). Lowell's position attracted distain from scientists independent of their belief in unity, because of his amateur approach to the scientific question. Norriss S. Hetherington, "Percival Lowell: Professional Scientist or Interloper?," *Journal of the History of Ideas* 42, no. 1 (1981).

207. Alfred Russel Wallace, *Is Mars Habitable? A Critical Examination of Professor Percival Lowell's Book "Mars and Its Canals," with an Alternative Explanation* (London: Macmillan & Co., 1907).

208. Even at the time many people recognized the rhetorical power of science to persuade. George Cornewall Lewis wrote in 1849, "No species of imposture is so captivating, so well-suited to the present time, and consequently so likely to meet with temporary success, as that which assumes the garb, and mimics the phraseology, of science." George Cornewall Lewis, *An Essay on the Influence of Authority in Matters of Opinion* (London: J. Parker, 1849), 55. The Darwinian revolution combined with rhetorical maneuvers of scientific advocates only cemented this credibility. Thomas F. Gieryn, *Cultural Boundaries of Science: Credibility on the Line* (Chicago: Chicago University Press, 1999), 62-64.

209. Fichman, *An Elusive Victorian*, 310-11.
210. Dick, *The Biological Universe*, 50-51.
211. *Man's Place in the Universe* went through seven editions in five years. Fichman, *An Elusive Victorian*, 293.
212. Alfred William Benn, *History of English Rationalism in the Nineteenth Century*, vol. 1 (London: Longmans and Green, 1906), 84-86.
213. "The Silence of Scripture," *National Magazine* 13 (1858).
214. "Plurality of Worlds," *Littell's Living Age* 11, no. 596 (1855).

Chapter Four

Quantum Unity

> "What should one make of this quartet of WAP, SAP, PAP, and FAP? In my not so humble opinion I think the last principle is best called CRAP, the Completely Ridiculous Anthropic Principle."—Martin Gardner, 1986[1]

The role of science in Western societies of the twentieth century weakened the foundations for grounding traditional philosophical, religious, and ethical appeals in unity.[2] The nineteenth century saw the success of a movement to split the scientific method from natural theology and philosophical concerns.[3] The scientific community largely promoted an "objective," materialist methodology that treated philosophical concerns like teleology and morality as irrelevant and occasionally antithetical to science.[4] This culture of "objectivity" manifested itself most clearly in its disregard for religious and spiritual ideas as tools for explaining how the world works.[5] Even absent outright hostility, the scientific social structure delimits thought outside the confines of accepted scientific methods, leaving spiritual and philosophical concerns to other disciplines.[6] This new understanding of "science," stripped of its past associations with religion, emerged as one of the definitive rhetorical frames of the twentieth century, such that appeals to science became a replacement for religious appeals.[7]

As science strived toward ever more "objectivity," the humanities moved in the opposite direction. The 1970s and 1980s saw the rise to prominence of postmodernist academics in western universities whose work many accused of undermining the idea of objective truth in disciplines like history, philosophy, and ethics.[8] With science disinterested in anything outside of its realm, and postmodernism positioned in opposition to objective searches for truth, the kind of natural philosophy that wedded science with philosophy and religion was all but extinct. The gulf between science and philosophy created a difficult environment for a rhetor to reintroduce the scientific myth of unity because of the way it depends on fluidity between myth, philosophy, religion, and science.

Tipler and Barrow's 1986 book *The Anthropic Cosmological Principle* managed to revive both the unity cosmology and its connection to absolute values, without surrendering scientific legitimacy. Their work moves smoothly from the latest thinking on cosmology and quantum mechanics to a deep review of the history of teleology. No longer in the days of natural philosophy, this combination had the potential to undermine the book's claim of scientific legitimacy.[9] Tipler and Barrow successfully walk a fine line, however, only hinting at the philosophical implications of their work, all the while adopting a technical style reminiscent of a scientific journal.[10] The successful balancing act made their work acceptable within the scientific community, despite the book's deviation from the normal practice of excluding the philosophical (and especially teleological) implications from discussions.[11] This scientifically legitimated philosophy helped ensure Tipler and Barrow's ideas took hold well beyond the hard sciences.[12] Philosophers and theologians made heavy use of the book to defend teleology with a "new" scientific justification, as will be discussed below. Beyond academia, *The Anthropic Cosmological Principle* became a best seller and even found its way into the political speeches of Vaclav Havel, the President of the Czech Republic.[13]

Despite the updated science found in the book, Tipler and Barrow present a version of the unity myth that builds on the work of Plato, Aquinas, and others. Brian Aldiss, writing in *Nature*, referred to the Tipler and Barrow's *Anthropic Cosmological Principle* as "a powerful sequel to Whewell's argument."[14] Aldiss makes the connection between the two works based solely on their shared rejection of the possibility of alien life. This chapter reads *The Anthropic Cosmological Principle* rhetorically, and claims that it serves as a sequel to the works of Plato, Aquinas, Whewell, and Wallace not only because of its denial of plurality but also in how it presents a modern-day instantiation of the scientific unity myth.[15] Tipler and Barrow face new constraints based on the preferences of their audience, which shape the way they frame the issue, but the connection of unity with absolute values remains. Tipler and Barrow present metaphysics in the guise of physics, which creates a broadbased rhetorical appeal grounded in the modern day respect afforded to science.

Before examining the arguments of Tipler and Barrow it is important to understand the philosophical climate at the time of their book's publication. Just as the earlier unity myths responded to forms of relativism in their own times, *Timaeus* to sophism, *Plurality of Worlds* to utilitarianism, and *Man's Place in the Universe* to Social Darwinism, Tipler and Barrow's work *The Anthropic Cosmological Principle* (1986) is in conversation with contemporary relativism. The next section examines postmodernists as contemporary defenders of relativism. Following this, the chapter examines the state of the scientific unity/plurality debate. It then examines how Tipler and Barrow advance their own unity myth in *The Anthropic Cosmological Principle* and details its public reception. Finally, the chapter examines later works by Tipler in order to help highlight the sometimes opaque scientific myths found in *The Anthropic Cosmological Principle.*

POSTMODERNISTS AS THE NEW SOPHISTS

Postmodernism, like many intellectual schools of thought, lacks clear boundaries of inclusion. Many scholars commonly associated with postmodernism, like Judith Butler and Michel Foucault, reject the label.[16] Jean-Francois Lyotard, in his book *The Postmodern Condition* (1978) says at the most basic and simplified level "I define postmodern as incredulity toward metanarratives," but the term "postmodernist" was widely used and few shared ideas connect its varied adherents.[17]

Because of the ambiguity of the term, this book uses the terms postmodernist and postmodernism to refer to various academics and academic philosophies in the late 20th and early 21st century that directly attack or indirectly weaken hierarchy, teleology, and the possibility of objective truth. The chapter uses this definition in part because Tipler and many of those that use Tipler and Barrow's cosmology to oppose relativism label their opponents postmodernists and treat relativism as their defining characteristic. The classification of thinkers who defended varying degrees of relativism or engaged in practices that promoted relativism under the label of postmodernist is also consistent with the broader critiques of postmodernism that circulated at the time, like this one relayed in the *Washington Post.*

> It is in academe—in the study of philosophy, literature and history—that postmodernism has had its most profound, and some say dangerous, impact. Under the influence of French philosophers Jacques Derrida, Michel Foucault and Jean-Francois Lyotard—who preach the debunking, dismantling and reexamination of accepted beliefs—scholars and students have evolved a new analytical approach that, according to conservatives, threatens to eviscerate the whole notion of "truth" and the practice of scholarship.[18]

Not all individuals that self-identified or were identified by critics as postmodernists advocated relativism or advanced ideas that promoted relativism (just as not all Sophists were relativists in their thinking despite the common association). This chapter echoes the poorly defined and homogenizing definition of postmodernism used by its critics, because it set the terms of the public debate and because there for the purposes of this limited study there is no better way to designate the broad group of thinkers in question.

Prior to the 1980s postmodernist thought had a series of high profile adherents, whose work acted to undermine the current academic landscape. Foucault argued that history emerges from contingent events and attempts by historians to order past events into inevitable historical trends were projections of our own society.[19] Derrida disrupted English, among many other disciplines, when he argued that texts lack a stable meaning.[20] The Critical Legal Studies movement pointed out how a country's system of laws is inherently indeterminate, such that statutes are applied in contradictory ways to similar cases. This legal indeterminacy becomes a tool to maintain the interests of those in power at the expense of minorities, the poor, women, and other disadvantaged groups.[21] In the late seventies and early eighties postmodern ideas had gained a stronghold in American universities and the incredulity toward metanarratives manifested itself in efforts to change or eliminate the cannon of books, calls for cultural relativism and identity politics, and critiques of capitalist culture.[22]

Like past groups that questioned the existence of absolute immutable values, critics were quick to label postmodernists as sophists. In *The Closing of the American Mind* (1987), Harvard professor of English Allan Bloom compares postmodernists like Derrida and Foucault to the "sophists."[23] He was hardly alone in doing so; prominent postmodernists like Richard Rorty, Derrida and many others, openly embraced the connection and moved to rehabilitate philosophy's ancient nemeses.[24]

Plato, Whewell, and Wallace's cosmologies all respond to a relativist movement in their own time. Whether the rise of postmodernism inspired *The Anthropic Cosmological Principle* is difficult to say. Barrow flatly denies the idea, though he does seem to acknowledge that *Anthropic Cosmological Principle* played an important role in the debate over postmodernism.[25] Tipler did not respond to any emails or voicemail messages, and thus his thoughts on the subject are unknown. What is known is that significantly after the publication of *Anthropic Cosmological Principle*, Tipler publically attacked postmodernism, calling it a form of "irrationality."[26] He also penned an article titled "Postmodern Physics: Colleges Fail to Teach Basics—Even in Physics" that weighed in favor of the traditional cannon. In the article he suggests that postmodernists "no longer believe there are fundamental truths in their own disciplines" and bemoans that this has contributed to a state where the "corruption of education is probably universal across all disciplines."[27]

Whether or not Tipler and Barrow intended *The Anthropic Cosmological Principle* as a specific intervention against postmodernism, it is useful to read it in conversation with postmodernism. As the chapter later demonstrates, the reception of the book clearly shows that many individuals interpreted the book as a response to postmodernism. Also the themes of the book directly respond to the anti-teleological, anti-metanarrative perspective that many associated with postmodern thought. While their book attacks relativism, not postmodernism by name, it is clear to anyone paying attention that postmodernists were the contemporary thinkers most closely associated with relativism. Of course the unity myth depends on unity science and just as relativism had gained a hold in academia so had pluralism made a steady advance in popularity from the time Wallace wrote *Man's Place in the Universe*.

STATE OF THE SCIENTIFIC DEBATE

Following Wallace, the 1920s and 1930s saw some other scientists dismiss the possibility of alien life outside the solar system. They argued that stellar collisions rendered much of the universe uninhabitable. This, however, was a minority position, rejected by astronomer royal Harold Spencer Jones's *Life on Other Worlds* (1940), which became the standard text on the subject for a quarter of century.[28] Jones put forth an image of a vast universe, in which our solar system did not have a special location. He wrote, "With the universe constructed on so vast a scale, it would seem inherently improbable that our small Earth can be the only home of life."[29] The book that would replace Spencer's as the dominant text on the plurality question, Carl Sagan and Iosef Shklovskii's *Intelligent Life in the Universe* (1966), proved even more supportive of the probability of alien life, estimating between 50 thousand and a million technologically advanced alien civilizations within our galaxy.[30]

Frank Drake created a formula (the Drake Equation) for determining the probable number of intelligent alien civilizations, which had the technology to send signals of their existence. The formula "$N = R^* \cdot f_p \cdot n_e \cdot f_l \cdot f_i \cdot f_c \cdot L$" took a variety of factors relevant to the development of an alien civilization and multiplied them together to estimate how many advanced alien civilizations exist, where:

N = The number of civilizations in the Milky Way Galaxy whose electromagnetic emissions are detectable.
R^* = The rate of formation of stars suitable for the development of intelligent life.
f_p = The fraction of those stars with planetary systems.
n_e = The number of planets, per solar system, with an environment suitable for life.
f_l = The fraction of suitable planets on which life actually appears.
f_i = The fraction of life bearing planets on which intelligent life emerges.
f_c = The fraction of civilizations that develop a technology that releases detectable signs of their existence into space.

L = The length of time such civilizations release detectable signals into space.

Early readings of the Drake equation by scientists of the time placed the number of advanced alien civilizations in the Milky Way galaxy between one thousand and one hundred million.[31] In no small part due to Drake's equation, the plurality thesis maintained a comfortable place with the astronomy community.[32]

By 1980, plurality had been the dominant scientific paradigm for decades and this influence spilled beyond academia to the broader culture.[33] Carl Sagan rocketed to fame by speaking and writing about the likelihood of extraterrestrial life and even promoted plurality on his immensely popular scientific program *Cosmos*.[34] ET (1982) became the highest-grossing movie of all time (a title it held for ten years, until the release of *Jurassic Park*).[35] Public interest in alien life pushed the US government to make it a priority for NASA. In fact, scientists attributed NASA's ability to acquire public funding in the economic downturn of the 1970s to Sagan's promise of discovering intelligent aliens.[36] NASA made the search for alien life one of its top priorities during the exploration of Mars, which helped gain the mission enthusiastic public support.[37] In 1983 the United States government provided $1.5 million to the Search for Extraterrestrial Intelligence (SETI) to scan for radio messages from alien civilizations.[38] SETI represented one of the few areas of agreement between the United States and the Soviet Union: both invested resources into exploring the question of alien life and scientists from both countries met to discuss the question.[39]

Despite the broad support for plurality some scientists still held strong in their support for unity beliefs. Many biologists never accepted the plurality consensus, with important scientists like Theodosius Dobzhansky, George Gaylord Simpson, J. Francois, Francisco Ayala, and Ernst Mayr defending unity.[40] These biologists argued for unity on the basis of the statistical improbability of the evolution of intelligence on Earth.[41] For them, Earth had won the equivalent of a cosmic lottery. The chances for intelligent life emerging on any given planet were so small that for it to occur twice was improbable even in a vast universe.[42]

The 1970s also contained hints of the emergence of a vocal unity contingent within the field of astronomy, the most ardent enclave of pluralist thought. In 1975, Michael Hart made the case against alien life in a major astronomy journal by invoking Fermi's paradox.[43] Fermi's paradox gains its name from physicist Enrico Fermi (although the argument's origin is much older).[44] In 1950, while working at the Los Alamos National Laboratory, Fermi asked the simple question, if intelligent aliens exist, "where are they?"[45] Fermi's paradox suggests that if alien life were plentiful, then at least one of those species should have colonized/explored the universe, providing evidence of their existence. Only a few years after Hart's article, in 1979, scientists opposed to belief in alien intelligence organized a conference focused on Fermi's paradox.[46] Although voices sympathetic to the plurality spoke, like Jill Tatar of the SETI institute, the conference participants came down firmly against the probability of intelligent aliens.

Still the consensus of scientists supported plurality in 1980 when Frank Tipler, then a professor of mathematics at U.C. Berkeley, entered the debate with his article "Extraterrestrial Intelligence Does Not Exist."[47] Tipler was born February 1, 1947 in Andalusia, Alabama. He claims that as a child he was inspired by "Werner von Braun, and decided then that [he] wanted to be an astrophysicist."[48] His parents raised him as a Christian fundamentalist, but he became an atheist at sixteen because he felt that science clearly contradicted the claim that the Earth was only 6,000 years old.[49] After finishing his PhD at MIT, he became a post-doc for the famous Princeton physicist John A. Wheeler, who Tipler notes on his website was the PhD advisor of the even more famous R. P. Feynman.[50] Before his article on alien life, Tipler published several times on general relativity in prestigious journals.[51] Even George Ellis, Professor of Applied Mathematics at the University of Cape Town, who became one of Tipler's harshest critics, claimed that Tipler did "nice work" in these early articles.[52]

"Extraterrestrial Intelligence Does Not Exist" provided a defense of Fermi's paradox and answered one of the common solutions to the paradox: that no aliens have visited Earth because space colonization would prove prohibitively expensive. He theorized that Von Neumann probes (intelligent machines that have the ability to self-replicate) could quickly and relatively cheaply explore the entire universe. Tipler argues that given the effectiveness of Von Neumann probes, some alien race interested in exploring the universe would use them. While one can object that Tipler would have little knowledge of what an *alien* race would or would not do, he makes a compelling case for why probes are a more efficient—and thus more likely—method of exploration than the radio waves that many in the SETI community believed aliens would send.[53] Tipler takes the absence of any signs of probes as proof of the lack of intelligent extraterrestrial life. More importantly for this book, Tipler makes the philosophical implications of the unity v. plurality debate a focus of his article.[54] Tipler entered the fray ready to combat both the scientific support for plurality and the prevalent scientific myths attached to plurality.

TIPLER'S RHETORICAL ANALYSIS

Tipler published his argument against unity in three articles in the *Quarterly Journal of the Royal Astronomical Society.*[55] The first article focuses primarily on Tipler's updates to the Fermi paradox. The other two articles provide a history of extraterrestrial life debate. In these articles, Tipler explores the extra-scientific implications of unity by means of examining the motivations of scholars engaged in the debate. Unsurprisingly, given his position as an active defender of unity, Tipler only recognizes the plurality defenders as putting forth a scientific myth in support of a broader philosophical and religious agenda. He concludes his section on the history of the plurality debate by saying, "I contend that, as has been the case for 2000 years, these philosophical and theological beliefs are the main motivations for the belief in [extraterrestrial intelligence]."[56] In contrast, Tipler portrays unity defenders as objective scientists (although he does make an exception for unity thinkers in the Medieval period). The dichotomy of objective scientists vs. ideologues becomes evident in the way Tipler narrates historical events. He describes Whewell as "the first person to take a critical *scientific* look at the empirical, as opposed to the philosophical or theological, evidence of extraterrestrial intelligence (author's italics)."[57] Tipler criticizes defenders of plurality for their lack of "a sense of history,"[58] meaning that plurality proponents fail to recognize that they merely recycle the arguments of previous plurality defenders, without any critical reflection.

Tipler acts to undermine the ethos of the plurality defenders further by connecting them to three discredited scientific theories: spontaneous generation, Laplace's nebular hypothesis, and the great chain of being (a teleological, hierarchical, anti-evolutionary worldview that preceded Darwin).[59] These links work to discredit the idea of plurality through guilt by association; the pluralists believed these wrong ideas, so their support of plurality is equally suspect. Leaving aside the "association fallacy" at the foundation of his argument, Tipler does not mention unity thinkers' belief in these same discredited concepts.

Tipler's article selectively engages the links between discredited science and participants in the plurality debate. Proponents of unity often supported these doctrines as well, just as many plurality proponents opposed them. Whewell actually defended a modified version of the nebular hypothesis in his book *Plurality of Worlds*, while many plurality defenders (like Brewster) not only rejected the nebular hypothesis, but actively castigated Whewell for relying on it.[60] Nor does Tipler mention that Whewell fiercely opposed Darwinism and supported a version of the great chain of being.[61] Similarly, Tipler points to Wallace's book *Man's Place in the Universe* to suggest that Darwinism demands unity, with no recognition of Wallace's significant back-tracking from orthodox Darwinism or the support of other Darwinians for plurality, including, most notably, Charles Darwin himself.[62] Tipler narrates events in a way that denies plurality defenders scientific credibility, while elevating the credentials of unity proponents. This sets the stage for his characterization of plurality advocates as motivated primarily by "philosophical and theological beliefs."[63]

Beyond Tipler's one-sided reporting of the scientific beliefs of the debate participants, he portrays defenders of plurality as religious fanatics. He accuses Brewster of going after Whewell with a "missionary zeal," by which he means Brewster behaved irrationally.[64] Tipler then attributes this same "missionary zeal" to Frank Drake's defense of SETI.[65] He argues that Carl Sagan, Fred Hoyle, Frank Drake, A.G.W. Cameron, and other contemporary scientists that advocate plurality practice a new religion based on salvation by an advanced alien race.[66] Tipler presents himself as part of a scientific corrective to the thousands-of-years-old mystical religion of plurality.[67] In this respect, he positions himself as another Whewell or Wallace, whom he champions as scientific purists. The truth of Tipler's representation of modern-day SETI defenders aside, he replicates the closed-minded commitment to his cosmology he accuses his opponents of having. His one-sided reading of the plurality debate reflects either his own lack of "a sense of history" or a purposeful slanting of events.

Although the *Quarterly Journal of the Royal Astronomical Society* did not have much readership outside a specialized audience, Tipler demonstrates an interest in reaching the wider public. He points to the lack of popular dissenting voices as reason for plurality's ubiquity at the end of the eighteenth century and compares it to the modern day. "If there is no opposition to a view, it will become generally accepted whatever the evidence for it (witness the contemporary situation of [Extraterrestrial Intelligence] in the popular press)."[68] Tipler's concern for the public perception of alien life points to his opinion of the stakes of the debate. Rather than approach the existence of alien life as a technical question best left to private debate between scientists, he inhabits a role as a public intellectual, countering the efforts of Sagan and others. It was not long before Tipler, now a professor of mathematical physics at Tulane University in New Orleans, got the opportunity to expose his ideas to a larger audience.

In 1981, the popular science magazine *Physics Today* published an abridged version of his article.[69] The adapted article, also titled "Extraterrestrial Intelligent Beings Do Not Exist," presents the argument for unity in a more accessible format, with much of the metaphysical undertones absent. The article condenses Tipler's original thirty-six pages of arguments into three and replaces his myriad of references with a suggestion that readers interested in learning more view his original articles in the *Quarterly Journal of the Royal Astronomical Society.* A comic of the universe with a speech bubble coming from one tiny spec of space (presumably Earth) saying "Where is everybody?" accompanies the piece.[70]

Tipler's initial forays into the alien life debate had a major impact on the debate. Senator William Proxmire drew on Tipler's arguments when he made

Figure 4.1. The comic accompanying Tipler's *Physics Today* article.

the case that Congress should cut funding for SETI.[71] More importantly, Tipler reintroduced the idea that philosophical beliefs intertwined with the question of alien life. Tipler argues that pluralists historically have had religious and quasireligious motivations for their scientific conclusions. He indicates that he wants to enter the public debate to correct the mistaken public belief in alien life. At this stage, he does not articulate the stakes of the debate beyond scientific accuracy. Tipler's next endeavor would begin to build the case for the philosophical values associated with unity, providing both a philosophical and scientific rebuttal of the position of the pluralists.

THE ANTHROPIC PRINCIPLE

Six years after the initial publication of Tipler's article "Extraterrestrial Intelligent Beings Do Not Exist," Tipler again saw his unity ideas presented in a popular venue. He co-wrote the *Anthropic Cosmological Principle* with John D. Barrow, a professor of astronomy at Sussex University. Like Tipler, Barrow had impressive scientific credentials. By the year 1986, when the *Anthropic Cosmological Principle* was released, Barrow had published seventy-four articles, many in top-tier academic journals. In 1983, he co-wrote *The Left Hand of Creation*, a popular science book explaining the origins of the universe with Joseph Silk, an astronomy professor at the University of California Berkley. Tipler and Barrow brought an impressive scientific pedigree to their work that bolstered its credibility.

Their book merges the titular concept of the Anthropic Principle with the unity cosmology and uses this fusion as a defense of teleology. Before one can understand how they deploy the unity cosmology, it is important to understand the Anthropic Principle. Despite having many similarities with unity, the Anthropic Principle exists as a distinct concept. The Anthropic Principle comes in a variety of forms, which Tipler and Barrow explore in the book.

The authors identify four meanings of the Anthropic Principle. The first, proposed in modern times by Princeton physics professor Robert Dicke, they refer to as the weak Anthropic Principle (WAP). The WAP claims that the universe has to be amenable to life; otherwise, no life would exist. The WAP amounts to little more than a tautology, and Tipler and Barrow suggest it is "in no way either spectacular or controversial."[72] The strong Anthropic Principle (SAP) goes further than the WAP and says that not only do the conditions of the universe allow for intelligent life, but "the Universe *must* have those properties which allow life to develop within it as some stage in its history" (italics added).[73] Tipler and Barrow acknowledge that the SAP often gets equated with intelligent design by a creator god, wherein a divinity creates the universe specifically for life.[74] Tipler and Barrow, however, defend a third version the Anthropic Principle, Princeton professor of astrophysics John Wheeler's participant Anthropic Principle (PAP), which says that intelligent observers, like humans, are necessary for a universe to exist.

The PAP derives from one of the theorized quirks of reality according to some accounts of quantum mechanics: nothing can exist in a physical form before an observer witnesses it. Tipler and Barrow argue that many universes exist, but universes without intelligent observers can only exist in a quantum (but not physical state).[75] The likelihood for a universe to allow the evolution of observers, thus, selects for universes amenable to life to become physically manifest. The PAP has the incredibly counterintuitive implication that life that postdates the existence of the universe is responsible for the physical manifestation of the universe.[76] Tipler and Barrow argue that the condition of future observers for making a universe physically manifest itself explains the large number of physical laws and properties of our universe that seem tailor made for life (like water's unique characteristics and the overall composition of atomic matter), because a universe hostile to life will not develop observers and thus not move from a quantum to a physical state.[77]

The elevation of life to cause, rather than effect, is compatible with the human-centric theme of the unity cosmology, although not identical to it. PAP is compatible with the unity philosophy, because it makes life central to the universe, rather than a result of chance. Without life the universe could not exist. The PAP alone does not point to a teleological purpose for humanity, however, because an alien living on a distant planet could play the role of observer that makes the universe physically present.

Tipler and Barrow vigorously deny the possibility of alien life, however. They rehash Tipler's version of Fermi's paradox as well as provide a series of biological arguments against the likelihood of intelligence developing.[78] At first the authors do not make clear the relevance of the absence of alien intelligence to the Anthropic Principle. One can easily get the impression that the authors had simply added Tipler's original article on the lack of aliens as a chapter to an otherwise unrelated book. The sections on alien life seem out of place with the broader discussion of the anthropic principle and philosopher J. J. C. Smart even suggests that unity contradicts PAP's contention that life selects for biofriendly universes.[79] After all, if the potential for observers selects which universes become physically manifest, then one could rightfully expect the existence of universes hospitable to observers. This brings us to the fourth form of the Anthropic Principle discussed by Tipler and Barrow, the Final Anthropic Principle (FAP), in which they marry the life-centric elements of the PAP with the human-centric elements of unity.

The FAP states that once "intelligent information-processing" (by which Tipler and Barrow mean humanity to be succeeded by artificial intelligence) emerges, it "will never die out."[80] The FAP builds on the framework of the PAP, because for the FAP to be true, intelligent life must be necessary for the universe to exist. The FAP also relies on the absence of alien life, because if extraterrestrial intelligent life existed then humanity could be replaced as the quantum observers necessary for the universe. Tipler later sums up the FAP bluntly in a TEDX talk saying, "We humans do not have the death option… the laws of physics will not permit us to do that."[81]

Early in the *Anthropic Principle* Tipler and Barrow take a guarded stance on what the FAP means for ethics.

> Although the FAP is a statement of physics and hence *ipso facto* has no ethical or moral content, it nevertheless is closely connected with moral values, for the validity of the FAP is the physical precondition for moral values to arise and to continue to exist in the Universe: no moral values of any sort can exist in a lifeless cosmology. Furthermore, the FAP seems to imply a melioristic cosmos.[82]

They acknowledge the idea of scientific myths by saying that physical principles cannot have moral content, but immediately reverse course and say that the FAP is closely connected with moral values and implies a "melioristic" universe. Tipler and Barrow only briefly mention the idea that the FAP implies "meliorism," which is the idea that the universe is evolving inevitably toward the Good. This kind of brief connection between their scientific ideas and teleology allows them to maintain the appearance of disinterested scientists, all the while signaling to readers that believe in teleology that the cosmology of the Anthropic Principle supports their view.

THE SCIENTIFIC FRAME

Before one can examine the argument for teleology and absolute values embedded in their cosmology, it is important to understand the rhetorical frame within which Tipler and Barrow operate. Unlike Plato, Aquinas, and Whewell, Tipler and Barrow wrote at a time when natural theology no longer existed as a major academic discourse and in fact often elicited hostility in scientific circles. Scientists treated natural theology as biased and fundamentally antagonistic to what they viewed as the disinterested, objective process at the heart of modern science. While scientific myths in support of modern ideologies circulated easily, scientific myths associated with traditional values, teleology, or religion risked the discredited label of natural theology or worse. This created a difficult rhetorical environment for the unity myth, which depends on the fusion of science with traditional values. In the beginning of their book, Tipler and Barrow act to forestall any claims against their objectivity. They write,

> The authors are cosmologists, not philosophers. This has one very important consequence which the average reader should bear in mind. Whereas many philosophers and theologians appear to possess an emotional attachment to their theories and ideas which requires them to believe them, most scientists tend to regard their ideas differently. They are interested in formulating many logically consistent possibilities, leaving any judgment regarding their truth to observation. Scientists feel no qualms about suggesting different but mutually exclusive explanations for the same phenomenon. The authors are no exception to this rule and it would be unwise of the reader to draw any wider conclusions about the authors' views from what they read here.[83]

Tipler and Barrow seek to maintain the aura of objective scientists crunching the data, rather than philosophizing or sermonizing. The passage amounts to an anti-rhetorical, rhetorical gambit, downplaying the importance of rhetoric in favor of hard science, at the same time making a rhetorical move to bolster their credibility.

Mary Midgley, a professor of philosophy at Newcastle University, examines this passage and writes, "*What these authors hope to do is to import into metaphysics the kind of impartiality that comes naturally in physical science, simply by handling it with scientific methods*" (author's italics).[84] The passage goes beyond describing the author's intent, however. It tells the audience that they should read the text as scientific, rather than a work of philosophy or even a book separated into distinct sections on science and philosophy. Tipler and Barrow paint a picture where they felt compelled to their conclusions by the raw data, rather than their own beliefs and biases. This establishes a framework requiring the reader to confront and refute the science behind their argument or risk the kind of "emotional attachment to their theories and ideas" that Tipler and Barrow renounce. In effect, Tipler and Barrow demand that the audience refute their science, before it assails their scientific myth/metaphysics.

The standard of scientific refutation as the metric of argument puts criticism of *The Anthropic Cosmological Principle* beyond the ability of much of the general public who purchased the book, as well as many academics unschooled in advanced physics. For example, Tipler and Barrow write,

> The absolute minimum amount of energy required to process a given amount of information is determined by the Second Law of thermodynamics. If ΔI is the information processed in bits, then the Second Law requires
>
> $$\Delta I \leq \Delta E / k_B T \ln 2 = \Delta E/T(\text{ergs/K})(1.05 \times 10^{16}) \quad (10.39)$$
>
> where kB is Boltzmann's constant, T is the absolute temperature in degrees K, and AE is the amount of free-energy expended. The inequality (10.39) is due to Brillouin.[85]

Certainly this passage and many others like it are outside the comprehension of the average reader. Nevertheless, studies have shown that even individuals with postgraduate degrees and experience reading research papers rate papers higher when they include mathematical formulas, even when these formulas are nonsensical.[86] This is to say that complicated formulas likely prime readers to accept Tipler and Barrow's conclusions, even if readers have no idea what the equations mean. In addition to the numerous equations, the length of the book alone, at over 700 pages, makes a close reading of *The Anthropic Cosmological Principle* a daunting task.[87] The authors include an enormous number of footnotes, both citations and further explanations, a fact noted in popular and academic reviews.[88] Tracking down all of the sources cited and reading them would take more effort than all but the most dedicated reader could sustain. Given the complicated nature of the material, even if someone did read all of the extensive citations it would be no guarantee of deeper understanding. The book often reads as series of scientific journal articles meant for specialized audiences, rather than a work that would have public success. The standard of scientific refutation forecloses the avenues of response to all but the most educated readers, despite *The Anthropic Cosmological Principle* finding a wide audience of non-specialists.[89]

The difficulty of the text creates a structural authority that places the reader in a dependent relation to the authors. Of course difficulty alone does not guarantee ethos, much less readership. Tipler and Barrow employ a strategy where most of the philosophical implications of their work occur at the beginning and the end of the text in highly readable prose, which draws in the lay reader. The difficult to follow sections occur in the middle of the book. Most readers will likely skip over the math sections and not follow up on the footnotes, despite the central role they play in the book's argument. The text arouses a desire for a knowledgeable authority to do the work of interpreting the scientific findings, which the authors then satisfy with their impressive credentials.[90]

The scientific ethos that Tipler and Barrow bring to the table makes readers more likely to trust their conclusions even if they cannot figure out exactly how they reached them.[91] Both men hold doctorates in scientific fields and work at respected universities.[92] The fact that famous Princeton scientist John Wheeler endorsed the book in the foreword also heavily contributes to their aura of objectivity. As Robert Klee, professor in the Department of Philosophy and Religion at Ithaca College, remarks, "The foreword was by none other than John Wheeler, [is] as sure a sign of the book's scientific legitimacy as anything."[93] The prominent academic publisher Oxford University Press adds another layer of intellectual sanction. The difficulty of the text combined with the qualifications of the authors means that most readers probably take for granted the accuracy of Tipler and Barrow's scientific claims.[94]

Tipler and Barrow's scientific credentials, the complexity and depth of their arguments, and their ethos of scientific neutrality all add to the persuasiveness of *The Anthropic Cosmological Principle*. Too explicit a defense of the unity myth, with its associations with natural theology, would undermine these persuasive elements. Tipler and Barrow's ethos depends on being objective, modern, and scientific, whereas natural theology was often treated as biased, outdated, and religious. Whewell and other earlier figures could all offer explicit and extended explanations of the implications of unity, because they wrote at times where science was seen as deeply intertwined with questions of religion and philosophy. In the late twentieth century, associating science with religion or philosophy risked the charge of bias.[95] This is precisely the charge that Tipler levels against plurality's defenders in his earlier article.[96]

The brilliance of *The Anthropic Cosmological Principle* is that it uses a variety of rhetorical techniques to hint at or imply support for the unity myth, without giving the appearance the book operated outside the scientific paradigm of the time. One strategy was to outline a history of teleological thought without explicitly endorsing teleology. The next section details the anthropocentric philosophical background for the supposedly objective and value-free Anthropic Principle.

THE UNITY MYTH AND THE ANTHROPIC COSMOLOGICAL PRINCIPLE

Because of the desire to appear objective, the metaphysical implications of their scientific inquiry emerge only sporadically throughout the 700-page book and often through the words of other thinkers. The first two major sections of the book detail the history of teleology, a concept known primarily from philosophy and theology. Tipler and Barrow's history begins with the Old Testament and the ancient Greeks, a time when "philosophy and science were conjoined and 'metaphysics' was concerned with the method as well as the meaning of science."[97] The connections they draw between the Anthropic Principle and religion and science come through clearly in the history, which touches on figures already discussed in this book like Plato, Democritus, and Aquinas, as well as a myriad of other important individuals (including advocates for the Anthropic Principle in nonwestern cultures).

Tipler and Barrow claim that the purpose of the history is to show that, despite what many scientists think, the Anthropic Principle is not a "new and revolutionary idea."[98] The eighty-two-page background on teleology, however, goes beyond establishing an argumentative precedent for their theory. It serves the function of establishing the stakes of the debate, because the historical figures that Tipler and Barrow chronicle do not treat teleology as a neutral fact of existence. Aquinas, Schelling, Whewell, and many of the other figures examined view teleology in nature as a justification for religion and absolute values, a fact that is made clear repeatedly throughout the history. Drawing a connection between their own work and these earlier figures acts to philosophize in precisely the way the authors disavow in their earlier claim to scientific objectivity. The form of a historical primer helps obfuscate the break with their earlier statement by allowing Tipler and Barrow to draw out the philosophical implications of their own theory by means of association with the ideas of others.[99]

The clearest example of how Tipler and Barrow use others ideas to attach value to their scientific arguments comes at the conclusion of the section on history, with nine whole pages of discussion on Teilhard de Chardin, the Christian mystic, philosopher, and scientist.[100] Tipler and Barrow focus on Teilhard's concept of the Omega point, a concept whose name references the Biblical passage where God says, "I am Alpha and Omega, the beginning and the ending."[101] Teilhard argues that life evolved purposefully and inevitably toward intelligence.[102] The emergence of intelligence brought about the "noosphere," which Teilhard describes as the collective and integrated thought of humanity. Over time the noosphere evolves as technology and socialization bring about more knowledge. Teilhard claims that eventually humanity will inevitably reach the Omega point, where the sum total of knowledge coalesces into a super-intelligent collective that cannot be undone.

Tipler and Barrow go beyond summarizing Teilhard's view, however, and actually build on his theories. Tipler and Barrow acknowledge that some scientists attacked Teilhard as a mystic masquerading as a scientist, but Tipler and Barrow defend his work as scientific.[103] Tipler and Barrow do point to areas where new science has definitively disproved Teilhard, such as his understanding of information processing theory. Rather than discredit Teilhard, Tipler and Barrow suggest that these refutations prove that Teilhard made refutable and thus scientific claims. The authors then suggest ways that one can save Teilhard's overall theory with a few changes. While Tipler said the Omega point will occur on Earth, Tipler and Barrow claim that it will occur at the singularity of the universe (the point in time right before the universe collapses back into the state it existed before the big bang, an idea to which they return more fully later in the book and is examined later in this chapter).[104] The authors indicate their own theory represents a modernized version of Teilhard's general thesis of evolution to totalizing intelligence.

Tipler and Barrow's move to modernize Teilhard's thesis does not represent a neutral scientific observation. Tipler and Barrow say plainly, "The basic framework of [Teilhard's] theory is really the only framework wherein the evolving Cosmos of modern science can be combined with an ultimate meaningfulness to reality."[105] An ultimate meaningfulness, described at various times in the language of purpose, teleology, absolute values, etc, is precisely what has animated philosophical, ethical, and religious debates going back to ancient Greece. Tipler and Barrow's move to provide scientific proof for Teilhard's position and thus provide an argument for meaning makes *Anthropic Principle* as much as a work of philosophy as a work of science and as a perfect example of Toulmin's idea of scientific myth. Teilhard's Omega point (and by extension their revised version of Teilhard's theory) represents a cosmos of meaning. In order to draw out this point, Tipler and Barrow contrast the meaningful existence inherent in the Omega point, with the possibility of "heat death," which they describe as a prevailing cosmology among scientists.

Tipler and Barrow cite individuals that claim heat death obliterates the possibility for a meaningful universe. Heat death is a phenomenon that some scientists have predicted, where far in the future energy becomes evenly dispersed throughout the universe. Because the universe is so massive, equal distribution of heat would create freezing conditions inhospitable to any form of life. The most famous reaction to heat death comes from Bertrand Russell, who Tipler and Barrow quote at length:

> The world which science presents for our belief is even more purposeless, more void of meaning... Amid such a world, if anywhere, our ideas hencefor-

> ward must find a home. That Man is the product of causes which had no prevision of the end they were achieving; that his origin, his growth, his hopes and fears, his loves and his beliefs, are but the outcome of accidental collocations of atoms; that no fire, no heroism, no intensity of thought and feeling, can preserve an individual life beyond the grave; that all the labours of the ages, all the devotion, all the inspiration, all the noonday brightness of human genius, are destined to extinction in the vast death of the solar system, and that the whole temple of Man's achievement must inevitably be buried beneath the debris of a universe in ruins—all these things, if not quite beyond dispute, are yet so nearly certain, that no philosophy which rejects them can hope to stand. Only within the scaffolding of these truths, only on the firm foundation of unyielding despair can the soul's habitation be safely built.[106]

For Russell, no human achievement, however spectacular, can escape the universe's ultimate fate of frozen silence. Tipler and Barrow claim that the idea of heat had a troubling effect on those that believed in teleology, including Charles Darwin, E. W. Barnes, and Teilhard.[107] People find heat death depressing not because it will adversely affect any human now alive, but because it robs humanity of a future. Inevitable extinction denies a *telos* or purpose to our existence, which many feel renders humanity insignificant and the universe meaningless.

Of course, Darwin, Barnes, and Teilhard's views that heat death renders humanity insignificant are no less scientific myths than the claim that the Omega point makes humans significant. Individuals may interpret heat death as depressing, but the scientific fact of a frozen universe does not require a particular attitude toward life today, an argument Toulmin makes forcefully in his initial essay on scientific myth.[108] To their credit Tipler and Barrow quote Russell saying that despite his famous quote he thinks the idea that heat death is "depressing" is "nonsense," which reflects how different individuals can interpret the same scientific facts as having different extra-scientific meanings.[109] This willingness to leave heat death open to interpretation fades later in the book, where Tipler and Barrow explicitly set it up as a foil to the teleology of the Omega point. For now they simply establish that Teilhard found the idea of heat death deeply troubling for his prediction of the Omega point.

It is important to note that heat death and the Omega point (understood as the emergence of an interconnected super intelligence) do not represent mutually exclusive possibilities. The future events' meanings for our present time, however, do conflict. This prompted Teilhard to posit the Omega point as mutually exclusive with heat death because if the Omega point did not exist forever, "it would not be Omega."[110] If a human created supercomputer inevitably collapsed in a universal heat death, then it lacked any meaning as far as Teilhard was concerned, just as the FAP only has meaning if humanity never goes extinct.

Tipler and Barrow end their discussion of the history of teleology with Teilhard and heat death and return to it in their final chapter. Rather than simply reviewing the various historical opinions on teleology this chapter solidifies the extra-scientific implications of nihilism and teleology with heat death and the Omega point respectively. In the first pages of their final chapter, they reference Russell's connection of heat death with "insignificance" and quote Nobel prizing-winning physicist Steven Weinberg claiming that it rendered life "pointless."[111] They argue these views reflect the feelings of "most philosophers and scientists over the past century that mankind is forever doomed to insignificance."[112] The FAP, however, offers hope to avoid this form of cosmic despair and give humanity significance,

> Although mankind [*sic*]—and hence life itself—is at present confined to one insignificant, doomed planet, this confinement may not be perpetual...once space travel begins, there are, in principle, no further physical barriers to prevent *Homo sapiens* (or our descendants) from eventually expanding to colonize a substantial portion, if not all, of the visible Cosmos. Once this has occurred, it becomes quite reasonable to speculate that the operations of all these intelligent beings could begin to affect the large scale evolution of the Universe... *Present-day life would then have cosmic significance because of what future life may someday accomplish* [Italics mine].[113]

No longer mediating their discussion of the implications of the Omega point through other individuals, like Teilhard, Tipler and Barrow appear to offer their own argument that the FAP makes humanity significant. In doing so Tipler and Barrow seem to accept Russell and Weinberg's central thesis that heat death does render human behavior pointless, but they challenge the science that heat death inevitably awaits humanity and offer an alternative vision of the future where the actions of intelligence can shape the cosmos. The ability to enact change on a universal scale gives "cosmic significance" to the lives of present-day humans. This follows in the tradition of Whewell and Wallace, both of whom accepted the scientific myths of their ideological opponents, only to dispute the myths' scientific foundations and thus reverse their mythic implications.

Tipler and Barrow lay forth the cosmic changes they imagine saving future intelligence from destruction and infusing present-day human life with value. In their model, human intelligence expands to encompass the entire universe and creates a super computer so powerful that it represents a final "Omega" point in a truer sense than Teilhard's belief in super-intelligence on Earth.[114] In the last paragraph of their book they write,

> Life will have gained control of *all* matter and forces not only in a single universe, but in all universes whose existence is logically possible; life will have spread into all spatial regions in all universes which could logically exist,

> and will have stored an infinite amount of information, including *all* bits of knowledge which it is logically possible to know. And this is the end.[115]

This picture of the future defies imagination, but it's clear what it symbolizes: total domination of the universe by intelligence. As "life spread[s] into all spatial regions in all universes which could logically exist," the universe becomes synonymous with intelligence (the intellectual offspring of humanity today). For Tipler and Barrow intelligence infuses the universe with meaning; nature and the desires of intelligence become one. Aesthetically the universe becomes unity, as the Omega point incorporates all of its constituent parts.

Buried in a footnote they further unpack the meaning of this final passage, "a modern-day theologian might wish to say the totality of life at the Omega point is omnipotent, omnipresent, and omniscient!"[116] The Omega point represents humanity becoming god through science.[117] This technological singularity provides the perfect foundation for transcendent ethics throughout all time, in some ways reminiscent of how the orderly, crystalline, celestial objects of Plato's *Timaeus* provided a model for human ethics.[118] Tipler and Barrow argue the FAP, with its Omega point, provides a scientific cosmological foundation for meaning and purpose in the lives of humanity.

ABSENT ALIENS

The previous sections demonstrate how Tipler and Wallace maintain a scientific style and present the FAP as a justification for teleology. This has obvious connections to the cosmological arguments of thinkers like Plato, Aquinas, Whewell, and Wallace, who connected their own teleological ideas to the structure of the universe. The FAP, however, is not the same as the absence of alien life central to the unity cosmology. One can read the fine-tuning of the universe discussed in *The Anthropic Cosmological Principle* as the only proof necessary of teleology and absolute values. Under this interpretation the discussion of the absence of aliens is a superfluous indulgence of Tipler's past work. If one views *The Anthropic Cosmological* as an argument for human value, however, the unity cosmology represents a central component of Tipler and Barrow's argument.

Without unity, the FAP and the Omega point fail to necessitate meaning and purpose to the lives of *humans*. First, the Omega point by definition must be singular, even absent the technical question of whether two universal super computers could exist simultaneously. Only by achieving total knowledge does that computer gain the status of the Omega point, Omega being finality. The existence of two such computers may raise Teilhard's objection that absent singularity, "it would not be Omega."[119] This means that to achieve the Omega point, any intelligence would have to join in the effort.

The addition of aliens to the cosmos forces Tipler and Barrow to explain what about humanity specifically, among all the life on earth, makes it the *telos* of the universe. Without confronting the alien question, beyond denying alien existence, Tipler and Barrow can rely on age-old anthropocentrism to explain why humans alone (and not, for example, dolphins, octopi, or worms) represent the central focus of the cosmos. Aliens blur the line between humanity and other species by forcing the question of what, if anything, makes humanity unique among animals.[120] Without intelligent aliens in the equation, Tipler and Barrow are free to assert that of the known life forms only humans and their robotic offspring can fulfill the demands of the FAP.

The existence of intelligent aliens also raises the possibility that humanity need not survive to create the Omega point, because some other race of intelligent observers could achieve the goal. In other words, the FAP becomes compatible with human extinction as long as some alien species survives to create the Omega point. This not only matters for our distant, distant offspring (be they organic or mechanical), but also for all humans living today. Tipler and Barrow assign our present-day lives meaning based on their contribution to the Omega point. Absent a clear connection to this final goal, humanity lacks "cosmic significance." Without future participation in the Omega point, humanity again faces the bleak prospects of extinction, just as in the scenario of universal heat death (even if it occurs earlier through other means). The prospect of human extinction returns the cosmic despair and pointlessness that Weinberg and others discuss, not for our distant descendants, but for all humans today.

Tipler and Barrow's connection of the absence of alien life to the FAP also provides some of the easiest-to-measure proofs of the theory. While they provide a substantial amount of arguments for the fine-tuned nature of the universe, what constitutes fine-tuning is often a matter of interpretation. The absence or presence of intelligent alien life represents a much easier thing to determine objectively, given the proper technology.[121] By conflating human singularity and the FAP, Tipler and Barrow provide a more persuasive scientific proof of humanity's centrality.

Beyond the absence of alien life's persuasion value, the very structure of Tipler and Barrow's argument for unity speaks to their broader concern for the special nature of humanity. They depend heavily on Fermi's paradox, which denies the existence of aliens because they have not contacted humanity, but Fermi's paradox presupposes, not proves, their teleological universe. The Soviet SETI scientists, in contrast to their American counterparts, highlighted the problem of radical difference when attempting to contact or understand the motivations of aliens.[122] Even though many Soviet scientists believed aliens certainly existed, they had doubts that we would have a means to communicate with them or recognize their attempts to communicate with us.[123] Furthermore, a focus on the complications of discovery and communication assume aliens have a desire to find and communicate with humanity in the first place. The idea that alien races could live their lives totally independent of any concern for humanity is an answer to Fermi's paradox that also calls into question our species' cosmic importance. Tipler and Barrow's faith in the fact that aliens would have both the interest and means to contact us has parallels to European attitudes prior to "discovering" the Americas. These attitudes had difficulty in accepting the fact that two continents of humans had lived for millennia unaware of Europe. If one allows for the possibility that humanity occupies a peripheral, rather than central, place in the cosmos, than it should come as no surprise that alien life has not contacted humanity.

These four reasons show that the unity cosmology represents an essential component of Tipler and Barrow's version of the Anthropic Principle. Including aliens in the cosmos raises questions about the anthropocentric nature of Tipler and Barrow's claim that humans, rather than, say, dolphins, are the purpose of the universe. Intelligent aliens could construct an Omega point independent of humanity, which means that humans need not be the species that survive until the end of the universe. The absence of aliens represents an easily understood proof of human specialness in contrast to that derived from mathematical formulas, which makes it useful for defending the philosophical implications Tipler and Barrow attribute to the Anthropic Principle. Finally, Fermi's paradox fits within the larger narrative of humanity serving as the metaphorical center of the universe. Just like Plato and Aquinas to whom they acknowledge their deep intellectual debt, Tipler and Barrow connect unity to teleology.[124]

ACADEMIC RECEPTION

An examination of the reception of *The Anthropic Cosmological Principle* helps to reveal the continued interest generated by the unity myth. Given Tipler and Barrow's connection of philosophy with science, one could have imagined a harsh reaction to their book, especially among scientists. Instead the response proved to be very positive overall. This chapter examines a variety of articles, books, and speeches inspired by Tipler and Barrow's book in order to gauge its rhetorical effectiveness and dissemination.

A JSTOR search reveals six reviews in peer-reviewed academic journals, a relatively high number.[125] Four reviews spoke of the book in an overall positive manner.[126] Berkley astronomy professor Joseph Silk called the book "a marvelous treasure trove" in *Science*, one of the most respected science journals in the world.[127] Philosopher Michael Heller appeared equally enamored with the book and wrote, "Nobody interested in philosophical aspects of modern cosmology, or modern science in general, could easily be excused from reading this book."[128] *The Anthropic Cosmological Principle* even received a positive review in the art journal *Leonardo*, with James Goldman highlighting the book's aesthetic connections, before declaring it "a definitive and indispensable resource."[129] Of the remaining two reviews, one opposed the book and the other gave a relatively neutral account.[130] The negative review came from a biologist, who was deeply unsettled by Tipler and Barrow's defense of teleology. He went so far as to question the sincerity of Tipler and Barrow's arguments and stated, "Assuming that the book was intended to be provocative, it is an unequivocal success."[131] Given the book's coverage of so many controversial topics one would expect many more such comments than actually occurred.

The academic presence of *The Anthropic Cosmological Principle* goes well beyond its reviews, however. *The Anthropic Cosmological Principle* made its way on the book lists of philosophy, science and philosophy of science journals.[132] The influence of the book becomes apparent when one examines the way other scholars cite it. Often academics cite *The Anthropic Cosmological Principle* for scientific purposes absent any reference to its broader teleological theme, typically Tipler and Barrow's description of how the universe emerged from nothing.[133] This speaks to Tipler and Barrow's success in passing their book as a legitimate work of science, despite their transgressing scientific norms. Historians cite Tipler and Barrow for anecdotes of important scientific figures or their defense of Whig history.[134] JSTOR results also reveal exposure of the book within the academic artistic community. The book inspired a poem, a new theme for science fiction, and an innovative approach to music.[135] And, of course, authors cited the book to reference the claim that the universe is finely tuned for human existence.[136] The achievement of so many citations, much less citations across a plethora of disparate disciplines, sets *The Anthropic Cosmological Principle* apart from all but the most successful academic works. This is especially true since a significant number of the citations occur upward of a decade after the book's original publication.

Most interesting for this study, however, academics began using the book as proof of absolute truths. B. L. Hebblethwaite cites the book's arguments for teleology as a proof of the God hypothesis.[137] James Applewhite, unhappy with the "postmodern" state of art," uses the Anthropic Principle to call for a return to transcendent universal values within aesthetics.[138] Stuart N. Hart begins his case to extend more rights to children with an appeal to the Anthropic Principle,

> Recent studies of the origins and progress of the universe have led cosmologists to postulate the "Anthropic Principle" to respect accumulating evidence that the universe embodies a design and fine-tuning mechanisms directed to-

> ward the achievement of human life (Barrow 1998). Evidently we are not the insignificant product of a chance occurrence in the backwater of evolution. These facts suggest that there is something quite special about human beings, something that deserves respect.[139]

Given the passion of Hart's plea for children's rights in the rest of his article, it is difficult to imagine him changing his mind based on new scientific data about the origin of the universe. One has to wonder if cosmology informs Hart's (and the other previous examples') opinion on child welfare or whether he latched onto Tipler and Barrow's cosmology as a defense of his preexisting beliefs.[140] Independent of these authors' actual beliefs on the universe's relation to values, they all felt that Tipler and Barrow's version of the Anthropic Principle lent credence to their argument.

Errol E. Harris argues that the PAP and SAP represent a natural argument for absolute values and design (in his second book he makes it clear that by design he means the truth of Christianity).[141] M. A. Corey wrote another defense of God, teleology, and absolute values that depends on Tipler and Barrow's book.[142] Unsuprisingly, given their religious conclusions, both Harris and Corey reject Tipler and Barrow's secular Omega point and replace it with a religious event.[143] Despite the move to desecularize (or resacralize) Tipler and Barrow's arguments, however, the same argumentative framework undergirds their positions.

EXPLAINING THE ACADEMIC SUCCESS

Given the suprising success of *The Anthropic Cosmological Principle* across disciplines, its worth exploring why it was so popular. John Leslie had made the connection between the God hypothesis and the Anthropic Principle several years earlier with little fanfare.[144] A JSTOR search for Leslie's article returns only 17 results, including the original article and three other articles by John Leslie.[145] Of the remaining 13 articles, nine overlap with citations of Tipler and Barrow's book, which suggests interest in Leslie's work did not grow until after the publication of the *Anthropic Cosmological Principle*.[146] Five of the thirteen share a place in the same footnote as Tipler and Barrow's book.[147] Removing trivial results leaves only one article that cites Leslie's article without reference to Tipler and Barrow, and this article is in a journal that focuses on religion.[148] Leslie's 17 responses contrasts sharply to 131 responses for Tipler and Barrow's book.

Given the similarity of their arguments, the difference cries out for explanation. The medium of the two pieces, certainly, represents a contributing factor. Leslie's argument came in the form of a journal article meant for specialized audiences; Tipler and Barrow's book became a popular success, making it more visible across academic disciplines. Tipler also proselytized his views, attending philosophy conferences and presenting primers specifically designed for academics not in the hard sciences to incorporate the Anthropic Principle in their work.[149] The length allowed Tipler and Barrow to cover a range of topics, reflected in the diversity of the ways others cite them. These factors alone cannot explain such a great gulf in success, however. The articles and books that use the Anthropic Principle to defend absolute moral values cite Tipler and Barrow and not Leslie, unless these works explicitly defend religion.[150]

Tipler and Barrow's book epitomizes a more rhetorically palatable defense of teleology. For starters Tipler and Barrow's work rhetorically performs the scientific objectivity that they claim in their introduction, as detailed above. The fact that so many scientific journals cite the book for its discussion of physics, independent of any reference to teleology suggests that scientists take the work seriously. This in turn gives those that cite the book in defense of absolute moral values or teleology, scientific authority. The "objective," "hard science" of Tipler and Barrow makes their work appear more legitimate than Leslie's philosophizing.

The non-religious nature of the book may have even been a boon to religious academics. William Lane Craig, a theist, takes issue with Tipler and Barrow because "the thrust of the book's argument is in the end anti-theistic," but proclaims, "I have already seen this book cited by two prominent philosophers of religion in support of the teleological argument."[151] Craig, like these unnamed "prominent philosophers" and many others, uses Tipler and Barrow's work as a starting point to defend religion. Craig does not need to spend much time responding to Tipler and Barrow's opposition to the design argument, because said opposition barely exists in the book.[152] As this chapter later demonstrates, the lack of opposition may well be by design. To put things another way, Tipler and Barrow present metaphysics in the guise of physics, which appeals to the religious and nonreligious alike. As Earman notes, "my concern is with attempts to wrap PAP [the belief that observers create the physical universe] in the cloak of scientific respectability. These attempts amount to no more than hand waving."[153] What Earman dismissively refers to as "hand waving" actually constitutes a successful rhetorical strategy to co-opt the respectability of science. Evidence of its success: law professor Jeffrey F. Addicott even proposes the Anthropic Principle as an acceptable compromise to the ciriculum debate over evolution and intelligent design in public schools. Addicott's argument: "[the Anthropic Principle] is recognized as a legitimate scientific axiom within the scientific community."[154] While many saw science and religion in conflict, Tipler and Barrow provided a justification for theism grounded in scientific discourse. Theists were quick to use it to their advantage.

Beyond the credibility gained by a scientific approach, Tipler and Barrow's version of the Anthropic Principle likely also performed better than

Leslie's because it catered to human vanity. Tipler and Barrow acknowledge that any intelligent observer could represent the catalyst that causes a universe to become actual; however, throughout the book they vigorously deny the likelihood of intelligent alien life. The privileging of intelligent observers, combined with the belief that humans are the only intelligent observers, positions humanity as critical to the cosmos. In a quote that reflects both the need for scientific credibility and the absence of aliens, William Gairdner writes,

> For it was the hardcore scientists like Copernicus and Galileo who inadvertently drove us out of the centre of the universe as it was then conceived, and now, as we shall see, it is the hardcore scientists [Tipler and Barrow] suggesting we belong back at the centre again.[155]

The connection to scientific credibility should be obvious. "Hardcore scientists" engineered the Copernican revolution, therefore a counter paradigm must also emerge from scientists. The quote does not mention aliens specifically, but the desire for a return to the "centre" speaks to a unity cosmology. While humanity cannot literally hold a place in the center of the universe, it can metaphysically through unity, as Whewell and Wallace suggest.

Leslie, by contrast, goes out of his way to disconnect *Anthros* (humanity) from the Anthropic Principle. He even suggests that because the universe has characteristics befitting intelligent life generally, rather than just humans, the Anthropic Principle should be renamed the "Psychocentric Principle."[156] Unlike the Anthropic Principle, the Psychocentric Principle would not metaphorically return humanity to the center of the universe and thus likely lacked the anthropocentric appeal of Tipler and Barrow's book.

The Anthropic Cosmological Principle's accessibility, both in writing style and availability, scientific credibility, and focus on humanity, made it a player across disciplines in academia. The enduring success of any academic book written in the last three decades remains a rarity. For Tipler and Barrow's work to achieve such attention across disciplines over such a long period of time represents a remarkable feat. *The Anthropic Cosmological Principle* did not only succeed within academia, it spilled over into the public consciousness.

PUBLIC RECEPTION

The Anthropic Cosmological Principle had an impact beyond the confines of academia. The book became a bestseller, a difficult task for any scientific text, much less one with so many opaque concepts.[157] The popularity of the book helped expose many new people to the idea of a human-focused cosmology. While scholars debated the Anthropic Principle going back to the middle of the twentieth century, Tipler and Barrow's book brought it into the public realm. As a writer for the *Hamilton Spectator* put it, "The Anthropic Cosmological Principle, co-authored with Frank J. Tipler in 1986, took 'anthropic principle' from the pages of obscure journals and introduced it to popular culture."[158]

The idea of the Anthropic Principle had obvious appeal to defenders of religion outside the confines of academia. As previously mentioned, William Craig wrote that the Anthropic Principle represents a proof of God in academic journals. Craig also makes the same argument, in more generally accessible language, on his website.[159] The self-proclaimed purpose of his website is to inform the "public arena," in order "to provide an articulate, intelligent voice for biblical Christianity in the public arena."[160] Craig not only talks about the Anthropic Principle on the website; he travels the country lecturing at colleges or debating atheists on the question of God. The Anthropic Principle represents a core of his argument in these public presentations.[161] While Craig represents a player in the academic debate about the Anthropic Principle, one should have little doubt that his larger interest lies in using the idea to spread the Christian faith.

Beyond a religious audience, it is difficult to imagine the success of *The Anthropic Principle* not being in part a reaction to postmodernism. Certainly the postmodern connection to relativism and directionlessness runs directly counter to the teleology and purpose central to *The Anthropic Principle*. Many viewed postmodernists as joined by a shared rejection of metanarratives. *The Anthropic Principle*, with its treatment of all human history as building to one particular technological "Omega point," may hold the distinction of the most meta of metanarratives. Its role as a counter to postmodernism provides a lens with which to understand the success of a rather difficult academic book, an argument bolstered by a comparison to a more explicitly anti-postmodernist book.

A year after the release of the *Anthropic Cosmological Principle*, Allan Bloom published *The Closing of the American Mind* (1987), a book that accused postmodern relativists of destroying academia. Like the *Anthropic Cosmological Principle*, Bloom's book went on to exceed publishing expectations, selling half a million copies in hardback and maintaining the number one spot on the *New York Times* Non-Fiction Best Seller list for four months.[162] Andrew Ferguson, a writer for the *Weekly Standard*, wrote, "The best minds in American publishing were boggled. Never in their experience had a book about Jean-Jacques Rousseau, Friedrich Nietzsche, and Martin Heidegger outsold memoirs by Patty Duke, Shirley MacLaine, and Sam Donaldson—combined."[163] The book served as a focal point for a public backlash against the postmodern thought emanating from the academy. Social critic Camille Paglia later called it "the first shot in the culture wars," the prolonged conflict over values between liberal and conservatives that extends to the present.[164] Given the analysis above, however, Paglia's distinction of "first shot" may belong to *The Anthropic Cosmological Principle*. Two books attacking relativism becoming best sellers within two years cannot be separated from the cultural zeitgeist of the time, which was driven by backlash to what people saw as the relativism of postmodernism. Even though Tipler and Barrow never mention postmodernism directly, several prominent examples exist of others using Tipler and Barrow's ideas to attack postmodernism.

The president of the Czech Republic, Vaclav Havel, made the Anthropic Principle a central part of his speech when he received the prestigious Liberty Medal on July 4th 1994 at Independence Hall in Phildelphia. He received the medal for his inspiration of the Velvet Revolution in Czechoslovakia, which peacefully overthrew the Communist regime. At the time of the speech, Havel held the position of president in the newly created Czech Republic. He had served as last president of Czechoslovakia, but had stepped down because he did not want to preside over the split of the country into Slovokia and the Czech Republic. In the early nineties other former Communist countries like Yugoslavia were also being torn apart by ethnic conflicts, although admittedly more violently. It's clear from the content of his Liberty Medal speech that a concern for social fragmentation occupied his mind.

Rather than place the blame for fragmentation on the material reality of the collapse of the Soviet Union, which had used force to maintain national unity, Havel posits a philosophical culprit. He begins his speech by claiming that postmodernism had broken down stable foundations for shared meaning and absolute ethics:

> And thus today we find ourselves in a paradoxical situation. We enjoy all the achievements of modern civilization that have made our physical existence on this earth easier so in many important ways. Yet we do not know exactly what to do with ourselves, where to turn. The world of our experiences seems chaotic, disconnected, confusing. There appear to be no integrating forces, no unified meaning, no true inner understanding of phenomena in our experience of the world. Experts can explain anything in the objective world to us, yet we understand our own lives less and less. In short, we live in the postmodern world, where everything is possible and almost nothing is certain.[165]

Havel argues that this lack of grounding has resulted in "cultural conflicts" and "tribal cults" that threaten the "survival of a civilization." He likely refers to the ethnic conflicts occurring in the former Yugoslavia and elsewhere. Given postmodernism's focus on cultural pluralism, Havel singles it out as a cause of this ethnic fragmentation.

After putting forth a philosophical sickness at the heart of the world's problems, he offers a philosophical cure. He uses the Anthropic Principle as a 'transcendent" grounding that allows for a universal ethic that can prevent "tribal" conflicts.

> I think the Anthropic Cosmological Principle brings to us an idea perhaps as old as humanity itself: that we are not at all just an accidental anomaly, the microscopic caprice of a tiny particle whirling in the endless depth of the universe. Instead, we are mysteriously connected to the entire universe, we are mirrored in it, just as the entire evolution of the universe is mirrored in us. Until recently, it might have seemed that we were an unhappy bit of mildew on a heavenly body whirling in space among many that have no mildew on them at all. This was something that classical science could explain. Yet, the moment it begins to appear that we are deeply connected to the entire universe, science reaches the outer limits of its powers. Because it is founded on the search for universal laws, it cannot deal with singularity, that is, with uniqueness. The universe is a unique event and a unique story, and so far we are the unique point of that story. But unique events and stories are the domain of poetry, not science. With the formulation of the Anthropic Cosmological Principle, science has found itself on the border between formula and story, between science and myth. In that, however, science has paradoxically returned, in a roundabout way, to man, and offers him—in new clothing—his lost integrity. It does so by anchoring him once more in the cosmos.[166]

Havel uses the singularity inherent in the Anthropic Principle to push for human singularity. Although he does not state it explicitly, a plurality cosmology would justify the dangerous separatism he fears will engulf the globe in violence. The Anthropic Principle not only elevates humanity above an "unhappy bit of mildew," but also connects all people together.

Havel clearly grasps the rhetorical power of the unity cosmology as an argument against relativism. He recognizes the ancient roots of the argument and claims that it can "anchor" humanity. He appreciated the rhetorical power of the Anthropic Principle so much that he delivered a very similar speech in Washington on October 3, 1997, after he received the Fulbright Prize.[167] Beyond those watching the initial speech, several major newspapers published the text of Havel's speech, giving a wide platform to Tipler and Barrow's ideas.[168]

William Gairdner references *The Anthropic Cosmological Principle* as a proof of absolute values against what he sees as the current postmodern agenda of politics and philosophy.[169] On the importance of the cosmic structure to philosophy he writes,

> Some highly respected physicists and mathematicians go much further. They are persuaded that *the entire cosmos came into existence so that human consciousness would eventually arise to observe it!* We may have gotten bumped from the centre of our own solar system over recent centuries, but such new and fascinating modern views argue that we belong right back at the centre again.... [T]he general direction of today's scientific community and, there-

> fore, of the public mind—and certainly of public education, which we must assume to reflect that mind—has strongly run against this idea. We increasingly describe ourselves in definitely *materialistic*, *atheistic*, and therefore, *relativistic* terms.... The bleakness of this view could hardly be better expressed than by Havard astronomer Margaret Geller, who asks: "Why should the universe have a point? What point? It's just a physical system, what point is there?" ... when there is no point to anything, relativism rules. [author's italics][170]

Gairdner recognizes the power of the Anthropic Principle to influence social values. Its ability to return humanity to the "centre" of the universe reverses what he sees as moral decline rather than scientific error. Similar arguments are made by James Applewhite, who, unhappy with the "postmodern" state of art, uses the Anthropic Principle to call for a return to transcendent universal values within aesthetics.[171]

The academic and popular uptake of Tipler and Barrow's work demonstrates the staying power of the unity cosmology even in modern times. Theists, philosophers, artists, and politicians all found the Anthropic Principle a useful tool to persuade others of their beliefs in absolute values. Perhaps the most bizarre and forthright deployment of Tipler and Barrow's Anthropic Principle comes from Tipler himself, not in the original book *The Anthropic Cosmological Principle*, but in his later works.

TIPLER'S LATER WORK

Tipler and Barrow present, at least the fascade, of disinterested scientists in *The Anthropic Cosmological Principle*. They leave the implications of the Anthropic Principle rather vague beyond its connection to teleology generally. In Tipler's book *Physics of Immortality* (1994), he much more clearly defines the stakes of the debate over cosmology.[172] The reception of *The Anthropic Cosmological Principle* appears largely unaffected by Tipler's *The Physics of Immortality*, as very few of the sources cited above mention it. Nor does it appear that the book proved as big a commercial success as *The Anthropic Cosmological Principle*.[173] A brief examination of the *The Physics of Immortality*, will, however, help elucidate Tipler's purpose in writing the original book and give credence to my reading of his unity cosmology. In the *The Physics of Immortality,* Tipler makes explicit many of the ideas that required a subtle reading in *The Anthropic Cosmological Principle*.

Tipler returns to the connection between cosmological ideas and political and ethical thought. This time, instead of speaking mainly through the citation of other authors, Tipler explicitly interjects his own views on the subject. He begins this conversation by using the cosmology of eternal return, as a foil for his own beliefs. In a way this is similar to Tipler and Barrow's use of heat death as a foil to Teilhard's Omega point. Eternal return is the idea that the universe expands and collapses infinitely, which means that all events repeat themselves infinitely. Tipler argues that the eternal return cosmology had an enormous effect on both politics and philosophy within the twentieth century.[174]

Specifically, Tipler draws a series of connections between eternal return and national socialism. He begins by arguing that eternal return "is the basic foundation for the entire corpus of Nietzsche's philosophy," an idea that he says holds true for German philosopher Martin Heidegger as well.[175] Tipler portrays both of these philosophers as critical ideological justifications for Nazism. He then claims that the Nazis chose the swastika as a symbol based on its ancient meaning as a symbol for the cosmology of eternal return. Thus cosmology stands at the heart of Nazism. As Tipler writes, "The political consequences of Nietzsche's Eternal Return philosophy have been catastrophic."[176] Given the boldness of the attribution of such an important historical consequences to the cosmology of eternal return, Tipler provides scant evidence to support his argument. He quotes a few passages from Nietzsche, Heidegger, and some leading Nazi officials talking about eternal return. While it is interesting that these figures talk about cosmology, Tipler fails to prove anything beyond a correlation.

Much more significantly, however, Tipler appears prepared to grant that if the Nazis had been correct about the eternal return, then their beliefs would be justified.

> The Nazis who were aware of [the nonantisemitic] attitudes of Nietzsche emphasized that, in so rejecting racism, Nietzsche was rejecting the implications of his own Eternal Return idea. In this one and only aspect of philosopy the Nazis are correct; I thus reject the Eternal Return.[177]

This passage goes beyond Tipler's previous argument that the eternal return enormously influenced the Nazis. Here he suggests that Nazism logically makes sense in a universe of eternal return, a rather clear repudiation of his claim in *The Anthropic Cosmological Principle* that "statement[s]s of physics … [have] no ethical or moral content.[178]

In *The Anthropic Cosmological Principle*, Tipler and Barrow position heat death as the anti-teleological cosmology, but they do not go so far as to say that if heat is true it merits existential despair. Tipler pretty straightforwardly argues that we have no room to criticize Nazi behavior in a universe of eternal return. The ethical implications of cosmology in the background of *The Anthropic Cosmological Principle* come to the foreground in *The Physics of Immortality.* Of course, Tipler rejects eternal return and instead advocates the FAP, which does provide the basis for proper absolute ethics.

In addition to his renewed focus on philosophy, Tipler does more to draw out the religious importance of the Anthropic Principle. *The Anthropic Cosmological Principle* kept the Omega point a vague construction, and relegated its religious connotation to a footnote. In *The Physics of Immortality*, Tipler begins to explore what he meant by the cryptic remarks about the Omega point in *The Anthropic Cosmological Principle*. He predicts the giant super computer will have the capacity to replicate everyone who has ever existed in the form of data. The personalities of every person who has ever lived will be uploaded into a simulation, which Tipler labors to argue is equivalent to actually being reborn. The result:

> The Omega Point Theory allows the key concepts of the Judeo-Christian-Islamic tradition now to be modern physics concepts; theology is nothing but physical cosmology based on the assumption that life as a whole is immortal. A consequence of this assumption is the resurrection of everyone who have ever lived to eternal life. Physics has now absorbed theology; the divorce between science and religion, between reason and emotion, is over.[179]

Tipler's futuristic technoheaven draws the connections between the Anthropic Principle and religion further than any of the religious scholars who cited his earlier book. Cosmology no longer represents a sign of God, but literally creates God.[180]

None of these claims conflict with the arguments Tipler presented in his earlier book. In fact, one can easily read Tipler's new arguments as clarifications of his old positions. Although it is worth nothing that when Barrow heard of Tipler's new ideas about the Omega point he said they were best left as "science fiction."[181] Still, Tipler and Barrow originally wrote the Omega point would be "omnipotent, omnipresent, and omniscient!"[182] Why not imagine it capable of resurrecting all human life? They discussed dangerous philosophical implications of heat death; why not imagine the wrong cosmology at the heart of Nazism? The foundations of these new arguments rested within *The Anthropic Cosmological Principle*; *The Physics of Immortality* simply brings them to the fore.[183]

The change from the subtle to the explicit alerted the critics of Tipler's rhetorical strategy, resulting in a set of reviews very different from the reviews of *The Anthropic Cosmological Principle.* The three reviews on JSTOR all spoke negatively of the book, in contrast to the generally positive reviews of his previous book.[184] Brian Rotman wrote,

> I suspect physicists will dismiss Tipler's claims and project as those of an isolated raver and move to distance themselves from what they'll see as sheer bad publicity for their science. Others, especially those concerned to question the rhetorical maneuvers and metaphysical presuppositions of contemporary

> science, will see it as a symptom of a near-suffocating hubris on the part, not just of Tipler, who is merely its vehicle, but of forces within physics itself.[185]

This quote indicates that Rotman understands that most of Tipler's argument represents rhetorical appeals to science as the ultimate authority. While the extremity of Tipler's claims may cause scientists to "dismiss" them as an aberration, Rotman says the really interesting takeaway is how Tipler's book represents a common practice (appeal to scientific authority) taken a step too far beyond its traditional rhetorical acceptance. Varadaraja V. Raman comes to a similar conclusion.

> Readers of E. T. Bell's Men of Mathematics may remember the questionable anecdote in which Euler reportedly told an unbelieving Diderot in Catherine the Great's court, "Sir, $(a + b^n)/n = x$, hence God exists." The mathematically untutored Diderot (so the story goes) could not reply and promptly went back to France. This book may be regarded as a more serious, and considerably expanded, version of Euler's quip.[186]

Raman recognizes the practical effect of the scientific complexity of *Physics of Immortality* is that most readers have to take Tipler at his word. Many of the reviews of *The Anthropic Cosmological Principle* acknowledge its scientific difficulty, but they did not read this difficulty as a strategy of obfuscation. The more explicit nature of Tipler's conclusions in the *Physics of Immortality* appears to have ruptured the protective shield of "scientific objectivity." As a result, the reviewers clearly identify the rhetorical strategies deployed by Tipler. Effective deployment of the unity cosmology as an argument in modern times requires a much more subtle approach, better exemplified by *The Anthropic Cosmological Principle*, than the *Physics of Immortality*.

CONCLUSION

In the *Anthropic Principle* Tipler and Barrow combine the idea that the universe appears tailor made for intelligent life, with the unity cosmology. They claim that humanity exists at the center stage of universal events, so important that it is literally incapable of going extinct. This species wide importance trickles down to each human life, as humanity moves toward the end goal of the Omega point. This magnificent *telos* of the universe gives meaning to lives of all humans who live, have lived, and will live and thus staves off the relativistic, nihilistic, and genocidal philosophies that Tipler claims a purposeless cosmos justifies.

While the idea of a futuristic super computer dominating the universe sounds like science fiction or a new age religion, Tipler and Barrow don the cloak of scientific discourse, which gives legitimacy to their ideas. They write in a style with enough accessibility that their philosophical implications are understandable to the average reader, but the scientific justifications for these principles have to be taken on faith. The Omega point becomes a mathematical inevitability rather than a bold, largely unsubstantiated prediction.

The unity cosmology represents a central tenant of the rhetorical appeal of the *Anthropic Cosmological Principle*. Because of humanity's solitary place in the universe, we move to the center of the stage in the cosmic drama Tipler and Barrow outline. The Omega point does not represent an abstract end of the universe created by alien beings; the Omega is built by *our* descendants, retroactively granting purpose and value to our own lives.

Tipler and Barrow's book, the *Anthropic Cosmological Principle*, had an enormous impact. Presses publish tens of thousands of academic books every year. Only an elite few works capture the imagination of both the academic and wider audiences, spawning over a hundred articles and being referenced by heads of state. The success of the book only seems unusual when viewed as a random occurrence, however. It would be wrong to take the book's popularity as an improbability that demands explanation, much as the authors argue humanity's existence demands explanation. If one looks historically, the unity cosmology has often struck a chord with audiences and elevated scientific works to best sellers.

Although in many ways vastly different, strains of the unity mythology in Tipler and Barrow's work owe their genesis to ancient Greece. The extensive history of teleology at the start of the book demonstrates that Tipler and Barrow recognize this philosophical genealogy. Something about the connection of unity with perfection resonates powerfully in western societies as an answer to variations of the ancient claim that "man is the measure of all things."

NOTES

1. Martin Gardner, "WAP, SAP, PAP, & FAP," *The New York Review of Books* 33, no. 8 (1986).

2. Scientific myths still abounded as Toulmin demonstrates through his examples, but the more traditional natural theology lost its prominence. Stephen Edelston Toulmin, "Contemporary Scientific Mythology," in *Metaphysical Beliefs: Three Essays*, ed. Alasdair MacIntyre (London: SCM Press, 1957).

3. David Cahan, "Looking at Nineteenth-Century Science: An Introduction," in *From Natural Philosophy to the Sciences: Writing the History of Nineteenth*, ed. David Cahan (Chicago: University of Chicago, 2003), 4-5.

4. Of course the situation is more complicated than some scientists importing the structure of religion into their scientific beliefs. Yet it is easy to go too far in the direction of conflating science and religion. For example, Lessl uses Carl Sagan as a case of one who merges science with morality and philosophy, but Sagan was highly criticized by many professional scientists

for his methods of popularizing science. Lessl, "Science and the Sacred Cosmos."; Lessl, "The Culture of Science and the Rhetoric of Scientism."

5. Lessl, "The Culture of Science and the Rhetoric of Scientism," 126, 44; David J. Tietge, *Rational Rhetoric: The Role of Science in Popular Discourse* (West Lafayette, IN: Parlor Press, 2008), 123-31.

6. William J. Kinsella, "Rhetoric, Action, and Agency in Institutionalized Science," *Technical Communication Quarterly* 14, no. 3 (2005).

7. Wander, "The Rhetoric of Science," 226; Kenneth Burke, *A Grammar of Motives* (Berkeley: University of California Press, 1969), 351-54.

8. Postmodernist is a nebulous term that has been used to include many who are not relativists and whose work does not support relativism. This chapter offers a clearer definition and justification of how it defines postmodern in the following section.

9. Richard Weaver, "Dialectic and Rhetoric at Dayton, Tennessee " in *Landmark Essays on the Rhetoric of Science: Case Studies* ed. Randy Allen Harris (Mahway, NJ: Lawrence Erlbaum, 1997), 107-26

10. Lawrence J. Prelli, "The Rhetorical Construction of Scientific Ethos " in *Landmark Essays on the Rhetoric of Science: Case Studies* ed. Randy Allen Harris (Mahway, NJ: Lawrence Erlbaum, 1997), 107-26.

11. Advertisements for the book appeared in philosophy journals, science journals, and religion journals which suggests Oxford University Press understood the potential cross over nature of the work.

12. Even in the Creationism vs. Evolution debate a prime example of religion vs. science, modern defenders of creationism put forth the scientific justification of Intelligent Design.

13. Vaclav Havel, "The New Measure of Man," *The New York Times*, July 8, 1994; Julian Barbour, "Why We Should Fear Vast Change," *The Times Higher Education Supplement*, April 18 2003.

14. Brian W. Aldiss, "Desperately Seeking Aliens," *Nature* 409 (2001): 1081.

15. Marie George argues that the Anthropic Principle generally represents a reformatted version of Aquinas' arguments. Marie I. George, "On the Tenth Anniversary of Barrow and Tipler's Anthropic Cosmological Principle," *American Catholic Philosophical Quarterly* 72, no. 1 (1998).

16. Paul Harrison, "Post-Structuralist Theories," in *Approaches to Human Geography*, ed. Stuart Aitken and Gill Valentine (London: Sage Publications, 2006), 122; Gérard Raulet, "Structuralism and Post-Structuralism: An Interview with Michel Foucault," *Telos*, no. 55 (1983): 205; Judith Butler, "Contingent Foundations: Feminism and the Question of Postmodernism," in *Feminists Theorize the Political*, ed. Judith Butler and Joan W. Scott (New York: Routledge, 1992).

17. Jean-François Lyotard, *The Postmodern Condition: A Report on Knowledge*, trans. Geoff Bennington and Brian Massumi (Minneapolis, MN: University of Minnesota Press, 1984), xxiv.

18. Jo Ann Lewis, "It's Postmodern: And if you don't get it, you don't get it," *The Washington Post*, March 27, 1994.

19. Michel Foucault, *Archaeology of Knowledge and Discourse on Language*, trans. M. Sheridan Smith (New York: Pantheon Books, 1972).

20. Jacques Derrida, *Of Grammatology*, trans. Gayatri Chakravorty Spivak (Baltimore, MD: John Hopkins University Press, 1997).

21. Mark Tushnet, "Critical Legal Studies: A Political History," *The Yale Law Journal* 100, no. 5 (1991).

22. Barbara Epstein, "Postmodernism and the Left," *New Politics* 6, no. 2 (1997).

23. Allan Bloom, *The Closing of the American Mind* (New York: Simon and Schuster, 1987), 352.

24. Susan C. Jarratt, *Rereading the Sophists: Classical Rhetoric Refigured* (Carbondale, IL: Southern Illinois University, 1991), 7-10.

25. John D. Barrow, Email, April 24 2013.

26. Frank J. Tipler, *The Physics of Christianity* (New York: Doubleday Broadway, 2007), 124.

27. Frank Tipler, "Postmodern Physics: Colleges Fail to Teach Basics - Even in Physics!," http://www.popecenter.org/commentaries/article.html?id=1843.
28. Dick, *The Biological Universe*, 53-54.
29. Jones Spencer, H., *Life On Other Worlds* (New York: Macmillan, 1940), vii.
30. Shklovskii and Sagan, *Intelligent Life in the Universe*, 418.
31. Frank Drake and Dava Sobel, *Is Anyone Out There?* (New York: Delacorte Press, 1992), 62.
32. Dick, *The Biological Universe*, 437-38.
33. Ibid.
34. Lynda Walsh, *Scientists as Prophets: A Rhetorical Genealogy* (Oxford: Oxford University Press, 2013), 136-38.
35. Although Voltaire, Kepler, and many others wrote stories about aliens Dick argues the "the true extraterrestrial alien, replete with its own physical and mental characteristics, is a relatively recent invention." Dick, *The Biological Universe*, 223. Whether or not Dick is correct in this assessment its clear that aliens took an especially prominent place in popular culture beginning in the late 19th century with the works of H. G. Wells and Kurd Lasswitz and remaining strong through the present day. Ibid., 222-66.
36. Basalla, *Civilized Life in the Universe*, 112-13.
37. Dick, *The Biological Universe*, 141-46.
38. Basalla, *Civilized Life in the Universe*, 165.
39. Mark A. Sheridan, "SETI's Scope: How the Search for Extraterrestrial Intelligence became Disconnected from New Ideas about Extraterrestrials" (Dissertation, Drew University, 2009), 67-93. The two groups of scientists did disagree on some of the issues relating to the science of SETI, as I will discuss later in the chapter.
40. Frank J. Tipler, "Extraterrestrial Intelligent Beings do Not Exist," *Quarterly Journal of the Royal Astronomical Society* 21 (1980): 267.
41. The use of probable as opposed to certain claims by many of the unity and plurality proponents reflects a general trend in the sciences. Lorenz Krüger et al., eds., *The Probabilistic Revolution* (Cambridge, MA: MIT Press, 1987).
42. Tipler, "Extraterrestrial Intelligent Beings do Not Exist," 267.
43. Michael H. Hart, "An Explanation for the Absence of Extraterrestrial Life on Earth," *Quarterly Journal of the Royal Astronomical Society* 16 (1975).
44. David Grinspoon points to Fontenelle's *Conversations on the Plurality of Worlds* as containing a precursor to Fermi's paradox. Grinspoon, *Lonely Planets*, 23.
45. Eric Jones, "Where is Everybody," *Physics Today* 38, no. 8 (1985): 11-13.
46. Ben Zuckerman and Michael H. Hart, eds., *Extraterrestrials: Where Are They?* (Cambridge: Cambrige University Press, 1995).
47. Tipler, "Extraterrestrial Intelligent Beings do not Exist."
48. Frank J. Tipler, "Biography," *Frank J. Tipler Home Page*, http://129.81.170.14/~tipler/biography.htm.
49. Megan Tressider, "The Megan Tressider Interview: Meaning of Life is, Er, God and Omega, Physicist Frank J Tipler, an atheist, says he has found God," *The Guardian* 1995.
50. Tipler, "Biography."
51. Frank J. Tipler, "Rotating Cylinders and the Possibility of Global Causality Violation," *Physical Review* D9, no. 8 (1974); Frank J. Tipler, "Causality Violation in Asymptotically Flat Spacetimes," *Physical Review Letters* 37 (1976); Frank J. Tipler, "Energy Conditions and Spacetime Singularities," *Physical Review* D17 (1978); Frank J. Tipler, "General Relativity and Conjugate Ordinary Differential Equations," *Journal of Differential Equations* 30 (1978); Frank J. Tipler, "General Relativity, Thermodynamics, and the Poincare Cycle," *Nature* 280 (1979).
52. George Ellis, "Piety in the Sky," *Nature* 371, no. 8 (1994): 115; Tressider, "Meaning of Life is, Er, God and Omega."
53. Tipler's predictions of what aliens would do bear some resemblance to Plato and Aquinas' discussion of how God designed the universe.

54. Tipler's original article was divided into three parts, because of a miscommunication between him and the editor. The three parts were released in the same journal within the same year. I treat these three articles as a unit for purposes of analysis.

55. The journal divided Tipler's initial manuscript because of a miscommunication, which meant that what was a unified document was released in three parts. The journal released the parts as independent articles over the course of the same year.

56. Frank J. Tipler, "A Brief History of the Extraterrestrial Intelligence Concept," *Quarterly Journal of the Royal Astronomical Society* 22 (1981): 133; Tipler, "Extraterrestrial Intelligent Beings do Not Exist," 143.

57. Tipler, "A Brief History of the Extraterrestrial Intelligence Concept," 139.

58. Ibid., 133.

59. Tipler, "Extraterrestrial Intelligent Beings do Not Exist," 279.

60. Brewster, *More Worlds Than One*, 232.

61. Shermer, *In Darwin's Shadow*, 204.

62. Kottler, "Alfred Russel Wallace," 154-56; Crowe, *Extraterrestrial Life Debate*, 223-24.

63. Tipler, "A Brief History of the Extraterrestrial Intelligence Concept," 143.

64. Frank J. Tipler, "Additional Remarks on Extraterrestrial Intelligence," *Quarterly Journal of the Royal Astronomical Society* 22 (1981): 289.

65. Ibid.

66. Ibid., 288-89.

67. In his defense, Tipler does provide quotes that make a compelling case that alien salvation does play a role in the way Sagan and others see the cosmos. Twenty-six years later George Basalla provides a much more in-depth version of this argument. Tipler also details the troubling history of his manuscript, as a justification for his claim that plurality defenders religiously guard their theory from attack. Tipler says that Carl Sagan refereed and rejected his article from two previous journals in a manner indicative of academic censorship. Ibid., 289. Like Whewell and Wallace before him, the problem emerges not from Tipler correctly pointing out the flaws of his opponents, but of his engaging in the same practices he criticizes.

68. Tipler, "A Brief History of the Extraterrestrial Intelligence Concept," 139.

69. Frank J. Tipler, "Extraterrestrial Intelligent Beings Do Not Exist," *Physics Today* 34, no. 4 (1981).

70. Ibid., 9.

71. Dick, *The Biological Universe*, 498.

72. Barrow and Tipler, *The Anthropic Cosmological Principle*, 16.

73. Ibid., 21. This represents a twist on the way academics typically define SAP. SAP is usually defined to say that if the universe was substantially different than it is now that it would be impossible for intelligent life to emerge. Gardner, "WAP, SAP, PAP, & FAP."

74. The authors do not take a firm stand on the SAP justified by God. Instead they write that it "does not appear to be open either to proof or to disproof and is religious in nature." Barrow and Tipler, *The Anthropic Cosmological Principle*, 22.

75. Ibid., 472-96.

76. It would be wrong to dismiss the PAP solely on the grounds of its counter-intuitiveness, as many of the foundational ideas from quantum physics seem incredibly strange.

77. Barrow and Tipler, *The Anthropic Cosmological Principle*, 505, 06, 10, 24-41.

78. Ibid., 556-70, 76-600.

79. J.J.C. Smart, "The Anthropic Cosmological Principle," *The Philosophical Quarterly* 37, no. 149 (1987).

80. Barrow and Tipler, *The Anthropic Cosmological Principle*, 23. The idea that intelligent machines will replace humanity represents an obvious difference from earlier manifestations of the unity cosmology. One could read Tipler and Barrow's depiction of the future as one that denies the importance of humanity, because of our inevitable replacement by machines. This reading clashes with the way Tipler and Barrow view personhood, however. For them information represents the foundation of personhood, which means the creation of intelligent machines differs little from bearing new human children. Both represent extensions of human information and thus one should view them as equally children of humanity. As I will discuss later, Tipler believes a human can become functionally resurrected by having all of the data on its

brain placed onto a computer. Alien intelligence does not exist as an extension of human knowledge and thus is threatening to the unity philosophy in ways that human created machines are not.

81. Frank TIpler, "TEDx Brussels 2010 - Frank Tipler - The Ultimate Future " (YouTube, 2011).

82. Barrow and Tipler, *The Anthropic Cosmological Principle*, 23.

83. Ibid., 15.

84. Mary Midgley, *Science as Salvation: A Modern Myth and its Meaning* (New York: Routledge, 1992), 23.

85. Barrow and Tipler, *The Anthropic Cosmological Principle*, 661.

86. Kimmo Eriksson, "The Nonsense Math Effect," *Judgment and Decision Making* 7, no. 6 (2012).

87. Even the authors recognize the unusual length of their book and begin with a joke, quoting the Duke of Gloucester's quip of, "Ah Mr. Gibbon, another damned, fat, square book. Always scribble, scribble, scribble, eh?" when he was presented with the second volume of *The Decline and Fall of the Roman Empire.* Barrow and Tipler, *The Anthropic Cosmological Principle*, v.

88. Smart, "The Anthropic Cosmological Principle," 466; Timothy Ferris, "I Think, Therefore the Universe Is," *New York Times*, February 16, 1986; Robert Klee, "The Revenge of Pythagoras: How a Mathematical Sharp Practice Undermines the Contemporary Design Argument in Astrophysical Cosmology," *British Journal for the Philosophy of Science* 53, no. 3 (2002): 337.

89. Wander, "The Rhetoric of Science," 227.

90. Burke discusses the way forms can create audience desires and than satisfy them as a way to impart narrative (or in this case messages). Kenneth Burke, *Counter-statement* (Berkeley: University of California Press, 1968), 29-32. Walsh explains how scientists gain ethos through means of having access to knowledge inaccessible to the general public. Walsh, *Scientists as Prophets*.

91. Tietge, *Rational Rhetoric*, 188.

92. Ibid.

93. Klee, "The Revenge of Pythagoras," 337. Of course that Klee feels the need to defend the legitimacy of the book in the first place reflects the fact that some prominent voices saw the book as deeply problematic.

94. This occurs frequently in scientific argument. Science exists as the dominant intellectual paradigm, but the average citizen lacks a strong scientific literacy. This means the ethos of the scientist stands in for actual understanding of the specifics of the science. Segal and Richardson, "Scientific Ethos."

95. This will be evident in the reception of Tipler's book *The Physics of Immortality*, which I will discuss towards the end of the chapter. Frank J. Tipler, *The Physics of Immortality: Modern Cosmology, God, and the Resurrection of the Dead* (New York: Doubleday, 1994).

96. Tipler, "A Brief History of the Extraterrestrial Intelligence Concept," 133; Tipler, "Extraterrestrial Intelligent Beings do Not Exist," 143.

97. Barrow and Tipler, *The Anthropic Cosmological Principle*, 28.

98. Ibid., 108.

99. Lessl argues that Carl Sagan uses historical symbolism to sneak metaphysical ideas into his "objective" science. Tipler and Barrow's rhetorical move is even cleverer, because they often do not need to invest history with symbolism as Sagan did, but rather they report the ideas of historical figures in a way that ultimately advances their own beliefs. Lessl, "Science and the Sacred Cosmos," 180.

100. Barrow and Tipler, *The Anthropic Cosmological Principle*, 195-203.

101. Rev. 1:8

102. Pierre Teilhard De Chardin, *The Phenomenon of Man* (New York: Harper Perennial Modern Thought, 2008), 276.

103. Barrow and Tipler, *The Anthropic Cosmological Principle*, 196.

104. Ibid., 203.

105. Ibid., 204.

106. Bertrand Russell, *Mysticism and Logic and Other Essays* (London: Longmans, Green and Co., 1918), 47-48.
107. Paul Davies independently made a similar assessment of heat death's effect and wrote that heat death had a "profoundly depressing effect on generations of scientists and philosophers." Paul Davies, *The Last Three Minutes: Conjectures about the Ultimate Fate of the Universe* (New York: Basic Books, 1994), 12.
108. Toulmin, "Contemporary Scientific Mythology."
109. Barrow and Tipler, *The Anthropic Cosmological Principle*, 167.
110. Teilhard quoted in ibid., 201.
111. Ibid., 613-14.
112. Ibid., 613.
113. Ibid., 614.
114. Ibid., 664-77.
115. Ibid., 677.
116. Ibid., 382.
117. Mary Midgley recognizes Tipler and Barrow's book as a form of techno-salvation myth. Midgley, *Science as Salvation*.
118. The figurative use of religious language in describing the Omega point gives scientific cover to the metaphysical ideas buried in the passage. The authors can always claim to have simply taken poetic license, which is bolstered by their claim that a "theologian" (not they) would describe the Omega point as "omnipotent, omnipresent, and omniscient!" Lessl, "The Culture of Science and the Rhetoric of Scientism," 181.
119. *Phenomenon of Man* 271, quoted in *Anthropic Cosmological Principle*, 201.
120. Wendt and Duvall, "Sovereignty and the UFO."
121. Of course, determining what constitutes "intelligent" or even "life" can be a tricky matter. Despite these concerns the endeavor still appears a much more straightforward task that reading purpose into the universal structure.
122. Sheridan, "SETI's Scope", 67-93.
123. Nicholas Rescher independently develops a similar argument focused on the likelihood that alien science differs radically from our own. Nicholas Rescher, "Extraterrestrial Science," in *Extraterrestrials: Science and Alien Intelligence*, ed. Edward Regis Jr. (Cambridge: Cambridge University Press, 1987).
124. It is worth noting that at the time Tipler and Barrow released their book, the American academy was embroiled in a "culture war." Postmodernism, poststructuralism, and identity politics had challenged many of the practices of the university. The literary canon came under attack by those interested in increasing the presence of works by underrepresented groups. Historians challenged the ability of their profession to provide objective analysis of the past. Professors influenced by continental philosophers, like Michel Foucault, Jean-François Lyotard, and Jacques Derrida attacked the possibility of "Truth." Whether or not they intended it Tipler and Barrow's work serves a rebuke to these modern forms of relativism by defending the possibility of absolute values, absolute knowledge, and an absolute future for humanity.
125. The analysis of academic reception comes primarily from a JSTOR search of the title "*The Anthropic Cosmological Principle.*" JSTOR is a premier database of over 500 academic research journals, which span across academic disciplines and track many journals back well before the mid-1990s. JSTOR provides a representative collection of the academic uses and criticisms of the book. Roger C. Schonfeld, *JSTOR: A History* (Princeton, NJ: Princeton University Press, 2003). Lots of recent scholarship justifies using book reviews to assess reception. Leah Ceccarelli, "A Hard Look at Ourselves: A Reception Study of Rhetoric of Science," *Technical Communication Quarterly* 14, no. 3 (2005): 257–65; James L. Machor, "The American Reception of Melville's Short Fiction in the 1850s," in *New Directions in American Reception Study*, ed. Philip Goldstein and James L. Machor (New York: Oxford University Press, 2008), 87-98; Steven Mailloux, "Judging and Hoping: Rhetorical Effects of Reading about Reading," in *New Directions in American Reception Study*, ed. Philip Goldstein and James L. Machor (New York: Oxford University Press, 2008), 23-32; Kenneth M. Roemer, "Placing Readers at the Forefront of Nowhere: Reception Studies and Utopian Literature " in

New Directions in American Reception Study, ed. Philip Goldstein and James L. Machor (New York: Oxford University Press, 2008), 99-119.

126. Joseph Silk, "New Series," *Science* 232, no. 4753 (1986); Michael Heller, *The Review of Metaphysics* 40, no. 3 (1987); Smart, "The Anthropic Cosmological Principle."; James A. Goldman, *Leonardo* 21, no. 3 (1988).

127. Silk, "New Series," 1036.

128. Heller: 565.

129. Goldman: 334.

130. The neutral account came from a literary journal, much later than the other reviews and situated the book among other books about scientific cosmology. Monroe K. Spears, "Review: Cosmology and the Common Reader," *Sewanee Review* 99, no. 1 (1991): 113-21.

131. Preston Cloud, *The Quarterly Review of Biology* 62, no. 1 (1987): 64.

132. "Recent Publications in Philosophy Source," *Noûs* 20, no. 4 (1986); "New Titles," *BioScience* 38, no. 1 (1988); "Recent Publications," *The British Journal for the Philosophy of Science* 37, no. 3 (1986).

133. I. Prigogine and J. Geheniau, "Entropy, Matter, and Cosmology," *Proceedings of the National Academy of Sciences of the United States of America* 83, no. 17 (1986): 6247; William Lane Craig, "The Origin and Creation of the Universe: A Reply to Adolf Grünbaum," *The British Journal for the Philosophy of Science* 43, no. 2 (1992): 239; Adolf Grünbaum, "Narlikar's "Creation" of the Big Bang Universe Was a Mere Origination," *Philosophy of Science* 60, no. 4 (1993): 638, 41-42; Robert J. Deltete and Reed A. Guy, "Hartle-Hawking Cosmology and Unconditional Probabilities," *Analysis* 57, no. 4 (1997): 304, 13; William Lane Craig, "A Swift and Simple Refutation of the "Kalam" Cosmological Argument?," *Religious Studies* 35, no. 1 (1999): 65; David S. Oderberg, "Temporal Parts and the Possibility of Change " *Philosophy and Phenomenological Research* 69, no. 3 (2004): 686; Adolf Grünbaum, "Creation as a Pseudo-Explanation in Current Physical Cosmology," *Erkenntnis* 35, no. 1/3 (1991); William Lane Craig, "The Caused Beginning of the Universe: A Response to Quentin Smith," *British Journal for the Philosophy of Science* 44, no. 4 (1993): 635, 29, 31-33.

134. Brett Clark, John Bellamy Foster, and Richard York, "The Critique of Intelligent Design: Epicurus, Marx, Darwin, and Freud and the Materialist Defense of Science," *Theory and Society* 36, no. 6 (2007): 526; Mary Jo Nye, "Temptations of Theory, Strategies of Evidence: P. M. S. Blackett and the Earth's Magnetism, 1947-52," *The British Journal for the History of Science* 32, no. 1 (1999): 89-90; Colin A. Russell, "Presidential Address: 'Rude and Disgraceful Beginnings': A View of History of Chemistry from the Nineteenth Century," *The British Journal for the History of Science* 21, no. 3 (1988): 291.

135. Barry Mazur, *Harvard Review*, no. 16 (1999): 145; David Rosenboom, "Propositional Music: On Emergent Properties in Morphogenesis and the Evolution of Music. Part I: Essays, Propositions and Commentaries," *Leonardo* 30, no. 4 (1997): 296; David N. Samuelson, "Review: Botching the Science in Science Fiction," *Science Fiction Studies* 19, no. 1 (1992): 104.

136. Mark Colyvan, Jay L. Garfield, and Graham Priest, "Problems with the Argument from Fine Tuning," *Synthese* 145, no. 3 (2005): 334-35; Neil A. Manson, "Anthropocentrism and the Design Argument," *Religious Studies* 36, no. 2 (2000): 171-72; D. J. Bartholomew, "Probability, Statistics and Theology," *Journal of the Royal Statistical Society* 151, no. 1 (1988): 140-41; M. A. B. Whitaker, "On Hacking's Criticism of the Wheeler Anthropic Principle," *Mind* 97, no. 386 (1988): 262; Timothy McGrew, Lydia McGrew, and Eric Vestrup, "Probabilities and the Fine-Tuning Argument: A Sceptical View," *Mind* 110, no. 440 (2001): 1027; B. L. Hebblethwaite, "Mellor's 'Bridge-Hand' Argument," *Religious Studies* 22, no. 3/4 (1986): 477; Peter Forrest, "Aesthetic Understanding," *Philosophy and Phenomenological Research* 51, no. 3 (1991): 538; Roger White, "Does Origins of Life Research Rest on a Mistake?," *Noûs* 41, no. 3 (2007): 457.

137. Hebblethwaite, "Mellor's 'Bridge-Hand' Argument," 477.

138. James Applewhite, "Postmodernist Allegory and the Denial of Nature," *Kenyon Review* 11, no. 1 (1989): 8.

139. Stuart N. Hart, "Making Sure the Child's Voice Is Heard," *International Review of Education* 48, no. 3/4 (2002): 251.

140. Tipler and Barrow make the claim that Herbert Spencer modeled his cosmology after his political beliefs. Barrow and Tipler, *The Anthropic Cosmological Principle*, 188.

141. Errol E. Harris, *Cosmos and Theos: Ethical and Theological Implications of the Anthropic Cosmological Principle* (Atlantic Highlands, NJ: Humanities Press, 1992); Errol E. Harris, *Cosmos and Anthropos: A Philosophical Interpretation of the Anthropic Cosmological Principle* (London: Humanities Press International, 1991). Importantly, Harris also denies the likelihood of intelligent alien life, citing Tipler and Barrow as his only source. Ibid., 9.

142. M. A. Corey, *God and the New Cosmology: The Anthropic Design Argument* (Lanham: Rowman & Little, 1993).

143. Harris, *Cosmos and Theos*, 22, 199; Corey, *God and the New Cosmology*, 187-88.

144. John Leslie, "Anthropic Principle, World Ensemble, Design," *American Philosophical Quarterly* 19, no. 2 (1982).

145. John Leslie, "Observership in Cosmology: The Anthropic Principle," *Mind* 92, no. 368 (1983); John Leslie, "Anthropic Explanations in Cosmology," *PSA: Proceedings of the Biennial Meeting of the Philosophy of Science Association* 2986 (1986); John Leslie, "Mackie on Neoplatonism's 'Replacement for God'," *Religious Studies* 22, no. 3/4 (1986).

146. Bartholomew, "Probability, Statistics and Theology," 141; John Jefferson Davis, "The Design Argument, Cosmic 'Fine Tuning,' and the Anthropic Principle," *International Journal for Philosophy of Religion* 22, no. 3 (1987): 141; Barry Whitney, "Divine Persuasion and the Anthropic Argument," *The Personalist Forum* 14, no. 2 (1998): 144; Murdith McLean, "Residual Natural Evil and Anthropic Reasoning," *Religious Studies* 27, no. 2 (1991): 185; Hebblethwaite, "Mellor's 'Bridge-Hand' Argument," 477; Rodney D. Holder, "Fine-Tuning, Multiple Universes and Theism," *Noûs* 36, no. 2 (2002); Patrick A. Wilson, "What Is the Explanandum of the Anthropic Principle?," *American Philosophical Quarterly* 28, no. 2 (1991); John Earman, "The Sap Also Rises: A Critical Examination of the Anthropic Principle," *American Philosophical Quarterly* 24, no. 4 (1987); Joseph M. Zycinski, "The Anthropic Principle and Teleological Interpretations of Nature," *The Review of Metaphysics* 41, no. 2 (1987).

147. Bartholomew, "Probability, Statistics and Theology," 141; Davis, "The Design Argument, Cosmic 'Fine Tuning,' and the Anthropic Principle," 141; Whitney, "Divine Persuasion and the Anthropic Argument," 144; McLean, "Residual Natural Evil and Anthropic Reasoning," 185; Hebblethwaite, "Mellor's 'Bridge-Hand' Argument," 477.

148. Two of the trivial JSTOR results are superficial references to the article in the same journal that published the article. One from the table of contents from the issue of *The American Philosophical Quarterly* containing the original article and another comes from a series of corrected errors from the original article in a later issue of the journal. "Corrigenda to Drengson and Leslie," *American Philosophical Quarterly* 19, no. 4 (1982); "Volume Information," *American Philosophical Quarterly* 19, no. 1 (1982).

149. Frank J. Tipler, "The Anthropic Principle: A Primer for Philosophers," *PSA: Proceedings of the Biennial Meeting of the Philosophy of Science Association* 1988, no. 2 (1988).

150. Corey, *God and the New Cosmology*; Hebblethwaite, "Mellor's 'Bridge-Hand' Argument."

151. William Lane Craig, "Barrow and Tipler on the Anthropic Principle vs. Divine Design," *The British Journal for the Philosophy of Science* 39, no. 3 (1988): 389.

152. Ibid., 395.

153. Earman, "The Sap Also Rises," 313.

154. Jeffrey F. Addicott, "Storm Clouds on the Horizon of Darwinism: Teaching the Anthropic Principle and Intelligent Design in the Public Schools," *Ohio State Law Journal* 63 (2002): 1595.

155. William D. Gairdner, *The Book of Absolutes: A Critique of Relativism and a Defense of Universals* (London: Mc-Gill-Queen's University Press, 2008), 101.

156. Leslie, "Anthropic Principle, World Ensemble, Design," 144.

157. Barbour, "Why We Should Fear Vast Change."

158. "The Anthropic Principle," *The Hamilton Spectator* March 17, 2006.

159. William Lane Craig, "The Teleological Argument and the Anthropic Principle," *Reasonable Faith*, http://www.reasonablefaith.org/site/News2?page=NewsArticle&id=5179.

160. "Our Mission," *Reasonable Faith*, http://www.reasonablefaith.org/site/News2?page=NewsArticle&id=5299.

161. William Lane Craig, "Question 97: Subject: Scepticism Quebecois," *Reasonable Faith*, http://www.reasonablefaith.org/site/News2?page=NewsArticle&id=6937.

162. William Goldstein, "The Story behind the Best Seller: Allan Bloom's *Closing of the American Mind*," *Publishers Weekly*, July 3 1987.

163. Andrew Ferguson, "The Book That Drove Them Crazy: Allan Bloom's 'Closing of the American Mind' 25 Years Later," *The Weekly Standard*, April 9, 2012.

164. Camille Paglia, "Ask Camille," *Salon.com*, http://web.archive.org/web/20080411071450/http://www.salon.com/july97/columnists/paglia2970722.html.

165. Havel, "The New Measure of Man."

166. Ibid.

167. Vaclav Havel, "From Plato to NATO," *The Herald*, January 31, 1998.

168. Vaclav Havel, "Looking at the World Through Post-modern Eyes," *Palm Beach Post*, July 14, 1994; Havel, "The New Measure of Man."; Vaclav Havel, "Our Changing Times, and the New Measure of Man," *Seattle Post-Intelligencer*, July 10, 1994.

169. Gairdner, *The Book of Absolutes*, 101-04. Gairdner unsuprisingly traces the genealogy of the modern day relativists back to Protagoras, whose ideas he claims, "evaporated under the withering criticisms first of Plato and then of his student Aristotle." Ibid., 6.

170. Ibid., 11.

171. Applewhite, "Postmodernist Allegory and the Denial of Nature," 8.

172. Tipler, *The Physics of Immortality*.

173. Although from what I have read *The Physics of Immortality* was on the bestseller list in Germany for 15 weeks. Tressider, "Meaning of Life Is, Er, God and Omega."

174. Tipler, *The Physics of Immortality*, 74-89.

175. Ibid., 79.

176. Ibid., 82.

177. Ibid., 83.

178. Barrow and Tipler, *The Anthropic Cosmological Principle*, 23. Admittedly, this section of Tipler's book has some ambiguities. Tipler sometimes refers to Nietzche's "Eternal Return philosophy," which could be read as distinct from the cosmology itself. After all, Tipler mentions the stoics shared the cosmology of eternal return, and while Tipler does not acknowledge it, their cosmology has little resemblance to Nazism. My reading of the text leads me to believe that Tipler uses these terms interchangeably such that the eternal return cosmology logically results in Nietzsche's philosophy of eternal return.

179. Tipler, *The Physics of Immortality*, 338. The idea that physics must encompass theology if theology is to survive occurs throughout the book. Ibid., 3, 6-11, 337.

180. Tipler foreshadows this conclusion again, in his response to an article by Martin Gardner when he writes, "It is fideism which is metaphysics and fantasy. Worse, fideism requires religion to be divorced forever from science. Gardner's real objection to our Omega Point theory is that it threatens to end the divorce." Frank J. Tipler, "Frank J. Tipler, reply by Martin Gardner," *The New York Review of Books* (1986).

181. Tressider, "Meaning of Life Is, Er, God and Omega."

182. Barrow and Tipler, *The Anthropic Cosmological Principle*, 382.

183. This highlights Booth's point that science often shares many of the characteristics of religion, but refuses to acknowledge these overlaps. Wayne C. Booth, "Rhetoric, Science, Religion," in *The Essential Wayne Booth*, ed. Walter Jost (Chicago: University of Chicago Press, 2006), 264-79.

184. Brian Rotman, *SubStance* 24, no. 3 (1995); ibid.; Varadaraja V. Raman, "Hope from Physics," *Science* 267, no. 5200 (1995); Albert B. Stewart, *The Antioch Review* 53, no. 3 (1995).

185. Rotman: 152-53.

186. Raman, "Hope from Physics," 1042.

Chapter Five

Unity in the Twenty-First Century

"Natural theology has a persistent habit of returning, even when its death notice has been extensively and repeatedly published."—Alister E. McGrath, 2011[1]

By the 1990s some postmodernists had expanded their critique beyond the softer academic disciplines like sociology, philosophy, and history to include the hard sciences. With the move to attack science, these thinkers now posed a major challenge to both elements of the scientific myth of unity; it questioned the possibility of absolute values, but it also challenged the foundation of the scientific project, in which unity grounds those absolute values. The burden of defending science did not fall solely on unity's defenders, however. In 1994, scientists Paul R. Gross and Norman Levitt published *Higher Superstition: The Academic Left and Its Quarrels with Science*, sparking what would become known as "the science wars."[2] The issue moved into the public sphere with articles discussing the science wars in *Newsweek*, *The Guardian*, and the *New York Times*.[3]

Many scientists viewed the postmodern critique of science as an unwanted intrusion upon their field by unqualified outsiders. Interestingly, the "science wars" helped cultivate a breed of public intellectuals with no compunction about using science as a springboard to comment on other disciplines, like ethics, philosophy, and religion. Scientists Daniel Dennett, Steven Pinker, and E. O. Wilson all attack postmodernism, poststructuralism, and deconstruction from the perspective of evolutionary science, but their worldview is no less hostile to the unity mythology.[4] While they accuse postmodernism of ignoring the foundation evolutionary theory provides for ethics, this is not a case of "the enemies of my enemies are my friends" for those interested in Platonic notions of the Good. Like Callicles in Plato's *Gorgias* or Francis Galton writing at the time of Alfred Russel Wallace, these thinkers embrace a form of ethical relativism grounded in the natural world.[5] They argue that evolution determines appropriate societal values, rather than the idea that correct values exist independently from human or natural causes. The unity mythology with its reliance on both scientific objectivity and absolute ethical values was caught between these two hostile camps.

By the end of the decade, science had gained as strong a position in society as it had ever held. The postmodern criticisms of science (the science wars) faded into the background, though postmodern thought continued to hold a prominent place in many other disciplines.[6] The disappearance of the science wars from the larger culture left science the *de facto* winner as it got to return to its previous exalted status. Despite the end of the science wars, the scientific expeditionary force of Dennett, Pinker, Wilson, and others continued to make forays into the areas of philosophy and ethics. Not only did science defend its domain against postmodern questioning, but also it increasingly pushed back against religion. Only 7 percent of National Academy of Science members surveyed in 1998 indicated that they believed in a personal God.[7] The poll represents the tail end of a century-long downward trend in religious belief among scientists, probably influenced by science's ever increasing explanatory power. The direction of scientific discoveries also likely played a role in the increasing ethical materialism and atheism of the scientific community. The discovery of other planets and galaxies, the solidification of evolutionary theory, and materialistic explanations of consciousness were more easily reconciled with a nonteleological universe without humanity at its metaphorical center. The unity/plurality debate, however, was one scientific question that could still resolve itself in favor of human centrality.

Against the backdrop of the surge of scientific myths hostile to teleology, two books became academic and popular successes making the case for the unity myth. The first, *Rare Earth* (2000), came from Peter Ward and Donald Brownlee, two prolific scientists at the University of Washington. Simon Conway Morris, a well-known evolutionary biologist and professor of Earth Sciences at the University of Cambridge, wrote the second, *Life's Solution: Inevitable Humans in a Lonely Universe* (2003).[8] Both of these books adopt a primarily scientific style in the mold of Wallace, Tipler, and Barrow, but nevertheless, they expand their reach beyond the scientific question of unity. This chapter examines each of these books in turn. It begins by exploring the specific ideologies at which these books take aim. Then it shows how *Rare Earth* and *Life's Solution* deploy updated unity mythologies. Finally it examines the books' receptions and what those receptions imply about their rhetorical power.

RARE EARTH AND MEDIOCRE HUMANITY

Peter Ward and Donald Brownlee CVs prior to the release of their book *Rare Earth* do not paint the picture of particularly public intellectuals. Prior to the year 2000, they had amassed almost a hundred articles in prestigious academic journals, but even the most prestigious academic journals are not particularly well circulated beyond academia. Their articles addressed narrow scientific questions, like the periodicity of chamber formation in cephalopods. These papers were of interest to specialists, but not the kind of thing the public clamored to read.[9] *Rare Earth* radically changed this modus operandi by speaking to broad cultural questions as well as purely scientific ones. In *Rare Earth* Ward and Brownlee make a thoroughly scientific case for unity, and use this case to offer a unity myth opposed to the scientific myth of mediocrity.

Carl Sagan coined the term "principle of mediocrity" to refer to the idea that scientists should assume that nothing is special about humanity's situation: not the planet it inhabits, its location, humanity's emergence through evolution, nor any other factor. The idea emerged from the 1961 Green Bank conference, an informal meeting of scientists interested in discussing the question of alien life.[10] The principle of mediocrity bares a significant resemblance to the Copernican principle, but deals with much more than humanity's location in the universe. As early as 1983, Sagan and Newman argue that the scientific debate has almost been decided definitively in the favor of the principle of mediocrity.

> We are not at the centre of the Solar System; our planet is one of many; it is vastly older than the human species; the Sun is just another star, obscurely located, on among some 400 billion others in the Milky Way, which in turn is one galaxy among perhaps hundreds of billions. We humans have emerged from a common evolutionary process with all the other plants and animals on Earth. We do not possess any uniquely valid locale, epoch, velocity, acceleration, or means of measuring space and time.[11]

Of course the still open question of unity vs. plurality bears on the principle of mediocrity, because if Earth were the only planet with life it would not be mediocre. Sagan and Newman dismiss this possibility, writing that defenses of unity fit a pattern of "anthropocentric and self congratulatory pronouncements" more concerned with humanity's place in the universe than good science.

The idea that humanity has no special place in the universe appears rather frequently from scientists. *Rare Earth* quotes as an example of the prevalence of the principle of mediocrity Marcello Gleiser, a professor of physics and astronomy, who writes, "Our planet is not in a special place in the solar system, our Sun is not in a special place in our galaxy and our galaxy is not in a special place in the Universe."[12] In a book on the history of science since the Renaissance, released shortly after *Rare Earth*, veteran science writer John Gribbin begins by saying, "The most important thing that science has taught us about our place in the Universe is that we are nothing special."[13] He goes on to list many of the same discoveries Sagan and Newman did, like the existence of many planets and stars, but his word choice points to something deeper. It is not just our circumstances that are not special, but rather "we are not special."

Whether Gribbin meant to imply that scientific discoveries in support of the mediocrity render humanity itself unspecial is unclear, but one can certainly find other examples that leave little doubt. Stephen Hawking once said, "The human race is just a chemical scum on a moderate-sized planet, orbiting around a very average star in the outer suburb of one among a hundred billion galaxies. We are so insignificant that I can't believe the whole universe exists for our benefit. That would be like saying that you would disappear if I closed my eyes."[14] The term "scum" and "insignificant" are extra-scientific concepts, a nascent scientific myth, in the vein of Thomas Paine's *Age of Reason*, which links the mediocrity principle to humanity's value(lessness). In a similar vein, P. Z. Meyers, a biology professor at the University of Minnesota writing significantly after the release of *Rare Earth*, writes,

> The mediocrity principle simply states that you aren't special. The universe does not revolve around you, this planet isn't privileged in any unique way, your country is not the perfect product of divine destiny, your existence isn't the product of directed, intentional fate, and that tuna sandwich you had for lunch was not plotting to give you indigestion.[15]

Meyers uses the scientific idea that human circumstances are average to attack teleology and fate, even though these concepts operate in another realm entirely.

In perhaps its most famous usage, Tyler Durden, one of the main characters of the underground hit film *Fight Club*, uses the mediocrity principle to put his comrades in line, "You are not special. You are not a beautiful or unique snowflake. You're the same decaying organic matter as everything else."[16] The world Durden describes is entirely materialistic, where humans are no more than their decaying bodies. Although actively searching for more examples of the deployment of the principle of mediocrity to make claims about humanity's value did not bear much fruit, one gets the impression that this attitude could exist below the surface, informing scientific attitudes even if it is rarely explicitly articulated. Certainly Ward and Brownlee feel that the myth of mediocrity plays a large role in shaping culture, and this idea bookends their otherwise very technical work.

RARE EARTH OR UNIQUE EARTH?

Before one begins an analysis of how Ward and Brownlee deploy the unity mythology in *Rare Earth*, it is important to establish that the book actually defends the science of unity. While the definition of unity has fluctuated among the case studies in this book, they all have agreed that whatever distinguishes Earth, its status as a planet, its intelligent life, etc., is a unique feature. A "rare" Earth is not the same as a unique Earth. If the Rare Earth hypothesis allows that human-like intelligence is *rare*, but not *unique*, it would be in tension with the unity myth as deployed by those like Plato, Whewell, and Tipler.[17] The unity myth gives humanity metaphorical centrality on the basis of some unique property. Multiple Earths would disrupt this centrality, because uniqueness represents a clear bright line for those seeking to ground their philosophical, religious, political, and ethical claims. "Rareness" is much more ambiguous. Are five intelligent civilizations in a galaxy considered rare? A hundred? After all, from a certain perspective even the millions of world's supporting alien civilizations predicted by Sagan in our own galaxy are only a small percentage of the roughly three trillion planets he estimated existing.[18]

When Ward and Brownlee use the word "rare" do they actually mean "singular?" The language in the book appears to leave open the possibility. One of the key ways the book connects rare to singular is that it repeatedly argues that *any* complex form of life whatsoever will be rare: "In this book we will argue that not only intelligent life, but even the simplest of animal life, is exceedingly rare in our galaxy and in the Universe."[19] By the simplest animal life Ward and Brownlee mean anything more complex than single-celled bacteria. Of course if any animal with more than one cell is rare, than the most advanced forms of life should be exceedingly rare, or as Ward and Brownlee put it, "If animal life is so rare, then intelligent animal life must be rarer still."[20] The book leaves the question of how rare ambiguous, but Ward and Brownlee argue that intelligent life on Earth required many conditions and planets that meet these conditions are "extraordinarily rare." Even if a planet like Earth existed it would be no guarantee of life, much less complex life, and even less so intelligent complex life. Given this logic one can easily be left with the impression that Ward and Brownlee believe that Earth is the only planet with intelligent life.[21]

In a chapter titled "Assessing the Odds," Ward and Brownlee discuss many of the critical components necessary for complex life to come into existence. They examine the Drake equation, which Drake and others argue points to between a thousand and one hundred million intelligent alien species in the Milky Way galaxy. Ward and Brownlee claim that Drake made a major mistake by treating planets in his equation as interchangeable.[22] Ward and Brownlee argue that most planets are uninhabitable to life, which means that the Drake equation errs in treating Earth as typical. This has a huge impact, they maintain, because "any factor in the equation that is close to zero yields a near-zero final answer, because all the factors are multiplied together."[23] For example, there may be trillions of stars, but if the percentage of stars that are located in a portion of the galaxy capable of supporting life is close to zero, then the number of possible inhabited planets drops radically. If another factor is also close to zero, say the number of stars with planets, then the number of possible inhabited planets moves even closer to zero. They argue that the original Drake equation had too few variables, and thus did not properly assess the probability of life. In order to better reflect the multitude of preconditions necessary for the development of life, Ward and Brownlee offer their own version of Drake's famous equation,

> $N^* \text{ X } fp \text{ X } ne \text{ X } fi \text{ X } fc \text{ X } fl = N$
> where:
> N^* = stars in the Milky Way galaxy
> fp = fraction of stars with planets
> ne = planets in a star's habitable zone
> fi = fraction of habitable planets where life does arise
> fc = fraction of planets with life where complex metazoans arise
> fl = percentage of a lifetime of a planet that is marked by the presence of complex metazoans[24]

The "Rare Earth equation" replaces "intelligent civilization" with "complex metazoans" which emphasizes the difficulties involved in the creation of even the most basic forms of life.[25] While the Drake equation includes factors geared toward the production of radio signals that the SETI project could detect, the Rare Earth equation only focuses on elements necessary for complex life. If one multiplies all the variables of the Rare Earth equation it reveals the number of planets that have complex life. When explaining the variables of the Rare Earth equation, Ward and Brownlee reiterate, "When any term of the equation approaches zero, so too does the final result." After discussing all of the unlikely variables that need to be present in order for complex life to develop, they "flesh [the Rare Earth equation] out a bit by adding some of the other factors featured in this book."[26]

> $N^* \text{ X } fp \text{ X } fpm \text{ X } ne \text{ X } ng \text{ X } fi \text{ X } fc \text{ X } fl \text{ X } fm \text{ X } fj \text{ X } fme = N$
> where:
> N^* = stars in the Milky Way galaxy
> fp = fraction of stars with planets
> fpm = fraction of metal-rich planets
> ne = planets in a star's habitable zone
> ng = stars in a galactic habitable zone
> fi = fraction of habitable planets where life does arise
> fc = fraction of planets with life where complex metazoans arise

fl = percentage of a lifetime of a planet that is marked by the presence of complex metazoans
fm = fraction of planets with a large moon
fj = fraction of solar systems with Jupiter-sized planets
fme = fraction of planets with a critically low number of mass extinction events[27]

With these additions the Rare Earth equation becomes significantly longer than Drake's counterpart. After offering this long list of requirements, Ward and Brownlee repeat for the third time, in italics, "*as any term in such an equation approaches zero, so too does the final product.*"[28] Sagan repeatedly emphasized the starting number of stars and planets, billions and trillions, to lead his audience to the conclusion that intelligent alien life must be widespread.[29] Ward and Brownlee flip the script and repeatedly draw attention to the variables necessary to sustain intelligent life. Only one of the variables needs to be rare for the likelihood of alien intelligence to be very low. Their repetition of "zero" acts to give attention to the idea that humanity is alone.

The contextualization of the Rare Earth equation within the broader book reinforces the idea of human uniqueness. The equation follows 200 hundred plus pages of Ward and Brownlee arguing that every single variable on the list is likely close to zero. Even if a planet has all of these variables it only meets the requirements for multicellular life, which is no guarantee for the emergences of intelligent creatures like humans. If planets with multicellular life are around zero, then one can easily be left with the impression that the number of planets with intelligent life is one: Earth. In the section on the Rare Earth equation, Ward and Brownlee say that despite the evidence they provide, "we cannot rule out the possibility that Earth is not unique in the galaxy as an abode of life that has just recently developed primitive technologies for space travel and interplanetary radio communication." Still, this leaves the impression that the possibility of other intelligent life is slim, even if it cannot be ruled out completely.

At one point in the book Ward and Brownlee seem more optimistic about the possibility of intelligent alien life saying; "There probably are other civilizations in the galaxy that have radio telescopes."[30] This represents the lone ray of hope for intelligent alien life in an otherwise pessimistic book, however.[31] This sentence is out of place and not consistent with the tone of the rest of *Rare Earth*. Even as they admit that other intelligent civilizations exist, Ward and Brownlee make the argument that SETI's efforts to listen to their radio signals will likely prove futile.

Media coverage treated *Rare Earth* as a case for unity, despite Ward and Brownlee's suggestion that intelligent aliens may exist in the galaxy. Nigel Hawkes, writing for the *Times*, quotes passages of *Rare Earth* that emphasize its connection to unity: "'What if the Earth, with its cargo of advanced animals, is virtually unique in this quadrant of the galaxy?' they ask. 'What if it is utterly unique, the only planet with animals in this galaxy or even in the visible universe?' In that case, it is time for thanks and even a little pride, in having pulled off the greatest accumulator in the history of betting."[32] The *Independent* ran a review of *Rare Earth* titled "Oh No! Humans Are the Only Intelligent Lifeform."[33] The Associated Press put out a story that said, "For years people have been startled by the tantalizing thought, 'We may not be alone.' Ward and Brownlee trump that by saying, 'We may be alone, after all.'"[34] Some of this media spin, no doubt, is the result of the writers' and editors' desire to create sensational stories. "Humanity Is Alone in the Universe," makes for a better headline than, "Estimate of Expected Intelligent Alien Life Forms Reduced." The dominant narrative had been that humanity was mediocre, but *Rare Earth*'s focus on the privileged conditions of Earth encouraged journalists to flip the script and argue that humanity was unique.

Fault, if any is deserved, does not lie solely with the media, however. Ward and Brownlee's book invites a unity interpretation. They repeatedly emphasize that any complex life will be *extremely* rare. Only one line in *Rare Earth* indicates that people should expect other intelligent life to exist. Prior to the release of the book, Brownlee made comments like "the underlying theme [of *Rare Earth*] is that the Earth is a charmed place," which one could easily interpret to mean that Earth is unique.[35] Most importantly, Ward and Brownlee frame the science of their book within the unity mythology in both the introduction and conclusion.

RARE EARTH AND UNITY MYTHOLOGY

Of all of the case studies of unity mythology in this book, the *Rare Earth*'s takes the subtlest approach. Its focus on the rare conditions of Earth is not as explicitly connected to ethics, religion, politics, and philosophy as Wallace, Whewell, or even Tipler and Barrow's cosmologies. As such, some critics treat *Rare Earth* as a scientific text with no social implications. One writer for the *Associated Press* goes so far as to praise Ward and Brownlee precisely for avoiding any form of scientific mythology: "The authors make no venture into the social or religious implications of 'Rare Earth.' ... They write with the detachment attained from the knowledge we all were once stardust and that Earth will once again return to ash."[36] For most of the book this holds true, as they confine their version of the unity mythology entirely to the introduction and conclusion of the book, much as Wallace bookends most of the unity mythology in *Man's Place in the Universe*. But the unity mythology is still present, framing the importance of their scientific results.

In the end of the introduction of *Rare Earth*, Ward and Brownlee describe what they see as the prevalent myth associated with plurality.

> Ever since Polish astronomer Nicholas Copernicus plucked it from the center of the Universe and put it in orbit around the sun, Earth has been periodically trivialized. We have gone from the center of the Universe to a small planet orbiting a small, undistinguished star in an unremarkable region of the Milky Way galaxy—a view now formalized by the so-called Principle of Mediocrity, which holds that we are not the one planet with life but one of many.[37]

They argue that the Copernican revolution robbed humanity of its spot at the center of the universe and rendered it mediocre. Of course, neither Earth's physical location nor the percentage of planets with life necessitates a "trivialization" of our planet. The association of Copernicanism with plurality, and, therefore, with human insignificance, represents the complement to the unity scientific myth, the mediocrity myth advanced by Hawkings and others. Rather than debunk the mediocrity myth, however, Ward and Brownlee accept it as a "trivializing" force, much as Whewell accepted plurality as an argument against Christianity. Instead of attacking the unscientific plurality myth, Ward and Brownlee reject the plurality science. If plurality decentralizes, then unity must centralize or, as they put it, "the Rare Earth Hypothesis will reverse [the Copernican] decentering trend."[38] They pose the question, "What if [Earth] is utterly unique: the only planet with animals in this galaxy or even in the visible Universe, a bastion of animals amid a sea of microbe-infested worlds?" The implied answer: unity would be the ultimate recentering of Earth. With the Earth recentered by unity, Ward and Brownlee use another rhetorical question to suggest that all of the living creatures on Earth would gain in value on account of their rareness. "If that is the case, how much greater the loss the Universe sustains for each species of animal or plant driven to extinction through the careless stewardship of Homo sapiens?"[39] While this quote places emphasize on the loss of animal lives, this is only because they (not humanity) face the serious short-term risk of extinction. If the tragedy of losing animals rises with their rareness, then the unity cosmology lends humanity a special value as the unique example of intelligent life and the "steward" of the planet.

Ward and Brownlee return to the ideas of centrality and value in their conclusion. They offer a short history of humanity's "centrality."[40] At first they posit centrality as a purely locational term. They begin by discussing the debates among ancient Greek cosmologists like Anaximander and the Pythagoreans about Earth's location in the solar system. They briefly trace the idea of geocentrism through the end of the Medieval period, where Copernicus "shattered the notion of an Earth-centered Universe and put the sun at the center of all orbits."[41] They claim that decentralization continued with the Sun losing its spot in the middle of the universe and the discovery that our galaxy is only one of many, with no special claim to an important or unique position.

After discussing how Edwin Hubble proved the existence of many galaxies, Ward and Brownlee make an interesting move from scientific discoveries about Earth's physical location to discussing a variety of other scientific advances unrelated to astronomy.

> It was not only astronomers who changed the world view. Einstein showed that there is no preferred observer in the Universe, and quantum mechanics told us that chance is king. Charles Darwin and his powerful theory of evolution demoted humans from the crown of creation to a rather new species on an already animal-rich planet, the chance offspring of larger-scale evolutionary and ecological forces. Nothing special. And yet . . .[42]

This short passage is replete with meaning. For starters, what is the connection between Darwinian evolution and Earth's location? Ward and Brownlee suggest in the passage that both these discoveries affect the value of humanity by attacking its metaphorical centrality. Prior to the discovery of evolution, humanity was "the crown of creation," just like a central Earth conferred special status upon us. The Darwinian revolution rendered humanity the mere "chance offspring of larger-scale evolutionary and ecological forces" just as the Copernican revolution rendered our home insignificant. Quantum mechanics, Darwinism, and Einstein's theory of relativity challenged humanity's metaphorical/spiritual/philosophical "centrality." The net effect of all of these discoveries in and out of astronomy is to render humanity "nothing special." But the Rare Earth hypothesis attempts to reverse this "decentering trend."[43] In contrast to the decentering trend that renders humanity mere animals, products of chance, and nothing special, unity implies the opposite. Humanity is above its animal origin, fated to exist, and most definitely special.

Ward and Brownlee leave the work of explaining how the Rare Earth hypothesis reverses the "decentering trend" to the ellipses, a rather blatant use of enthymeme. They call upon the audience to provide the rationale, which saves them the trouble of traveling too far from the physical into the metaphysical. The passage also relies heavily on flipping the scientific myth of the unnamed individuals who advocate the decentering view. Ward and Brownlee list the scientific myths associated with decentering and then claim that unity recenters humanity, flipping the myths. This allows them to advance the unity myth without directly endorsing it themselves.

In fact Ward and Brownlee actively take steps to avoid the appearance of advocating anything but science. In a move reminiscent of Tipler and Barrow's scientific disclaimer, in the paperback edition Ward and Brownlee deny a desire to explore any extra-scientific implications of the Rare Earth hypothesis, "the discussion spurred by the book often left the realm of scientific discourse, where we'd intended it to take place, and entered the arenas of religion, ethics, and science fiction."[44] Despite this disclaimer, they leave in the passages that build the case for the scientific myth of unity. The ellipses remain as an invitation to the audience to connect the Rare Earth science to the unity myth, and it is without question that the metaphysical and extra-scientific implications of *Rare Earth* help explain the massive success the book obtained.

INTELLIGENT DESIGN AND RARE EARTH

Given the success of many of the books profiled here it is important to reemphasize that *Rare Earth* did not have many indicators of success, beyond its unity subject matter. The book was a science work, written by a paleontologist and astronomer, and initially released by Springer Verlag, a publisher that caters to specialist audiences in the sciences. Each one of these factors would ordinarily weigh heavily against the prospect of *Rare Earth* receiving much notice, but the book continued the successful trend of unity publications.

Despite these limitations, *Rare Earth* received an enormous amount of public exposure. A search of LexisNexis Academic reveals that eighteen different newspapers published reviews of *Rare Earth* during the year of its publication, including major papers like the *New York Times* and *The Times*.[45] This list is no doubt far from complete. Ward published an article on the book in the popular science magazine *New Scientist*, which summarized the book's arguments, and he appeared on both CNN news and the cable news talk show the *O'Reilly Factor* on Fox to talk about *Rare Earth*.[46] The book sold 100,000 copies in a year and reached as high as number six on Amazon's bestseller list.[47] *Rare Earth* became the kind of book that inspired newspaper readers to write to the editor singing its praises and listeners to call into radio shows to discuss its ideas.[48]

The success of *Rare Earth* inspired mixed feelings in at least one group: intelligent design advocates, one of whom had played an enormous role in shaping the book. Guillermo Gonzalez, at the time a research assistant professor at University of Washington, provided a significant amount of research for *Rare Earth*. In the words of Ward, "[Gonzalez was] a major influence about the importance of some features of the earth [sic] that are unique to Earth and that we believe are important in the rise of complex life."[49] Unbeknownst to Ward, however, Gonzalez had been living what journalist David Darling describes as "something of a double life."[50]

On the one hand, Gonzalez had achieved impressive success publishing scientific articles in some of the best peer-reviewed journals. On the other hand, he regularly contributed to the creationist pamphlet *Connections*.[51] In his *Connections* articles, Gonzalez would take ideas he had published in peer-reviewed journals and explain their religious implications. For example, in his article "Live Here or Nowhere," Gonzalez cites one of his articles in the respected scientific journal *Astronomy & Geophysics* about the rareness of our sun's "galactic co-rotation radius" to make the case for "divine design."[52] While Ward and Brownlee claim to have desired to bracket off the Rare Earth hypothesis from ethical and religious questions, Gonzalez makes it clear that he had no such desire.

In a review of *Rare Earth* for the religious magazine *Touchstone*, Gonzalez shares his own views on what the Rare Earth hypothesis means. He relays a story of watching Ward lecture on the book and how he felt disappointed when Ward declined to connect Earth's rare properties to a religious designer. Gonzalez writes,

> I would say that Rare Earth presents yet more evidence of divine design. I would go even further and claim that the evidence shows our uniqueness and God's great care and love for us by his creating a vast universe just to have one planet inhabited by beings capable of worshiping him. What could be more meaningful?[53]

Four years after the release of *Rare Earth*, Gonzalez published his own book, *Privileged Planet* (2004), with Vice President and Senior Fellow of the Discovery Institute Jay Richards. In the book they take the scientific arguments in *Rare Earth* and explicitly give them religious implications.

Despite the similar content, the books were differently received. In contrast to the LexisNexis search for articles on *Rare Earth*, a similar search for *Privileged Planet* brings up roughly a quarter of the results. Of the articles that mention *Privileged Planet*, only seven address the content of the book itself, and these articles tend to come from sources with an ideological interest in spreading religion.[54] This leaves one article in the *New York Times* that gives Richard and Gonzalez's ideas a serious hearing in the context of the broader debate about intelligent design.[55] Not only did *Rare Earth* achieve considerably more recognition upon its release, but a LexisNexis search reveals eleven examples of positive references to *Rare Earth* a full ten years after its initial publication, including prominent mentions in both the *New York Times* and on *National Public Radio*.[56] Less than ten years after its release *Privileged Planet* has been all but forgotten in the media.

Similarly, while *Rare Earth* catapulted the academic careers of Ward and Brownlee, *Privileged Planet* had the opposite effect for Gonzalez, who was seeking tenure at Iowa State University. After the release of the book, 120 ISU faculty members signed a petition against intelligent design that John Hauptman, professor of physics at ISU, described as "obviously aimed at Gonzalez."[57] A report in the *Des Moines Register* uses interviews with faculty members and faculty emails to make the case that *Privileged Planet* played a large role in torpedoing Gonzalez's chance at tenure.[58] Despite authoring 68 peer-reviewed science papers, Gonzalez was forced to leave ISU, and he later accepted a position at Grove City College, a small Christian school.[59]

Interestingly Gonzalez predicted precisely this divergent reaction in his review of *Rare Earth*. He wrote that Hugh Ross, founder of the journal *Connections*, made many of *Rare Earth*'s arguments in his book *The Creator and the Cosmos* (1993).[60] Despite predating *Rare Earth*, "Ross's book was not widely read outside the Christian ghetto, his ideas were stillborn as far as the secular media are concerned; during the time the rare Earth hypothesis spread in the Christian community, it was being either ignored or actively opposed outside it."[61] Ross and Gonzalez made the religious implications of *Rare Earth* too obvious, undermining their scientific credibility to a broader audience. Or as Ben Wiker, PhD in Theological Ethics who has taught at Thomas Aquinas College and the Franciscan University of Stubenville, wrote,

> In regard to Christian theology, the most helpful thing about the arguments of Ward and Brownlee is that they are not trying to be helpful at all. They show no evidence of being Christian, or even being vaguely theistic, but argue from the perspective of evolutionary-based scientific materialism. No one can accuse them of stacking the deck in our favor. The cards are dealt by nature itself.[62]

Certainly there was no evidence at the time to link Ward to intelligent design, and after the publication of *Rare Earth* he engaged in several debates arguing against it.[63] As for his feelings about *Privileged Planet*, in a debate with Stephen Meyer of the Discovery Institute he called it "a bunch of trash—it's just crap."[64]

What the success gap between *Privileged Planet* and *Rare Earth* tells us is that the unity myth and religion are not equivalent. This lesson was already demonstrated by the relative failure of Tipler's *Physics of Immortality*, compared to *The Anthropic Cosmological Principle*. The popular and critical success of *Rare Earth* outside the "Christian ghetto" also demonstrates that the unity myth has appeal beyond religious believers. Furthermore, it shows that one needs to maintain outward religious or spiritualist sympathies to effectively advance the unity myth.[65]

SIMON CONWAY MORRIS AND THE NEW DARWINISTS

Ward and Brownlee direct their book at the principle of mediocrity and its mythic implications. Morris's unity myth deals with the principle of mediocrity as well, but it additionally takes aim at the movement to ground philosophy in evolutionary theory. These individuals argue that evolution proves atheism and provides the foundation for a particular set of ethical principles grounded primarily in the idea of biological utility. For them, humanity is not special or privileged, but rather the result of a process no different than the process that created snails or squids. Ethics emerge from this same evolutionary process, rather than existing immutably and eternally.

Of course individuals have found philosophical implications in evolution since its inception, most notably purveyors of Social Darwinism. The Nazi's use of Darwinism to justify a program of genocide, however, cast a dark shadow over the use of evolution as a foundation for ethics and philosophy. For several decades academics took a highly skeptical attitude toward using evolutionary ideas to comment on society.[66] Richard Dawkins, professor of biology at Oxford University, set the stage for the reemergence of evolution as a cultural force in the 1990s. In the 1970s, he created the Darwinian metaphor "meme" (the equation of an idea with a gene) as an explanation for why and how certain ideas spread, which suggested that the same evolutionary principles of fitness could apply to ideas like ethics.[67] In the 1980s, he prominently criticized teleology by means of evolution.[68] The 1990s saw the broad application of Dawkins's ideas by other scholars.

A variety of best-selling books of that decade used Darwinism as an explanatory tool in ethics, politics, religion, and even art. Daniel Dennett, Professor of Arts and Sciences at Tufts University, wrote *Darwin's Dangerous Idea* (1995). In it Dennett argues that Darwinism will ultimately prove "toxic" to religious fundamentalism, but he goes beyond the old Darwinism vs. religion debate to examine how evolution impacts all facets of human society.[69] He criticizes philosophy for its failure to reach consensus.

> Philosophers since Plato have attempted to organize these imperatives into a single rationally defensible and universal system of ethics, so far without achieving anything approaching consensus. Mathematics and physics are the same for everyone everywhere, but ethics has not yet settled into a similar reflective equilibrium.[70]

Dennett argues that the philosophy fails because it does not take into account humanity's biological reality. For Dennett, Darwinism has the power to offer solutions to thousands of years of philosophical stagnation as well as many social problems: "Darwin's idea… eats through just about every traditional concept, and leaves in its wake a revolutionized world-view, with most of the old land-marks still recognizable, but transformed in fundamental ways."[71] Dennett attacks the idea of any sort of universal or absolute morality. Instead, leaning heavily on Frederick Nietzsche and Thomas Hobbes, he argues that morality evolves over time. Dennett offers Darwinism as a method to understand this evolution. Dennett's book was a finalist for both the National Book Award in non-fiction and the Pulitzer Prize for general non-fiction.[72]

Steven Pinker, director of the Center for Cognitive Neuroscience at MIT, traces modern morality primarily to evolution in his book *How the Mind Works* (1997), but he also says that art can only be understood through the lens of evolution. While many people throughout history have treated art as transcendent, Pinker argues that art does not reflect external beauty or truth so much as it takes advantage of evolutionary circuits developed in the youth of humanity as a species. Art taps into the parts of human brains that developed to orient humans to their surroundings, saying, "the bright, crisp, saturated, contrasty images, whether from an expensive television set or from a colorful painting, may exaggerate the click of pleasure we get when we have adjusted our eyes properly."[73] In this way art operates the same as a strawberry cheesecake or pornography; it acts as a concentrated activation of the pleasure centers humans developed to survive.

Edward O. Wilson, Professor of Biology at Harvard University, wrote *Consilience* (1998), in which he not only uses Darwinism to attack religion, but also offers an explanation for its existence. He says that the belief in an afterlife made religious societies adept at spreading and defeating rivals, because "enemies of the faith can be savaged and suicidal martyrdom praised."[74] This advantage had the evolutionary effect of encouraging more religiosity, such that "[religious] indoctrinability became an instinct."[75] In other words, because religious societies fared better, the evolutionary pressure was to create more religious individuals. Wilson argues this does not apply just to religion, but to all ethics.

> The same [biological/Darwinian] evidence, I believe, favors a purely material origin of ethics, and it meets the criterion of consilience: Causal explanations of brain activity and evolution, while imperfect, already cover the most facts known about moral behavior with the greatest accuracy and the smallest number of freestanding assumptions. While this conception is relativistic, in other words dependent on personal viewpoint, it need not be irresponsibly so. If evolved carefully, it can lead more directly and safely to stable moral codes than transcendentalism, which is also, when you think about it, ultimately relativistic.[76]

Wilson suggests that stable moral codes owe their existence to the evolutionary advantage they confer upon humans, rather than their status as universal and timeless principles of the Good.

These thinkers reduce all human art, science, ethics, and religion to materialist, specifically biological, origins. They advocate the very kind of unholy union between the humanities and Darwinism that terrified William Whewell and Alfred Russel Wallace in the 1800s. Eventually these materialists would become/set the stage for the new-atheists: academics that conducted a very public attack on the idea of religion aimed at the broadest possible audience.[77] Even prior to their new atheist turn (which seems to be defined primarily by an author devoting an entire book to promoting atheism), the materialist views of Dawkins, Dennett, Pinker, Wilson, and others had become a potent intellectual force.

With the postmodernists playing the role of Protagorean sophists, challenging any truth, and the new Darwinists taking up the mantle of Callicles, grounding truth in the harsh reality of nature, Simon Conway Morris entered the fray to play the role of modern-day Socrates defending absolute, universal, and immutable values. Morris's transition to unity myth advocate mirrors Wallace's in many ways. Morris's early career was defined by his promising discoveries related to evolution, but like Wallace he began to express concern over how others interpreted his findings on evolution in regard to ethics. Also like Wallace, Morris would eventually modify his views on evolution in ways concordant with his ethical beliefs and put forward his own version of the unity mythology. Morris would go on to use his unity myth to defend absolute values from a myriad of enemies in the modern world.

BURGESS SHALE: EVOLUTIONARY SHOWDOWN

In the 1980s, Simon Conway Morris had a career much like those of Ward and Brownlee prior to the release of *Rare Earth*. He conducted research on fossils from the Burgess Shale, publishing the results in specialized scientific journals. In the 1990s, however, Morris became embroiled in a major conflict over evolutionary theory with one of science's most public figures: Steven Jay Gould. Gould claimed that evolution acts in a random way, whereas Morris argued that evolution continually finds its way to a limited set of good ideas. What makes this debate interesting for the study of scientific myth is that Morris steps outside the traditional realm of scientific questions and speaks to the ethical and philosophical implications. It begins Morris's life as a public intellectual, who uses science to comment on broader social issues, and foreshadows his in-depth defense of the unity mythology in his book *Life's Solution.*[78]

Gould's *Wonderful Life* (1988) champions contingent evolution, the idea that the evolutionary trajectory on Earth represents only one of a huge set of possibilities. Gould argues that evolutionary adaptations that succeed owe much to luck and chance and, relying heavily on Morris's early work, argues that many fossils in the Burgess Shale have no connection to any currently living animals. In other words, if these strange fossils had survived, rather than their luckier contemporaries, the animals that exist today would be radically different. To describe this phenomenon, Gould coins the now famous metaphor about rewinding the clock of time.

> The unpredictability of evolutionary pathways asserts itself against our hope for the inevitability of consciousness...Rewind the tape into the early Cam-

> brian, and perhaps this time our modern reefs are built by archaeocyathids, not corals. Perhaps no Bikini, no Waikiki; perhaps, also, no people to sip rum swizzles and snorkle amidst great undersea gardens.[79]

Many people think that consciousness and human intelligence offer such an evolutionary advantage such that they would inevitably arise. Gould does his best to disabuse the audience of this notion through his focus on the importance of contingency.

Despite inspiring Gould, Morris disagrees strongly with his theory of contingent evolution and wrote the book *The Crucible of Creation* (1998), devoted largely to attacking Gould's ideas. In the book Morris claims that the Burgess Shale fossils are not as weird as Gould suggests and can, in fact, be directly linked to modern-day animals. In contrast to Gould's emphasis on contingent evolution, the idea that evolution could advance in an innumerable number of directions, Morris defends convergent evolution, the idea that certain biological designs are so effective that they will repeatedly and inde-

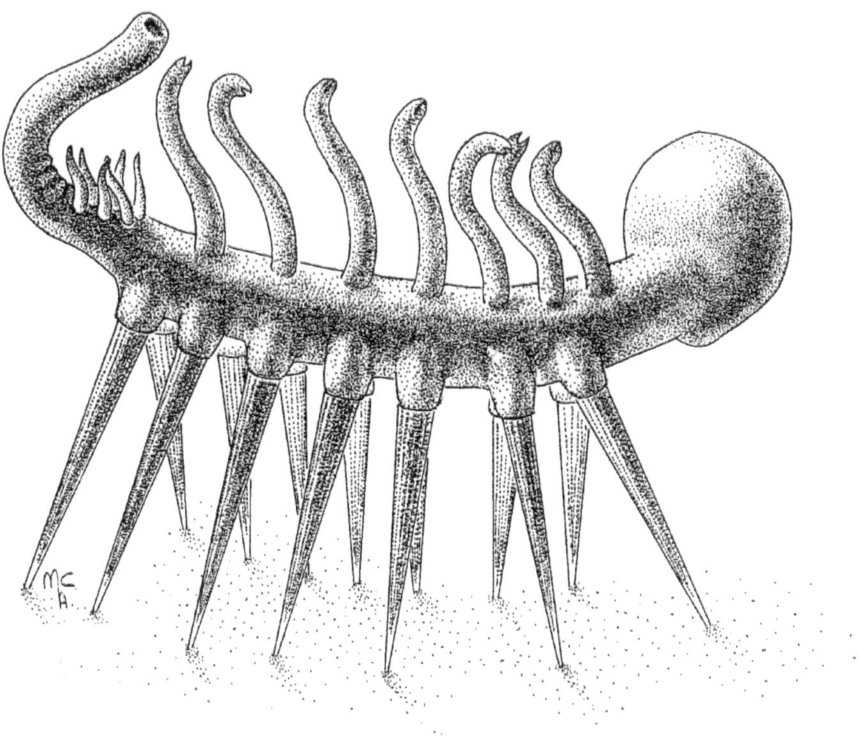

Figure 5.1. An illustration of one of the "weird" fossils from the Burgess Shale found in Gould's book *Wonderful Life*.

pendently evolve. For example, Morris points out that the eye has independently evolved multiple times.[80]

The charged language Morris uses to attack Gould's ideas suggests something beyond a mere clash of scientific paradigms. Morris claims that despite Gould's many attempts to radically rethink evolutionary theory, "little changed," an indication that Morris views Gould's work as lacking scientific rigor.[81] In reference to some of Gould's ideas Morris writes, "At first sight it is quite difficult to decide whether any of this needs to be taken seriously."[82] Morris treats Gould as a self-promoting master of "rhetoric" (used pejoratively), rather than a scientist worthy of the acclaim he had achieved.[83] Both academic and popular reviews of *The Crucible of Creation* note Morris's hostile tone toward Gould.[84] A reporter for *The Times* remarked, "In the polite world of academe, [Morris's language] is as close to a declaration of war as you get."[85]

What is particularly strange about the hostility underlying Morris's arguments against Gould is that Morris provided Gould with the evidence that the Burgess Shale fossils were significantly different from modern animals. In *Wonderful Life* Gould repeatedly singles Morris's work out for special acclaim.[86] One example of Gould's high praise: "Again, as for so much of this book, I owe this example to the suggestion and previous probing of Simon Conway Morris."[87] He went so far as to say that if paleontology had a Nobel Prize that Morris should be the first to receive it.[88] Not only did Morris's work inspire Gould, but Morris once shared Gould's interpretation of that data, saying in an interview on the question of contingent evolution "at one point I was very much in agreement with Stephen J. Gould."[89] No wonder that Gould points out the strangeness of Morris's hostile tone.

> Conway Morris has chosen…to be imperiously dismissive of my ideas, as if no sensible or experienced person could ever advocate such prejudiced nonsense. But he never tells us that Wonderful Life [*sic*] treats him, in his radical days as a graduate student, as an intellectual hero. I developed my views on contingency and the expanded range of Burgess diversity directly from Conway Morris's work and explicit claims, and I both acknowledged my debt and praised him unstintingly in my book… Conway Morris is certainly free to change his mind, as he has done. Indeed, such flexibility can only be viewed as admirable in science. But it is a bit unseemly never to state that you once held radically different opinions and to brand as benighted, in some obvious and permanent sense, a colleague who holds the views you once espoused.[90]

It is tempting to dismiss Gould's complaint as sour grapes, but he does raise an important point, especially given the widespread recognition of the strangeness of Morris's hostility. Why was Morris attacking Gould so strongly when Gould was advancing arguments grounded in Morris's work?

Morris's discomfort toward the extra-scientific implications that Gould draws (or that Morris believes Gould draws) from convergent evolution serves as a likely explanation.[91] Morris's book *Crucible of Creation* provides much evidence for thesis. In the introduction when discussing his disputes with Gould, Morris writes that "other issues, however, are at heart philosophical, and a resolution of differences need not rely solely on scientific data."[92] These philosophical implications that Gould draws from his contingent view of evolution may have prompted Morris's hostility. In the close of his introduction, Morris indicates that the philosophical debate over the implications of the Burgess Shale fossils makes ethical demands that scientists must address. "If science faces a particular challenge in the next few years, I believe it will not be from hewing away at the frontiers but from the necessity to reconcile its conclusions with other truths, many of which carry a particular ethical urgency."[93] Many scientists would find the idea of "ethical urgency" out of place in discussions over evolutionary events that occurred hundreds of millions of years ago. Morris sees his interpretation of the Burgess Shale fossils as reflective of the underlying forces of nature that shine light on various "truths" about the appropriate way to live. Gould's flawed science supports an opposing ethical system.

> Morris claims that Gould's commitment to Marxist ideology drives his scientific agenda, the underlying ideological agenda of Gould has always been fairly clear. Even where there has been a shift in thinking, it might be argued that in general the discussions were reflecting a particular world-view that at the least was sympathetic to the greatest of twentieth-century pseudo-religions, Marxism.[94]

Morris accuses Gould, rather directly, of twisting his scientific interpretation to advance his political beliefs, beliefs that the term "pseudo-religion" suggests are anathema to Morris. What is at stake to Morris is no longer just a correct view of evolution, but intellectual justification for Marxism. Furthermore, Morris claims that Gould's theory of contingency undermines the possibility of absolute morality.

> In brief, his assessment of Man as an evolutionary accident is to lead us into a libertarian attitude whereby, by virtue of a cosmic accident, we, and we alone have no choice but to take responsibility for our own destiny and mould it to our desire. At the very least, the activities of the last century as one of unrestricted political experimentation should give us pause for thought. The implications of an evolutionary process transcending the scientific evidence does indeed provide a metaphysic, albeit one that is etiolated and impoverished, but it should be decisively rejected.[95]

Morris suggests that Gould seeks to draw metaphysics from his contingent view of evolution, but that just as his evolutionary theory is incorrect, the metaphysics that would derive from it are "etiolated and impoverished." He indicates that the lack of teleological direction in contingent evolution justifies the kind of ethical relativism that resulted in "unrestricted political experimentation," which is likely a reference to fascism and communist totalitarianism, the two great failed political experiments of the 20th century.[96] Given contingency's support of such problematic metaphysics, Morris argues that it "should be decisively rejected."[97] Gould is not the only exemplar of problematic links between ideology and science in Morris's introduction (although he receives the most extensive treatment).

Morris sets up a dichotomy between Gould and a group he names the "hard Darwinists," of whom he singles out Dawkins and Dennett as prime examples.[98] While Gould seeks to challenge traditional tenants of Darwinism, the "hard Darwinists" focus myopically on them. Morris claims that the "hard Darwinists" are "strongly materialist," which leads them to offer a "severely reductionist" account of evolution.[99] Morris argues that these reductionist accounts miss the forest for the trees and says, "The world depicted by Dawkins has lost all sense of transcendence."[100] This transcendence could in part reflect the idea of universal, absolute, timeless, values (what Plato would call the Good). In a footnote he connects the "hard Darwinists" to postmodernist philosophy.

> Yet, I suspect that the creeping relativism seen in so much of the arts and philosophy, linked to the poisonous ideas of such individuals as Derrida, are far more influential than is widely acknowledged. Is it any coincidence that some of the most celebrated of the popularizers in science, such as Stephen Hawking and Richard Dawkins, offer us such arid manifestos?[101]

For Morris, the "arid manifestos" of the Darwinists lack any sense of transcendence and thus invite materialist and relativist philosophies.

Despite their different scientific approaches, Morris claims that both Gould and the hard Darwinists converge in facilitating relativism. Immediately after linking Hawking and Dawkins to Derrida, Morris accuses Gould of the same thing. "Even a closer reading of the essays of the last member of this triumvirate, Stephen Gould, beneath the hyperbole and enthusiasm, reveals a much bleaker message."[102] Presumably this "bleak message" functions the same as the "arid manifestos," excluding the possibility of transcendence and encouraging relativism.[103] This is bolstered by the fact that right after linking Gould, Hawking, and Dawkins, Morris calls for a reform of post-Saussurean literary theory, generally considered a mainstay of postmodern thought.[104]

These quotes early in *Crucible of Creation* show that Morris identifies his ideological opponents as deploying scientific myths. He accuses Gould of using contingency to advance Marxism and social constructivism and links the reductionist evolutionary theory of Dawkins and others with the "creeping relativism" of the postmodernists. Like Barrow, Tipler, Whewell, Ward, and Brownlee, however, Morris inverts the scientific myths of his philosophical opponents to advance his own philosophical views. Tipler, Whewell, Ward, and Brownlee accepted their opponents claim that plurality lessened human importance, but argued that we lived in a unity universe, which given their opponents' logic, must increase human importance.

Morris argues that if contingency advances social constructivism, then convergence must demonstrate absolute ethical values.

THE CONVERGENCE MYTH

Morris takes Dennett, Dawkins, and Gould to task for advancing a problematic philosophical agenda, but he does not attack them for seeking ideological and mythical implications from science. He ends his first chapter by saying, "We muddy the waters of the debate if we fail to acknowledge that the processes of evolution have metaphysical implications for us. This is because uniquely there is inherent in our human situation the possibility of transcendence."[105] Here Morris plainly states that science does matter for "metaphysics," presumably here "metaphysics" refers back to the ideological and philosophical debates already discussed in this chapter: Marxism, relativism, postmodernism, etc. While it is clear in the way Morris discusses these ideologies that he opposes them, the chapter also hints at his own worldview and metaphysics.

Although he never explicitly mentions it in the book, Morris is a Christian. A reporter for the *Times* says Morris's use of "transcendence" is "the giveaway," a term Morris admits is "not...commonly found among atheists."[106] Morris's Christian faith is in itself rather unremarkable; much like Gould's Marxism, scientists have all manner of philosophical and religious beliefs. But it becomes clear that Morris intertwines his religious beliefs with his scientific views, much as he accuses Gould of tying together his scientific beliefs and Marxism. Morris suggests that convergence opens up the possibility for "transcendence," because humans are the inevitable product of evolution, rather than an accident. Convergent evolution was destined to create creatures capable of appreciating religion.

> Perhaps independently [Neanderthals] developed some sort of sense of an afterlife, at least to judge from their practice of deliberate burial. Materialists will scoff at this as a shared delusion, but there are metaphysical alternatives that are perhaps more fruitful.[107]

The "metaphysical alternatives" to religion being a "shared delusion" go unstated, but it is implied that religion itself may be the alternative. In other words, convergent evolution reflects a divine plan to create creatures capable of appreciating the divine.

If Alfred Russel Wallace had in fact accepted and spearheaded the term Wallicism as a distinct approach to evolution, then it would be fair to call Morris a Wallicist. Both treat humans (or creatures sufficiently like humans) as the teleological end point of evolution, and both connect this belief to religion. Morris would even go on to suggest that evolution cannot explain the development of the human mind and posit the need for supernatural mechanism.[108] Given these similarities, Morris also runs into the problem that plagued Wallace. If humans are an inevitable part of evolution, is the universe full of planets with human-like creatures?[109] And if so, do not these worlds filled with these intelligent beings diminish humanity's importance? Morris addresses this issue in precisely the same way that Wallace does, by denying the possibility of life on other planets.

Morris closes *Crucible of Creation* with a short chapter about life on other planets. He claims that the "proposition, that we are alone, is worth thinking about seriously" and provides a footnote full of arguments about how Earth is uniquely suited for life.[110] Like Ward and Brownlee, Morris connects unity to environmental stewardship. He also says that unity means that humans should care more about their descendants and at least "leave the world a little better than we found it." Finally, he argues that unity calls upon humans to "understand a little of our history."

This last chapter has little to no connection to the rest of *Crucible of Creation*. First, Morris's book is supposed to be about the Burgess Shale; nothing in the previous chapters foreshadows the strange turn to discuss astrobiology. Second, one would expect that belief in convergent evolution would make one more likely to accept the possibility of intelligent life on other planets. After all, if evolution inevitably results in human-like creatures on Earth, then one would expect the same would also be true on another planet. The strange conclusion of *Crucible of Creation* foreshadows Morris's next book, *Life's Solution* (2003), where he explicitly links convergent evolution to the myth of unity.[111]

WHERE EVOLUTION AND UNITY CONVERGE

The concluding chapter of Ward and Brownlee's *Rare Earth* features prominently placed ellipses when discussing the implications of unity, inviting the audience to fill in much of the unity mythology for themselves. Morris's book instead begins with prominently placed ellipses. In the introduction to *Life's Solution*, Morris writes,

> There is, however, a paradox. If we, in a sense, are evolutionarily inevitable, as too are animals with compound eyes or tiny organelles that make hydrogen,

> then where are our equivalents, out there, across the galaxy? After all, the Milky Way has been available for colonization for at least a billion years, so in Enrico Fermi's famous words concerning putative extraterrestrials: 'Where are they?' To paraphrase much of this book, life may be a universal principle, but we can still be alone. In other words, once you are on the path it is pretty straightforward, but finding a suitable planet and maybe getting the right recipe for life's origination could be exceedingly difficult: inevitable humans in a lonely Universe. Now, if this happens to be the case, that in turn might be telling us something very interesting indeed. Either we are a cosmic accident, without either meaning or purpose, or alternatively . . .[112]

Morris says that Wallace's version of the unity mythology (humans are alone in the universe, but inevitable on Earth) could tell us something "very interesting." Like Ward and Brownlee, Morris uses an ellipsis in a way that avoids spelling out exactly what unity denotes. The passage makes it clear enough the general line the book takes on the issue, however. The "either" implies that if humanity is "evolutionarily inevitable" and thus not a "cosmic accident," it has "meaning" and "purpose." Morris does not take long before he more fully fills in this ellipsis for the audience.

For Morris, life is not an accident, but instead inevitable and purposeful: proof of metaphysics. As the first evidence of this claim Morris restates his argument from *Crucible of Creation* about the ethical distinction between convergent and contingent evolution. Morris says that contingent evolution makes humans a "cosmic accident, leaving us free to make the world as we will," a repetition of his attack on Gould.[113] Morris reflects on the unlikelihood of moral creatures (humans) evolving from an amoral process. Instead of viewing evolution as a random process, Morris suggests that evolution contains within it "deeper structures and potentialities" that allow us to counter the "materialism of much current thinking with its agenda of a world now open to limitless manipulation."[114] Morris accuses Darwinists like Wilson, Dennett, and Pinker of practicing "sophistry" and changes their label from "hard Darwinists" to the even more menacing sounding "ultra-Darwinists."[115] He accuses the "ultra-Darwinists" of attacking religion, while at the same time attempting to create their own pseudo-religion. Strangely, Morris spends three-and-a-half pages attacking Clarence Darrow for his "ruthless" behavior as an attorney during the Scopes Monkey trial.[116] Morris suggests "fundamentalist Darwinism" motivated Darrow and led him to the view that "'life is cruel, and mankind [*sic*] is unregenerate.'"[117] In short, Darwinism in its "ultra" form treats life in a reductionist fashion that renders life meaningless and denies the possibility for transcendence. Little has changed between the *Crucible of Creation* and *Life's Solution* with regard to the implications of contingent evolution or materialist Darwinism.

The second proof of teleology comes from unity, to which Morris gives different mythic implications than in *Crucible of Creation*. In *Crucible of*

Creation Morris said that unity provides incentives to be better stewards of the environment and seek knowledge. In *Life's Solution* he links human aloneness with the more traditional tenants of the unity cosmology. Like Ward and Brownlee, Morris argues that the Copernican revolution lead to the devaluation of humanity.

> Earlier I remarked on the almost gleeful abasement of humans, not least to inform us that we are insignificant worms in the cosmic drama. One powerful ingredient of this dreary world picture is the Copernican triumph, of Earth the Insignificant. Perhaps so, but it could be that our planet and its Solar System are both very much odder than is realized. Life may be a universal principle, but we can still be alone.[118]

The Copernican revolution robbed humanity of its centrality and thus rendered us no better than worms, but if we are alone we can regain our value. Morris claims that plurality represents one more push to render humans unremarkable.

> The scientific view is that the emergence of life is surely inevitable, and there is no shortage of planets on which the long climb from pond-scum to shopping must surely be taking place—or, happy thought, perhaps twinkling just out of sight is a planet now covered with shopping malls, parking lots, and internet cafes.[119]

Morris's ironic use of "happy thought" reflects his displeasure at a society currently guided by the wrong scientific myths. Belief in contingency and plurality promote the idea of human "insignificance," which reinforces the basest tendencies of humanity, like rampant consumerism, "shopping malls, parking lots, and internet cafes," at the expense of the transcendent. Implied is the idea that convergence and unity could reverse these social ills by providing scientific justification for more transcendent ways of living.

Morris claims that the ethical implications of unity are so powerful that regardless of whether unity "is literally true…it is far more prudent to assume that we are unique, and to act accordingly."[120] This idea bares significant resemblance to the argument from Plato's *Republic* that false myths with good messages are better than true myths with false messages. Morris's foregrounding of ethics before science exposes his priorities: transcendence over science. This is not to say that Morris conducted bad science or was deceitful in the way Plato advocated in the *Republic*; rather, it suggests that if the science of Morris's scientific myth (unity) were definitively disproven, he would not change his ethical beliefs. It also raises the serious possibility that Morris began his examination of the plurality/unity debate seeking to prove unity for the purpose of advancing his philosophical agenda.

RESPONSE TO LIFE'S SOLUTION

For an academic book about biology, *Life's Solution* made another major splash, selling an impressive number of copies and receiving its sixth reprint in 2009. It also gained prominence in the scientific world.[121] Even Dawkins, who Morris personally attacks in *Life's Solution* and who has no sympathy toward Morris's theological views, cites *Life's Solution* favorably in his book *The Ancestor's Tale* (2004).[122] Like Tipler and Barrow, the science of Morris's book caused the scientific community to embrace it, while the mythic undertones gave the book a broader audience. This meant that scientists like Dawkins could praise the book, while at the same time, Paul Davies, professor of physics and a public intellectual, could write that *Life's Solution* is "a welcome antidote to the bleak nihilism of the ultra-Darwinists."[123]

This scientific credibility gave Morris a platform to attack the ultra-Darwinists, materialists, and postmodernists. In 2005, Morris delivered the keynote address at the Boyle Lecture, an annual gathering "designed to address topics which explore the relationship between Christianity and our contemporary understanding of the natural world," where he decried the "sleight-of-hand" of "materialist philosophy."[124] Together with John Barrow, coauthor of *Anthropic Cosmological Principle*, Morris edited *Fitness of the Cosmos for Life* (2007).[125] *Fitness of the Cosmos for Life* presents some of the latest scientific evidence from a variety of scientific disciplines in order to make the case for teleology. In this sense it very much serves as a modern-day version of a *Bridgewater Treatises*, the scientific examinations of how nature reflects the goodness of God written in the 1800s. Both the *Bridgewater Treatises* and *Fitness of the Cosmos for Life* were commissioned by wealthy religious men with the goal of spreading Christian ideals: Reverend Francis Henry, Earl of Bridgewater and John Templeton respectively. Both contain articles from prominent scholars from some of the best universities. Both were supported by prominent organizations. Davies Gilbert, President of the Royal Society, organized the *Bridgewater Treatises*, and the Cambridge University Press published *Fitness of the Cosmos for Life*. It is no surprise that four separate chapters of *Fitness of the Cosmos for Life* reference the *Bridgewater Treatises*.[126]

The natural religion of the *Bridgewater Treatises* runs directly counter to the supposedly value neutral methodology of modern-day science. What was once cutting edge thinking appears on some level dated and obsolete, and yet, Morris's writings in the *Crucible of Creation, Life Solution,* and *Fitness of the Cosmos for Life* all, to varying degrees, embrace the spirit of the *Bridgewater Treatises* without appearing anachronistic. The popular success of his works and the public celebrity of Morris himself all suggest that natural theology and the unity myth have a place in the 21st century.

CONCLUSION

Despite being contemporaries and proponents of unity, Ward and Morris have some fundamental disagreements. Ward would likely accuse Morris of supporting intelligent design and religion, and one could easily see Morris describing Ward and Brownlee as overly reductionist and materialist. But like the divisions between Whewell and Wallace before them, these other disagreements do not stop them from espousing similar unity myths. Plurality renders humans mediocre and insignificant, whereas unity makes them special. Plurality decenters humanity, whereas unity places it metaphorically at the center of the universe. The common features of the unity myth not only connect Morris, Ward, and Brownlee but also connect them through time to the writings of Plato, Aquinas, Whewell, and Wallace.

The success of *Rare Earth* and *Life's Solution* in the 21st century shows the enduring power of the unity myth. In a time with cell phones, the Internet, quantum mechanics, psychotherapy, existentialism—in other words—a world completely alien to Plato's own—the unity myth of his ancient dialogues continues to prove rhetorically salient. Given the endurance of the myth, it is likely to continue into the indefinite future or until unity is scientifically disproven.

NOTES

1. Alister E. McGrath, *Darwinism and the Divine: Evolutionary Thought and Natural Theology* (Malden, MA: Wiley-Blackwell, 2011), 18. John Wiley & Sons Ltd, The Atrium, Southern Gate, Chichester, West Sussex, P019 8SQ, United Kingdom.

2. Paul R. Gross and Norman Levitt, *Higher Superstition: The Academic Left and Its Quarrels With Science* (Baltimore: John Hopkins University Press, 1994).

3. Janny Scott, "Postmodern Gravity Deconstructed, Slyly," *The New York Times*, May 18, 1996; Sharon Begley and Adam Rogers, "The Science Wars," *Newsweek*, April 21 1997; Tom Wakeford, "Too Far, Too Fast?," *The Guardian*, Mar. 5, 1997.

4. E. O. Wilson, *Consilience: The Unity of Knowledge* (New York: Vintage Books, 1999), 207; Steven Pinker, *How the Mind Works* (New York: W. W. Norton & Company, 1997), 57, 308; Daniel C. Dennett, *Darwin's Dangerous Idea: Evolution and the Meanings of Life* (New York: Simon & Schuster, 1995), 275 note 2, 344.

5. Wilson, Pinker, and Dennett would likely feel that this was an unfair comparison given the ruthlessness of Calicles and Galton, but this focuses on the wrong issue. Calicles, Galton, and Wilson, Pinker, and Dennett argue that morality derives from nature. The current scientific understanding of nature is simply less ruthless (eusociality, symbiotic relations, Gaia Hypothesis, etc.) than the view of nature when Calicles and Galton lived, which results in a less ruthless morality.

6. William A. Stahl et al., *Webs of Reality: Social Perspectives on Science and Religion* (Piscataway, NJ: Rutgers University Press, 2002), 161-62; Philip S. Baringer, " Introduction: The "Science Wars"," in *After the Science Wars: Science and the Study of Science*, ed. Keith M. Ashman and Philip S. Baringer (New York: Routledge, 2005), 1-2.

7. Edward J. Larson and Larry Witham, "Leading Scientists Still Reject God," *Nature* 394, no. 6691 (1998).

8. Brownlee, *Rare Earth: Why Complex Life Is Uncommon in the Universe*; Morris, *Life's Solution.*

9. Peter D. Ward, "Periodicity of chamber formation in chambered cephalopods: evidence from *Nautilus macromphalus* and *Nautilus pompilius*," *Paleobiology* 11, no. 4 (1985).
10. Sheridan, "SETI's Scope", 25.
11. Sagan Carl and William I. Newman, "The Solipsist Approach to Extraterrestrial Intelligence," *Quarterly Journal of the Royal Astronomical Society* 24 (1983): 113.
12. Ward and Brownlee, *Rare Earth*, 272. The quote originally comes from Marcelo Gleiser, *The Dancing Universe: From Creation Myths to the Big Bang* (New York: Dutton Books, 1997), 274.
13. John Gribbin, *The Scientists: A History of Science Told Through the Lives of Its Greatest Inventors* (London: Allen Lane, 2002), xvii. Interestingly, Gribbin would go on to write a book that makes the case for unity, which may reflect a change of heart or an acknowledgment of the financial dynamics of books about unity. Gribbin, *Alone in the Universe*.
14. Ken Campbell, "Beyond Our Ken," in *Reality on the Rocks* (1995).https://www.youtube.com/watch?v=jXBkF7CQaf8 5:10-50:35
15. P. Z. Myers, "The Mediocrity Principle," *The Edge*, http://www.edge.org/q2011/q11_12.html#myerspz.
16. David Fincher, "Fight Club," (Twentieth Century Fox Film Corporation, 1999).
17. The unity myth has changed throughout time, most notably from Plato suggesting no other planets existed with any life to Whewell later suggesting that intelligent life was the key concern.
18. Ward and Brownlee, *Rare Earth*, 267.
19. Ibid., XVIII.
20. Ibid., 272.
21. Ibid., 275.
22. Drake and Sobel, *Is Anyone Out There?* , 62.
23. Ward and Brownlee, *Rare Earth*, 270.
24. Ibid., 271-72.
25. Ibid., 272.
26. Ibid., 274.
27. Ibid., 274-75.
28. Ibid., 275.
29. Nigel Hawkes, "Earthlings, thank your uniquely lucky stars," *The Australian*, Jan. 31, 2000.
30. Ward and Brownlee, *Rare Earth*, 251.
31. In interviews Ward and Brownlee make it much more clear that Rare Earth does not imply a unity cosmology. John McLaughlin, in *John McLaughlin's "One on One"* (PBS, 2003).
32. Nigel Hawkes, "Why Earth is a lonely planet," *The Times*, Jan. 26, 2000.
33. Andrew Marshall, "Oh No! Humans are the only Intelligent Lifeform," *The Independent*, Feb. 10, 2000.
34. Doug Esser, "Authors say we're probably alone in the universe," *The Associated Press*, May 15, 2000.
35. Nigel Hawkes, "Profile of Earth may be quite unique in universe," *Windsor Star*, Feb. 26, 2000.
36. Esser, "Authors say we're probably alone in the universe."
37. Ward and Brownlee, *Rare Earth*, xxvii-xxviii.
38. Ibid., xviii.
39. Ibid., xxvii.
40. Ibid., 279.
41. Ibid., 280.
42. Ibid., 282.
43. Ibid., xxviii.
44. Ibid., xii.
45. This list is most likely far from exhaustive. Robin McKie, "When it comes to intelligent life, we're as good as it gets," *The Observer*, July 16, 2000; Michael LeGault, "Earth to ET: Is anyone out there?," *The Globe and Mail*, May 27, 2000; Esser, "Authors say we're probably alone in the universe."; "Books shatters space myths ", *St. John's Telegram*, May 28, 2000; Erik

Robinson, "Message of Earth Day Hits Home," *The Columbian*, April 22, 2000; Paul Davies, "Are We Still Alone?," *The Advertiser*, May 6, 2000; Lynda Hurst, "Finally the Unhappy Truth: ET has no Home to Phone," *The Toronto Star* March 25, 2000; Robin McKie, "Wake up, it's an alien on line one: British scientists are at the heart of a new project to answer the old question is there anybody out there?," *The Observer*, March 5, 2000; , *The Roanoke Times*, Feb. 14, 2000; Richard Woods and Christopher Goodwin, "In space, there may be no one there to hear us scream," *The Australian*, 2000; "Advanced Life Likely Only Here," *Wyoming Tribune-Eagle*, Feb. 20, 2000; Marshall, "Oh No! Humans are the only Intelligent Lifeform."; Ron Seely, "A Setback in the Eternal Hope for Alien Life?," *Wisconsin State Journal*, Feb. 10, 2000; "Can We Be Alone?," *The Columbian*, Feb. 9, 2000; William J. Broad, "Earthlings may be alone in the universe after all," *Austin American-Statesman*, Feb. 8, 2000; William J. Broad, "Is there anybody out there?," *Hamilton Spectator*, Feb. 8, 2000; Michael Bywater, "Alone at Last," *The Independent* Feb. 6, 2000; William J. Broad, "Maybe We Are Alone in the Universe, After All," *The New York Times*, Feb. 8, 2000; Hawkes, "Why Earth is a lonely planet."; Hawkes, "Profile of Earth may be quite unique in universe."; Nigel Hawkes, "A Very Lonely Planet: Charmed Place," *Edmonton Journal*, Feb. 13, 2000; Hawkes, "Earthlings, thank your uniquely lucky stars."; Nigel Hawkes, "The lonely planet: Maybe we ARE alone in the universe," *The Gazette*, 2000; Nigel Hawkes, "Are we alone? Quite probably," *The Vancouver Sun*, Jan. 27, 2000.

46. Peter Ward, "No place like home," *New Scientist*, March 18, 2000; Bill O'Reilly, in *The O'Reilly Factor* (Fox News Network, 2000); Andria Hall, "Rare Earth' Ponders Question of Life on Other Planets," (CNN, 2000).

47. Suzy Hansen, "We've got company," *Salon.com*, http://www.salon.com/2001/06/29/darling/; Holter, "Dead Reckoning."

48. Michael O'Driscoll, "Why life is more than an accident," *Irish Examiner*, Oct. 3, 2007; James Bakunas, "Letter to the editor," *The State Journal-Register*, Nov. 27, 2004; Albert L. Weeks, "Water on Mars? Don't take a cup," *Sarasota Herald-Tribune*, July 25, 2000.

49. Hansen, "We've got company."

50. David Darling, *Life Everywhere: The Maverick Science of Astrobiology*, First ed. (New York: Basic Books, 2001), 112.

51. In one article Gonzalez speculates openly on the possibility that "aliens" contacted by SETI may actually be fallen angels. Hugh Ross and Guillermo Gonzalez, "A Spectrum of Views on ETI," *Reasons.org*, http://www.reasons.org/articles/a-spectrum-of-views-on-eti.

52. Hugh Ross and Guillermo Gonzalez, "Live Here or Nowhere," *Reasons.org*, http://www.reasons.org/articles/live-here-or-nowhere.

53. Guillermo Gonzalez, "No Other Eden," *Touchstone Magazine*, http://www.touchstonemag.com/archives/article.php?id=13-05-036-b.

54. Most of the articles deal with Guillermo Gonzalez's fight for tenure at Iowa and the controversy about showing a film based on *Privileged Planet* at the Smithsonian. Two articles come from the "Religion and Faith," not the "Science" section, of the *Calgary Herald*, including one by a professor of theology and science at a Franciscan university. Another article comes from Philip Gold, Jay Richard's colleague at the Discovery Institute. A brief mention of *Privileged Planet* occurs in David Klinghoffer's article that summarizes recent books with relevance to religion; Klinghoffer would go on to become a senior fellow at the Discovery Institute. The book receives coverage in the *Deseret Morning News*. This is unsurprising given that the Church of Latter-Day Saints owns the paper, an organization with obvious interest in scientific narratives with religious implications. One article comes from the extremely conservative magazine *Human Events*, which is a sister company to the publisher of *Privileged Planet*. Two articles come from the authors themselves.

55. Dennis Overbye, "Hunting for Life in Specks of Cosmic Dust " *The New York Times*, July 19, 2005.

56. "Vast tides may explain life's origin," *The New Zealand Herald*, April 3, 2013; "Some Regions of the Universe are Not Favorable for Life," *The Daily Galaxy*, Dec. 24, 2012; "Super-Earth is only 600 light years away; Kepler-22b is the first planet outside our solar system confirmed as possibly habitable," *National Post*, Dec. 6, 2011; Dennis Overbye, "Hot on Trail Of 'Just Right' Far-Off Planet," *The New York Times*, Dec. 3, 2011; "Is A Moon Necessary For

A Planet To Support Life?," in *Talk of the Nation Science Friday* (National Public Radio, 2011); Gemma Lavender, "Living in the Galactic Danger Zone," *Space Daily*, Oct. 5, 2011; Lewis Dartnell and Chloe Darracott-Cankovic, "In the sky above, a guardian angel," *The Times Higher Education Supplement*, May 12, 2011; Lewis Dartnell, "Seeking out new worlds, new life," *The Times Higher Education Supplement*, April 8, 2010; Robin McKie, "Meet the neighbours?," *The Observer*, Feb. 7, 2010; "The Great Filter theory suggests humans have already conquered the threat of extinction," *io9*, Dec. 21, 2012; "What if the Earth had no moon?," *io9*, May 3, 2013.

57. John Hauptman, "Rights are intact: Vote turns on question, 'What is science?'," *Des Moines Register*, June 2, 2007.

58. Lisa Rossi, "Intelligent design theory influenced ISU tenure vote," *Des Moines Register*, Dec. 1, 2007.

59. Doug Schneider, "Flunking the tenure test," *The Washington Times*, April 23, 2008.

60. Hugh Norman Ross, *The Creator and the Cosmos: How the Greatest Scientific Discoveries of the Century Reveal God* (Colorado Springs, CO: NavPress Publishing Group, 1993).

61. Gonzalez, "No Other Eden."

62. Ben Wiker, "Billions of Planets, But Only One Earth," *National Catholic Register*, September 2000.

63. "Intelligent Design vs. Evolution—Stephen Meyer vs. Peter Ward (rematch)," http://www.youtube.com/watch?v=Sakmq5L3IiE&feature=youtube_gdata_player; Peter Ward and Stephen C. Meyer, "Stephen Meyer vs. Peter Ward Transcript," *The Dori Monson Show*, http://www.discovery.org/a/3421.

64. In the debate Meyer argued for intelligent design and Ward argued against it. Peter Ward and Stephen C. Meyer, "Hear the Great Debate on Intelligent Design!," http://www.discovery.org/a/3456. Ward's dismissive attitude to a former collaborator later work seems very similar to Barrow's disavowal of Tipler's *Omega Point*.

65. Ward has repeatedly declined to comment on his religious beliefs, but his role as Intelligent Design opponent in many debates leaves one with the impression that he is not particularly religious. Ward and Meyer, "Stephen Meyer vs. Peter Ward Transcript"; Ward and Meyer, "Hear the Great Debate on Intelligent Design!"

66. Geoffrey M. Hodgson, "Social Darwinism in Anglophone Academic Journals: A Contribution to the History of the Term," *Journal of Historical Sociology* 17, no. 4 (2004): 442-52.

67. Richard Dawkins, *The Selfish Gene* (Oxford: Oxford University Press, 1976).

68. Richard Dawkins, *The Blind Watchmaker* (New York: W. W. Norton & Company, 1986).

69. Dennett, *Darwin's Dangerous Idea*, 515.

70. Ibid., 495.

71. Ibid., 63.

72. "National Book Awards—1995," *NationalBook.org*, http://www.nationalbook.org/nba1995#.UmQ7SyQmzyc; "1996 Finalists," *Pulitzer.org*, http://www.pulitzer.org/finalists/1996.

73. Pinker, *How the Mind Works*, 527.

74. Wilson, *Consilience*, 268.

75. Ibid., 269-70.

76. Ibid., 263.

77. Richard Dawkins, *The God Delusion* (New York: Bantam Books, 2006); Sam Harris, *The End of Faith: Religion, Terror, and the Future of Reason* (New York: W. W. Norton, 2004); Victor J. Stenger, *God the Failed Hypothesis: How Science shows that God does not exist* (Amherst, New York: Prometheus Books, 2007); Daniel C. Dennett, *Breaking the Spell: Religion as a Natural Phenomenon* (New York City: Viking, 2006); Christopher Hitchens, *god is not Great* (New York: Twelve Books, 2007).

78. In another foreshadowing of Morris' unity views, *Rare Earth* cites him more than any other single author.

79. Stephen Jay Gould, *Wonderful Life: The Burgess Shale and the Nature of History* (New York: W. W. Norton & Co., 1989), 316.

80. Simon Conway Morris, *The Crucible of Creation: The Burgess Shale and the Rise of Animals* (Oxford: Oxford University Press, 1998), 171-208.
81. Ibid., 10.
82. Ibid., 12.
83. Ibid., 10.
84. Nigel Hawkes, "Whose life is it anyway?," *The Times*, Mar. 2, 1998; David Norman, "Curtains fall on the freak show," *The Times Higher Education Supplement*, Feb. 6, 1998; Andrew Berry, "Wonderful Crucible," *Evolution* 52, no. 5 (1998); Jeffrey Levinton, "Who Owns the Cambrian Explosion?," *The Quarterly Review of Biology* 74, no. 2 (1999).
85. Hawkes, "Whose life is it anyway?"
86. Stephen Jay Gould, *Wonderful Life: The Burgess Shale and the Nature of History* (New York: W. W. Norton & Co., 1989).
87. Ibid., 293.
88. Ibid., 84.
89. Simon Conway Morris, "Considering Convergence: Interview with Simon Conway Morris, part 1," http://www.astrobio.net/index.php?option=com_podcast&task=showall&id=10&msg=&pageNo=2.
90. Stephen Jay Gould and Charles E. Morris, "Showdown on the Burgess Shale," *Natural History* 107, no. 10 (1998).
91. Morris's contempt for Gould's style probably compounded the motivation for his harsh criticisms. In response to an interview question about his feelings on Gould, Morris replied, "The sense of [his] correct attitudes could almost be read as intolerance. When you start to simplify people's thoughts, and when you start to demonise the opposition, you end up with a parody, or a set of straw men." Stuart Wavell, "Dissent about descent," *The Sunday Times*, Jan. 5, 1997. Certainly Morris would not be alone in expressing distaste for Gould's style. Robert Wright, a journalist and author of a book on evolutionary on biology wrote, "among top-flight evolutionary biologists, Gould is considered a pest—not just a lightweight, but an actively muddled man who has warped the public's understanding of Darwinism." Steven Pinker, Werner Kalow, and Harold Kalant, evolutionary psychologists, call Gould "discourteous" and claim he misrepresented their ideas, saying the "ideas of evolutionary psychology are not as stupid as Gould makes them out to be. Indeed, they are nothing like what Gould makes them out to be." Ralph L. Holloway, an expert on human evolution at Columbia, calls Gould a "charlatan," but much of this hostility likely owes itself to the fierce debate Gould waged with these individuals over evolutionary psychology: the idea that evolution plays the primary role in shaping social structures. Robert Wright, "Homo Deceptus: Never trust Stephen Jay Gould.," *Slate* (1996), http://www.slate.com/articles/news_and_politics/the_earthling/1996/11/homo_deceptus.html; Steven Pinker, Werner Kalow, and Harold Kalant, "Evolutionary Psychology: An Exchange," *The New York Review of Books* (1997), http://www.nybooks.com/articles/archives/1997/oct/09/evolutionary-psychology-an-exchange/?pagination=false; Nicholas Wade, "Scientists Measure the Accuracy of a Racism Claim," *The New York Times*, June 12, 2011.
92. Morris, *The Crucible of Creation*, viii.
93. Ibid. viii
94. Ibid., 11.
95. Ibid., 14.
96. The view that Morris references these ideologies is bolstered further by Morris's disdain for Marxist theory generally.
97. I was unable to locate anywhere in *Wonderful Life* any suggestion that Gould thought this was an implication of his contingency theory. Neither could Mark Ridley, a lecturer in the department of zoology at the University of Oxford, who wrote, "I do not actually recall that Gould derived any libertarian do-our-own-thing conclusion from his lottery argument." Mark Ridley, "Rock of Ages," *The New York Times*, May 10, 1998. The suggestion that Gould is simultaneously libertarian and Marxist has its own sort of absurdist element, given the ongoing antagonism between these two viewpoints. A similar tension exists between describing Gould as a relativist and a Marxist. Marxism, in almost all its forms, puts forth an ethical system where certain behaviors and practices are wrong in an absolute and objective sense.

98. Morris, *The Crucible of Creation*, 7.
99. Ibid., 7, 9.
100. Ibid., 9.
101. Ibid., 15 note 3.
102. Ibid.
103. From one perspective this is a rather outlandish accusation, both because Gould and the "hard Darwinists" were bitter scientific rivals and because, as previously mentioned, the "hard Darwinists" were outspoken against postmodernism. It is clear though that Gould, the "hard Darwinists," and the postmodernists all oppose a platonic notion of truth and in this sense could be accurately described as agents of relativism. For more on the conflict between Gould and the hard Darwinists, including an examination of the extra-scientific issues at stake. See Walsh, *Scientists as Prophets*, 136-62.
104. Morris, *The Crucible of Creation*, 15 note 3.
105. Ibid., 14.
106. Hawkes, "Whose life is it anyway?."
107. Morris, *The Crucible of Creation*, 14.
108. Simon Conway Morris, "Darwin was right. Up to a point," *The Guardian*, Feb. 12, 2009.
109. Morris does not mention problem of evolution's waste in *Crucible of Creation* that Wallace focuses so heavily on, but he does address this point later in a lecture, which does suggest that it could have been a motivating factor. Simon Conway Morris, "Darwin's Compass: How Evolution Discovers the Song of Creation," http://www.stmarylebow.co.uk/#/boyle-lecture-2005/4535725162.
110. Morris, *The Crucible of Creation*, 223.
111. Morris, *Life's Solution*.
112. Ibid., xiii.
113. Ibid., xv.
114. Ibid., 2.
115. Ibid., 314-17. The term ultra-Darwinist bears some familiarity to Whewell's use of the term ultra-Lockean.
116. Ibid., 321.
117. Morris quotes Kevin Tierney to make this point ibid., 322.
118. Ibid., 105.
119. Ibid., 22.
120. Ibid., 328.
121. A JSTOR search reveals eleven articles citing *Life's Solution* as a (often the) source on convergent evolution. Anurag A. Agrawal and Mark Fishbein, "Plant Defense Syndromes," *Ecology* 87, no. 7 (2006); Zachary D. Blount, Christina Z. Borland, and Richard E. Lenski, "Historical Contingency and the Evolution of a Key Innovation in an Experimental Population of Escherichia coli," *Proceedings of the National Academy of Sciences of the United States of America* 105, no. 23 (2008); Derek E. G. Briggs and Richard A. Fortey, "Wonderful Strife: Systematics, Stem Groups, and the Phylogenetic Signal of the Cambrian Radiation," 31, no. 2 (2005); Graham E. Budd, "The Earliest Fossil Record of the Animals and Its Significance," *Philosophical Transactions: Biological Sciences* 363, no. 1496 (2008); Rosemary G. Gillespie, "The Ecology and Evolution of Hawaiian Spider Communities: The diversification of Hawaiian spiders illustrates universal principles behind community assembly on evolutionary and ecological scales," *American Scientist* 93, no. 2 (2005); Richard E. Glor, "Phylogenetic Insights on Adaptive Radiation," *Annual Review of Ecology, Evolution, and Systematics* 41 (2010); Ronald A. Jenner, "Challenging Received Wisdoms: Some Contributions of the New Microscopy to the New Animal Phylogeny," *Integrative and Comparative Biology* 46, no. 2 (2006); Bhavin S. Khatri et al., "Statistical Mechanics of Convergent Evolution in Spatial Patterning," *Proceedings of the National Academy of Sciences of the United States of America* 106, no. 24 (2009); Colin P. Osborne and David J. Beerling, "Nature's Green Revolution: The Remarkable Evolutionary Rise of C4 Plants," *Philosophical Transactions: Biological Sciences* 361, no. 1465 (2009); Geerat J. Vermeij, "Historical Contingency and the Purported Uniqueness of Evolutionary InnovationsAuthor(s): Geerat J. Vermeij," *Proceedings of the National Academy*

of Sciences of the United States of America 103, no. 6 (2006); Robert Woods et al., "Tests of Parallel Molecular Evolution in a Long-Term Experiment with *Escherichia coli*," *Proceedings of the National Academy of Sciences of the United States of America* 103, no. 24 (2006).

122. Richard Dawkins, *The Ancestor's Tale: A Pilgrimage to the Dawn of Evolution*, Paperback ed. (Boston: Houghton Mifflin Company 2005), 595.

123. Paul Davies, "Reviews," *Cambridge.org*, http://www.cambridge.org/sk/knowledge/isbn/printView/item1170875/?site_locale=sk_SK.

124. Morris, "Darwin's Compass."

125. John D. Barrow et al., eds., *Fitness of the Cosmos for Life: Biochemistry and Fine-Tuning* (Cambridge: Cambridge University Press, 2007).

126. Ibid., 7, 20, 34, 134.

Conclusion

Humanity as the Measure vs. the Unity of the World

> Sometimes I think we're alone in the universe, and sometimes I think we're not. In either case the idea is quite staggering. —Unknown Science Fiction Writer[1]

The existence of life on other planets has proved to be one of the most enduring questions in human history. This scientific debate has garnered such attention because its implications transcend science, having been intertwined with some of the longest lasting political, religious, ethical, and philosophical debates. Alfred North Whitehead once famously quipped that the European philosophical tradition consists of "a series of footnotes to Plato."[2] Tracing the unity myth throughout history, one sees elements of Plato's original argument in the *Timaeus*. The connection of unity and teleology is so prominent that it often goes without explicit articulation, acting as an enthymeme to combat relativist thinking. Just as the unity myth can bolster teleology, however, it also threatens it.

The unity argument has proved rhetorically efficacious throughout history, but that does not necessitate it will be so in the future. Perhaps the greatest danger to its effectiveness lies in the scientific foundation, "we are alone." A 2011 study recently suggested over a billion habitable planets in our galaxy alone.[3] For those that tie their values to the absence of alien life, these sorts of discoveries represent a recalcitrance that must be overcome. As Kenneth Burke writes, "Interpretations themselves must be altered as the universe displays various orders of recalcitrance to them."[4] In the Medieval period, the failure of scholars truly to come to terms with the problems of the Aristotelian cosmology reflects the power of cognitive dissonance as a tool to overcome recalcitrance. Scholastics did not seize upon mounting inconsistencies in the cosmology in order to fashion a new one, in large part because of the Aristotelian cosmology's connection to political, philosophical, and religious beliefs. Even the Aristotelian system, however, ultimately collapsed under the weight of the scientific evidence for heliocentrism.

The large unknowns about the question of alien life present a potential catastrophe for those that rely on the unity myth. I do not mean to suggest that alien life (intelligent or otherwise) exists for certain, or even if it did that humanity will ever find definitive proof of its existence. There is a very real possibility that this could occur, however. By tying their beliefs closely with a particular cosmology, defenders of teleology, Truth, and absolute morality risk a scientific paradigm shift that could undermine support for their beliefs. Steven J. Dick, NASA's chief historian, claims that the discovery of alien life has the potential to set off mass social change similar to the Copernican and Darwinian revolutions.[5]

THE PROBLEMS OF UNITY

The unity argument points to a disturbing tendency in humanity—the need to assign worth relationally. Western society typically finds value in centrality and singularity: exclusive rather than inclusive characteristics. Under this metric of value, the existence of alien life threatens humanity's sense of purpose and self-worth. This framing raises the specter of serious problems for the way that humans approach alienness.

The connection of unity with human value raises the stakes in identifying potential alien life. The National Research Council wrote an extensive report on the difficulties scientists could have in recognizing alien life.[6] The belief that the discovery of life will collapse teleology, Truth, and absolute morality will politicize these efforts in a way that further complicates them. Whewell, Tipler, and Barrow all allowed for the possibility of non-intelligent alien life in their versions of the unity argument. The line between intelligent and non-intelligent life remains hotly contested, even with animals like dolphins. If difficulties emerge in categorizing a species with which humanity has shared the planet since its evolution, one can imagine the enormous difficulty in trying to determine whether a truly alien species should be considered "intelligent." Similar debates occur over what would constitute sentient artificial intelligence. Anthropocentrism puts social barriers in front of the classification of other species as intelligent, because recognizing alien life as intelligent threatens to break down human "containment strategies, props supporting a dangerously brittle identity."[7] The threat to human identity creates incentives to depict any alien life we discover as unintelligent in order to eliminate the threat to the scientific myth of unity and the political, religious, philosophical, and ethical systems it buttresses.

One can see the stakes involved in such a (mis)calculation in the violence meted out on our planet to creatures deemed sentient, but unintelligent. Elephants, crows, dolphins, pigs, and many other animals denied the label of intelligent are harvested for resources, often in a truly horrifying manner. It may well be that humanity becomes the nightmarish aliens of our UFO nightmares, swooping down on less "intelligent" creatures of other worlds and snatching them up for scientific experimentation.

The problems of unity do not exist entirely in the future, however, as the powerful myth shapes behavior and forecloses pathways today. For starters, there may exist value in seriously contemplating plurality; as Grinspoon writes, "Certainly, efforts to communicate with intelligent extraterrestrials do not make much sense unless they are made on behalf of all humans. Merely contemplating the possibility of finding other life makes obvious our deep identification with all Earth's inhabitants."[8] Or as Primack and Abrams put it, "Dealing wisely with aliens or simply contacting them may be a distant goal, but understanding what it would mean can have an immediate and powerful effect. It makes clear what truly matters today: to *be* the kind of human beings we aspire to be in the long run, and to adopt this perspective *now*."[9] These authors embrace the decentering of plurality and hope that it brings with it a new scientific myth where humanity refuses to be content with the status quo and instead embraces a new, more noble existence.

The unity myth may also forestall a plurality revolution in the model of the Copernican or Darwinian revolutions. In both of these cases the collapse of a scientific myth that one can trace back to the *Timaeus* resulted in massive social change. What change could a plurality revolution bring? It is possible, as Whewell, Wallace, and others suggest, that a scientific myth grounded in plurality would usher in nihilism, but other more optimistic possibilities exist.

A plural, decentered universe could alter social relations for the better by opening space for alternative ways of being. Certainly there exists no shortage of groups marginalized by western society's standards of acceptable religion, appearance, gender, sexuality, race, ad infinitum. Western culture's long history of colonization and imperialism also reflect a similar, outwardly directed disregard for otherness. The existence of multiple loci of alien life may give credence to a Western ethic more accepting of "aliens" within and without the boundaries of the planet.

Just as the Copernican and Darwinian discoveries revolutionized scientific inquiry, a similar paradigm shift could emerge from a plurality revolution. It is only recently, for example, that psychology recognized that many of its fundamental principles of human behavior only apply to westerners.[10] Psychologists missed these "alien" ways of thinking by centering their study too closely on themselves, but change would not necessarily be limited to social science. Nicholas Rescher raises the possibility that human science is deeply mediated through our own physical bodies and environment.[11] Rescher argues that it is likely a mistake think that only one universal science exists, even if only one physical universe exists.[12] Alien creatures may have radically different environments and ways of interacting with the world, resulting in science that may be different from humanity's. A plurality revolution could point light on the parochial nature of humanity's scientific endeavors in a way that opens up new pathways for development.

A social revolution grounded in the scientific reality of plurality would be no less a myth than the social order grounded in unity. This is to say that a plurality myth would not be truer than the unity myth, regardless of the scientific truth of plurality, just as the new social order brought on by heliocentrism did not necessarily follow from Copernicus' discoveries. Science has little intrinsic connection to most questions of ethics, values, religion, and politics. The unity myth, however, has foreclosed avenues for exploration throughout history, just as geocentrism anchored a particular worldview. A new mythos based in plurality could help open up new epistemologies. Of course, predicting the precise course of the next Copernican revolution goes well beyond the scope of this project.

FINAL THOUGHTS

The unity argument has represented a rhetorical force throughout much of human history, and few signs exist that its use to defend teleology, Truth, and absolute morality will abate anytime soon. The universe as a foundation for argument has enormous appeal. The connection between singularity and Platonic values can serve as an unstated premise, which allows individuals to structure the unity argument as an enthymeme. The cosmological nature of the unity vs. plurality debate speaks to the structure and purpose of the universe. Unity, thus, undergirds a broader understanding of human ethos, understood as our dwelling place. The scientific nature of the defense of unity gains it the credibility society affords to science. All of these rhetorical strengths, combined with the unity argument's long track record, indicate that it will remain an argumentative force well into the future.

NOTES

1. Jeremy Bernstein, "How About a Little Game?," *The New Yorker*, Nov. 12 1966, 92. This quotation comes from an interview with Stanley Kubrick, he may have been slightly misquoting Arthur C. Clarke who once said, "Two possibilities exist: either we are alone in the universe or we are not. Both are equally terrifying." "Arthur C. Clarke," *Economist*, http://www.economist.com/node/10918055.

2. Whitehead, *Process and Reality: An Essay in Cosmology*, 39.
3. Michael G. Gowanlock, David R. Patton, and Sabine M. McConnell, "A Model of Habitability Within the Milky Way Galaxy," *arXiv.org* http://arxiv.org/abs/1107.1286.
4. Kenneth Burke, *Permanence and Change: An Anatomy of Purpose*, 3 ed. (Berkeley: University of California Press, 1984), 256.
5. House Science Committee, *Astrobiology: Search for Biosignatures in our Solar System and Beyond*, December 4, 2013.
6. Committee on the Origins and Evolution of Life Committee on the Limits of Organic Life in Planetary Systems, National Research Council, The Limits of Organic Life in Planetary Systems (Original Publication; repr., Washington: The National Academies Press, 2007; reprint, Reprint) http://www.nap.edu/catalog/11919.html (2010); ibid.
7. Peters, *Speaking into the Air: A History of the Idea of Communication*, 230.
8. Grinspoon, *Lonely Planets*.
9. Joel R. Primack and Nancy Ellen Abrams, *The View from the Center of the Universe: Discovering our Extraordinary Place in the Cosmos* (New York: Riverhead Books, 2006), 235.
10. Ethan Watters, "We Aren't the World," *Pacific Standard* (2013), http://www.psmag.com/magazines/magazine-feature-story-magazines/joe-henrich-weird-ultimatum-game-shaking-up-psychology-economics-53135/.
11. George, *Civilized Life in the Universe: Scientists on Extraterrestrials*, pg 177
12. Rescher, "Extraterrestrial Science."

Bibliography

"1996 Finalists." *Pulitzer.org*, http://www.pulitzer.org/finalists/1996.
Abrams, Joel R. Primack and Nancy Ellen. *The View from the Center of the Universe: Discovering our Extraordinary Place in the Cosmos*. New York: Riverhead Books, 2006.
Abulafia, David. *The Discovery of Mankind: Atlantic Encounters in the Age of Columbus*. New Haven: Yale University Press, 2008.
Addicott, Jeffrey F. "Storm Clouds on the Horizon of Darwinism: Teaching the Anthropic Principle and Intelligent Design in the Public Schools." *Ohio State Law Journal* 63 (2002): 1507.
"Advanced Life Likely Only Here." *Wyoming Tribune-Eagle*, Feb. 20 2000, A12.
Agrawal, Anurag A., and Mark Fishbein. "Plant Defense Syndromes." *Ecology* 87, no. 7 (2006): S132-S49.
Aldiss, Brian W. "Desperately Seeking Aliens." *Nature* 409 (2001): 1080-82.
Algra, Keimpe. "The Beginnings of Cosmology." In *The Cambridge Companion to Early Greek Philosophy*. Edited by A. A. Long. 45-65. Cambridge: Cambridge University Press, 1999.
Anderson, Elizabeth S. "John Stuart Mill and Experiments in Living." *Ethics* 102, no. 1 (1991): 4-26.
Annas, Julia. "Self-knowledge in Early Plato." In *Platonic Investigations*. Edited by Dominic J. O'Meara. 111-38. Washington, DC: Catholic University of America Press, 1985.
Applewhite, James. "Postmodernist Allegory and the Denial of Nature." *Kenyon Review* 11, no. 1 (1989): 1-17.
Aquinas, Thomas. *Contra Gentiles Book Three: Providence (Q. 1-83)*. Translated by Vernon J. Bourke. New York: Hanover House, 1956.
———. "De Aeterniate Mundi." *Medieval Sourcebook*, http://www.fordham.edu/halsall/basis/aquinas-eternity.asp - f2.
———. *The Heavens*. Translated by Fabian R. Larcher and Pierre H. Conway. Washington, DC: Priory of the Immaculate Conception.
———. *Quaestiones Disputatae De Potentia Dei*. Translated by English Dominican Fathers: The Newman Press, 1952.
———. *The Summa Theologica of St. Thomas Aquinas. Part I QQ. L.-LXXIV*. Translated by Fathers of the English Dominican Province. 2nd Revised ed. 10 vols. Vol. 1. London: Burns Oates and Washbourne, 1920.
"Are the Stars Inhabited." *Scientific American* 9, no. 44 (1854): 349.
Aristophanes. "Wasps." In *Aristophanes: Clouds Wasps Birds*. Edited by Peter Meineck Translated by Peter Meineck. 133-233. Indianapolis: Hackett, 1998.

Aristotle. "Metaphysics." In *The Basic Works of Aristotle*. Edited by Richard McKeon Translated by W. D. Ross. 689-934. New York: Random House, 1941.

———. *On the Heavens*. Translated by Stuart Leggatt. Warminster England: Aris & Phillips, 1995.

———. "Physics." In *The Basic Works of Aristotle*. Edited by Richard McKeon. Translated by R. P Hardie and R. K. Gaye. 218-397. New York: Random House, 1941.

———. "Politics." In *The Basic Works of Aristotle*. Edited by Richard McKeon. Translated by Benjamin Jowett. 1127-324. New York: Random House, 1941.

"Art. II." *British Quarterly Review* 20 (1855): 45-85.

"Arthur C. Clarke." *Economist*, Mar. 27. http://www.economist.com/node/10918055.

Augustine, and Thomas S. Hibbs. *Enchiridion on Faith, Hope, and Love*. Translated by Thomas S. Hibbs. Washington, DC: Regnery, 2002.

Ayala, Francisco José, and Theodosius Grigorievich Dobz, eds. *Studies in the Philosophy of Biology: Reduction and Related Problems*. Berkeley: University of California, 1974.

B. D. "Letter 1." *The Royal American Magazine* 1, no. 7 (1774): 241.

Bain, Alexander. *James Mill: A Biography*. London: Longmans, Green and Co., 1882.

Baker, Thomas. *Reflections Upon Learning*. London: A. Bosvile, 1699.

———. *Reflections Upon Learning: Wherein is Shewn the Insufficency Thereof, in its Several Particulars: In order to envince the Usefulness and Necessity of Revelation*. 4 ed. London: Dyal and Bible, 1708.

Bakunas, James. "Letter to the editor." *The State Journal-Register*, Nov. 27 2004.

Baldry, H. C. *Unity of Mankind in Greek Thought*. London: Cambridge University Press, 1965.

Barbour, Julian. "Why We Should Fear Vast Change." *The Times Higher Education Supplement*, April 18, 2003.

Baringer, Philip S. " Introduction: the "science wars"." In *After the Science Wars: Science and the Study of Science*. Edited by Keith M. Ashman and Philip S. Baringer. 1-12. New York: Routledge, 2005.

Barrow, John D. Email, April 24, 2013.

Barrow, John D., Simon Conway Morris, Stephen J. Freeland, and Charles L. Jr. Harper, eds. *Fitness of the Cosmos for Life: Biochemistry and Fine-Tuning*. Cambridge: Cambridge University Press, 2007.

Barrow, John D., and Frank J. Tipler. *The Anthropic Cosmological Principle*. New York: Oxford University Press, 1986.

Barthes, Roland. *Mythologies*. London: Paladin, 1973.

Bartholomew, D. J. "Probability, Statistics and Theology." *Journal of the Royal Statistical Society* 151, no. 1 (1988): 137-78.

Basalla, George. *Civilized Life in the Universe: Scientists on Extraterrestrials*. Oxford: Oxford University Press, 2006.

———. *Civilized Life in the Universe: Scientists on Intelligent Extraterrestrials*. Oxford: Oxford University Press, 2006.

Battaglia, Debbora. ""For Those Who Are Not Afraid of the Future": Raelian Clonehood in the Public Sphere." In *E.T. Culture: Anthropology in Outerspaces*. Edited by Debbora Battaglia. ix, 281 p. Durham, NC: Duke University Press, 2005.

Begley, Sharon, and Adam Rogers. "The Science Wars." *Newsweek*, April 21, 1997.

"Bellarmine Theological Forum." *Catholic Apologetics International Publishing*, http://www.catholicintl.com/.

Benn, Alfred William. *History of English Rationalism in the Nineteenth Century*. Vol. 1. London: Longmans and Green, 1906.

Bernstein, Jeremy. "How About a Little Game?" *The New Yorker*, Nov. 12 1966, 70-110.

Berry, Andrew. "Wonderful Crucible." *Evolution* 52, no. 5 (1998): 1528-32.

Bloom, Allan. *The Closing of the American Mind*. New York: Simon and Schuster, 1987.

———. *The Republic of Plato*. Second ed. New York: Basic Books, 1991.

Blount, Zachary D., Christina Z. Borland, and Richard E. Lenski. "Historical Contingency and the Evolution of a Key Innovation in an Experimental Population of *Escherichia coli*." *Proceedings of the National Academy of Sciences of the United States of America* 105, no. 23 (2008): 7899-906.

Blumenberg, Hans. *The Genesis of the Copernican World*. Translated by Robert M. Wallace. Cambridge, MA: MIT Press, 1987.
"Book Notices." *The National Magazine* 5 (1854): 282-83.
"Books Shatters Space Myths ". *St. John's Telegram*, May 28 2000.
Booth, Wayne C. "Rhetoric, Science, Religion." In *The Essential Wayne Booth*. Edited by Walter Jost. 264-79. Chicago: University of Chicago Press, 2006.
Borrell, Brendan. "Are Octopuses Smart?" *Scientific American*, Feb. 27 2009.
Bostrom, Nick. "Existential Risks: Analyzing Human Extinction Scenarios and Related Hazards." *Journal of Evolution and Technology* 9 (2002).
Braziel, Jana Evans. "Being and Time, Non-being and Space: Introductory Notes toward and Ontological Study of Women and Chora." In *Belief, Bodies, and Being: Feminist Reflections on Embodiment*. Edited by Deborah Orr. 103-26. Oxford: Rowman and Littlefield, 2006.
Brewster, David. *More Worlds Than One: The Creed of the Philosopher and the Hope of the Christian*. New York: Carter & Brothers, 1854.
Briggs, Derek E. G., and Richard A. Fortey. "Wonderful Strife: Systematics, Stem Groups, and the Phylogenetic Signal of the Cambrian Radiation." 31, no. 2 (2005): 94-112.
Brisson, Luc. *Plato the Myth Maker*. Translated by Gerard Naddaf. Chicago: University of Chicago, 1998.
Broad, William J. "Earthlings May Be Alone in the Universe After All." *Austin American-Statesman*, Feb. 8 2000, A1.
———. "Is There Anybody Out There?" *Hamilton Spectator*, Feb. 8, 2000.
———. "Maybe We Are Alone in the Universe, After All." *The New York Times*, Feb. 8, 2000.
Broadie, Sarah. "The Sophists and Socrates." In *The Cambridge Companion to Greek and Roman Philosophy*. Edited by D. N. Sedley. 73-97. Cambridge: Cambridge University Press, 2003.
Brogan, A. P. "John Locke and Utilitarianism." *Ethics* 69, no. 2 (1959): 79-93.
Bronowski, Jacob. *William Blake and the Age of Revolution*. London: Routledge, 1972.
Brooke, John Hedley. "Indications of a Creator: Whewell as Apologist and Priest." In *William Whewell: A Composite Portrait*. Edited by Menachem Fisch and Simon Schaffer. 149-74. Oxford: Clarendon, 1991.
———. "Natural Theology and the Plurality of Worlds: Observations on the Brewster-Whewell debate." *Annals of Science* 34, no. 5 (1977): 221-86.
Brownlee, Peter Ward and David. *Rare Earth: Why Complex Life is Uncommon in the Universe*. New York: Copernicus 2000.
Bruno, Giordano. *The Ash Wednesday Supper*. Translated by Edwards A. Gosselin and Lawrence S. Lerner. Toronto: University of Toronto, 1995.
———. "On the Infinite Universe and Worlds." In *Giordano Bruno: His Life and Thought*. Edited by Dorothea Waley Singer. Translated by Dorothea Waley Singer. New York: Henry Schuman, 1950.
Buchdahl, Gerd. "Deductivist versus Inductivist Approaches in the Philosophy of Science as Illustrated by some Controversies between Whewell and Mill." In *William Whewell: A Composite Portrait*. Edited by Menachem Fisch and Simon Schaffer. 311-44. Oxford: Clarendon, 1991.
Budd, Graham E. "The Earliest Fossil Record of the Animals and Its Significance." *Philosophical Transactions: Biological Sciences* 363, no. 1496 (2008): 1425-34.
Burke, Kenneth. *Counter-statement*. Berkeley: University of California Press, 1968.
———. *A Grammar of Motives*. Berkeley,: University of California Press, 1969.
———. *A Grammar of Motives*. Berkley: University of California Press, 1962.
———. *Permanence and Change: An Anatomy of Purpose*. 3rd edition. Berkeley: University of California Press, 1984.
Burkert, Walter. "The Logic of Cosmogony." In *From Myth to Reason?: Studies in the Development of Greek Thought*. Edited by Richard Buxton. 87-106. Oxford: Oxford University Press, 1999.
———. *Lore and Science in Ancient Pythagoreanism*. Translated by Edwin L. Minar Jr. Cambridge: Harvard University Press, 1972.

Burket, Walter. *Lore and Science in Ancient Pythagoreanism*. Translated by E. L. Minar. Cambridge: Harvard University Press, 1972.

Butler, Judith. "Contingent Foundations: Feminism and the Question of Postmodernism." In *Feminists Theorize the Political*. Edited by Judith Butler and Joan W. Scott. 3-21. New York: Routledge, 1992.

Bywater, Michael. "Alone at Last." *The Independent* Feb. 6, 2000.

Cahan, David. "Looking at Nineteenth-Century Science: An Introduction." In *From Natural Philosophy to the Sciences: Writing the History of Nineteenth*. Edited by David Cahan. 3-15. Chicago: University of Chicago, 2003.

Campanella, Thomas. *A Defense of Galileo the Mathematician of Florence: Which is an Inquiry as to Whether the Philosophical View Advocated by Galileo Is in Agreement with, or Is Opposed to, the Sacred Scriptures*. Translated by Richard J. Blackwell. London: University of Notre Dame Press, 1994.

Campbell, Ken. "Beyond Our Ken." In *Reality on the Rocks*, 1995.

"Can We Be Alone?". *The Columbian*, Feb. 9 2000, a10.

Cannon, Susan Faye. *Science in Culture: The Early Victorian Period*. New York: Dawson and Science History Publications, 1978.

Caputo, John D. *Against Ethics* . Bloomington: Indiana University Press, 1994.

Carl, Sagan, and William I. Newman. "The Solipsist Approach to Extraterrestrial Intelligence." *Quarterly Journal of the Royal Astronomical Society* 24 (1983): 113-21.

Carlyle, Thomas, and James Anthony Froude. *Reminiscences*. New York: Charles Scribner's Sons, 1881.

Carone, Gabriela Roxana. *Plato's Cosmology and its Ethical Dimensions*. Cambridge: Cambridge University Press, 2005.

Carpenter, Andrew Norris. "Kant's Earliest Solution to the Mind/Body Problem." PhD diss., University of California at Berkeley, 1998.

Ceccarelli, Leah. "A Hard Look at Ourselves: A Reception Study of Rhetoric of Science." *Technical Communication Quarterly* 14, no. 3 (2005): 257–65.

"Chalmer's Discourses on the Christian Revelation." *Eclectic Review*, no. 8 (1817): 205-19.

Chalmers, Thomas. *A Series of Discourses on the Christian Revelation, Viewed in Connection with the Modern Astronomy*. 10 ed. Glasgow: Robert Chapman, 1822.

Chambers, Robert. *Vestiges of the Natural History of Creation*. London: John Churchill 1845.

Cherniss, Harold. "The Philosophical Economy of the Theory of Ideas." *The American Journal of Philology* 57, no. 4 (1936): 445-56.

Cherniss, Harold F. *Aristotle's Criticism of Plato and the Academy*. New York: Russell & Russell, 1962.

Chitwood, Ava. *Death by Philosophy: The Biographical Tradition in the Life and Death of the Archaic Philosophers Empedocles, Heraclitus, and Democritus*. Ann Arbor: University of Michigan Press, 2004.

Claeys, Gregory. "The "Survival of the Fittest" and the Origins of Social Darwinism." *Journal of the History of Ideas* 61, no. 2 (2000): 223-40.

———. *Thomas Paine: Social and Political Thought*. Florence, KY: Routledge, 2004.

Clark, Brett, John Bellamy Foster, and Richard York. "The Critique of Intelligent Design: Epicurus, Marx, Darwin, and Freud and the Materialist Defense of Science." *Theory and Society* 36, no. 6 (2007): 515-46.

Clark, David L. "Kant's Aliens: The Anthropology and Its Others." *New Centennial Review* 1, no. 2 (2001): 201-89.

Cloud, Preston. *The Quarterly Review of Biology* 62, no. 1 (1987): 63-64.

Colyvan, Mark, Jay L. Garfield, and Graham Priest. "Problems with the Argument from Fine Tuning." *Synthese* 145, no. 3 (2005): 325-38.

Committee on the Limits of Organic Life in Planetary Systems, Committee on the Origins and Evolution of Life, National Research Council. Washington: The National Academies Press, 2007, http://www.nap.edu/catalog/11919.html.

Consolmagno, Guy. *Aquinas Lecture*. Atlanta: The Aquinas Center, Emory University, 2010. Audio Lecture.

———. *The Heavens Proclaim: Astronomy and the Vatican*: Vatican Observatory, 2009.

Cooper, Andrew R. "'Monumental Inscriptions': Language, Rights, the Nation in Coleridge and Horne Tooke." *ELH* 66, no. 1 (1999): 87-110.
Cooper, John. "Aristotle's Natural Teleology." In *Language and Logos: Studies in Ancient Greek Philosophy*. Edited by Malcolm Schofield and Martha Nussbaum. 197-222. Cambridge: Cambridge Univeristy, 1982.
Cooper, John M. "Letters." In *Plato: Complete Works*. Edited by John M. Cooper. 1634-35. Indianapolis: Hackett, 1997.
———, ed. *Plato Complete Works*. Indianapolis: Hackett, 1997.
Copernicus, Nicolaus, and Stephen Hawking. *On The Revolutions of Heavenly Spheres*. Philadelphia: Running Press 2002.
Corey, M. A. *God and the New Cosmology: The Anthropic Design Argument*. Lanham: Rowman & Little, 1993.
Corlett, J. Angelo. "Interpreting Plato's Dialogues." *The Classical Quarterly* 47, no. 2 (1997): 423-37.
Cornford, Francis MacDonald. *From Religion to Philosophy: A Study in the Origins of Western Speculation*. Princeton: Princeton University Press, 1991.
———. *Plato's Cosmology: The Timaeus of Plato*. Indianapolis: Hackett, 1997.
———. *Principium Sapientiae*. Cambridge: Cambridge University Press, 1952.
Cornish, Alison. "Dante's Moral Cosmology." In *Cosmology: Historical, Literacy, Philosophical, Religious, and Scientific Perspective*. Edited by Norriss S. Hetherington. 201-15. New York: Garland Publishing, 1993.
"Corrigenda to Drengson and Leslie." *American Philosophical Quarterly* 19, no. 4 (1982): 380.
Craig, William Lane. "Barrow and Tipler on the Anthropic Principle vs. Divine Design." *The British Journal for the Philosophy of Science* 39, no. 3 (1988): 389-95.
———. "The Caused Beginning of the Universe: A Response to Quentin Smith." *British Journal for the Philosophy of Science* 44, no. 4 (1993): 623- 39.
———. "The Origin and Creation of the Universe: A Reply to Adolf Grünbaum." *The British Journal for the Philosophy of Science* 43, no. 2 (1992): 233-40.
———. “Question 97: Subject: Scepticism Quebecois.” *Reasonable Faith*, http://www.reasonablefaith.org/site/News2?page=NewsArticle&id=6937.
———. "A Swift and Simple Refutation of the "Kalam" Cosmological Argument?" *Religious Studies* 35, no. 1 (1999): 57-72.
———. “The Teleological Argument and the Anthropic Principle.” *Reasonable Faith*, http://www.reasonablefaith.org/site/News2?page=NewsArticle&id=5179.
Crowe, Michael J. *The Extraterrestrial Life Debate, 1750-1900*. Mineola, NY: Dover Publications, 1999.
———. "A History of the Extraterrestrial Life Debate." *Zygon* 32, no. 2 (1997): 147-62.
———. *Theories of the World from Antiquity to the Copernican Revolution*. Mineola, NY: Dover, 2001.
Cudworth, Ralph. *The True Intellectual System of the Universe: The First Part Wherein All the Reason and Philosophy of Atheism is Confuted, and its Impossibility Demonstrated*. 2 ed. London: J. Walthoe, 1743.
"Current Periodical Articles." *The Review of Metaphysics* 35, no. 4 (1982): 919-58
Cusanus, Nicolas. *Of Learned Ignorance*. Translated by Father Germain Heron: New Haven, 1954.
Däniken, Erich von. *Chariots of the Gods? Unsolved Mysteries of the Past*. New York: Bantam Books, 1971.
Darling, David. *Life Everywhere: The Maverick Science of Astrobiology*. First ed. New York: Basic Books, 2001.
Dartnell, Lewis. "Seeking Out New Worlds, New Life." *The Times Higher Education Supplement*, April 8, 2010.
Dartnell, Lewis, and Chloe Darracott-Cankovic. "In the Sky Above, a Guardian Angel." *The Times Higher Education Supplement*, May 12, 2011.
Davies, Paul. "Are We Still Alone?" *The Advertiser*, May 6, 2000.
———. *The Last Three Minutes: Conjectures about the Ultimate Fate of the Universe*. New York: BasicBooks, 1994.

———. “Reviews.” *Cambridge.org*, http://www.cambridge.org/sk/knowledge/isbn/printView/item1170875/?site_locale=sk_SK.
Davis, John Jefferson. "The Design Argument, Cosmic 'Fine Tuning,' and the Anthropic Principle." *International Journal for Philosophy of Religion* 22, no. 3 (1987): 139-50.
Dawkins, Richard. *The Ancestor's Tale: A Pilgrimage to the Dawn of Evolution*. Paperback ed. Boston: Houghton Mifflin Company 2005.
———. *The Blind Watchmaker*. New York: W. W. Norton & Company, 1986.
———. *The God Delusion*. New York: Bantam Books, 2006.
———. *The Selfish Gene*. Oxford: Oxford University Press, 1976.
De Chardin, Pierre Teilhard. *The Phenomenon of Man*. New York: Harper Perennial Modern Thought, 2008.
de Santillana, Giorgio. *The Crime of Galileo*. Chicago: University of Chicago Press, 1955.
Deltete, Robert J., and Reed A. Guy. "Hartle-Hawking Cosmology and Unconditional Probabilities." *Analysis* 57, no. 4 (1997): 304-15.
Dennett, Daniel C. *Breaking the Spell: Religion as a Natural Phenomenon*. New York City: Viking, 2006.
———. *Darwin's Dangerous Idea: Evolution and the Meanings of Life* . New York: Simon & Schuster, 1995.
Derrida, Jacques. *Of Grammatology*. Translated by Gayatri Chakravorty Spivak. Baltimore, MD: John Hopkins University Press, 1997.
Dick, Steven J. *The Biological Universe: The Twentieth-Century Extraterrestrial Life Debate and the Limits of Science*. 1st pbk. ed. Cambridge: Cambridge University Press, 1999.
———. *Plurality of Worlds: The Origins of the Extraterrestrial Life Debate from Democritus to Kant*. Cambridge: Cambridge University Press, 1982.
———. "The Universe and Alfred Russel Wallace." In *Natural Selection and Beyond*. Edited by Charles H. Smith and George Beccaloni. 220-40. Oxford: Oxford University Press, 2008.
Dillon, John. "Timaeus in the Old Academy." In *Plato's Timaeus as Cultural Icon*. Edited by Gretchen J. Reydams-Schils. Notre Dame, Indiana: University of Notre Dame Press, 2003.
Drake, Frank, and Dava Sobel. *Is Anyone Out There?* New York: Delacorte Press, 1992.
Duhem, Pierre. *Medieval Cosmology: Theories of Infinity, Place, Time, Void, and the Plurality of Worlds*. Translated by Roger Ariew. Chicago: University of Chicago Press, 1985.
Dupre, Louis, and Nancy Hudson. "Nicholas of Cusa." In *A Companion to Philosophy in the Middle Ages*. Edited by Jorge J. E. Gracia and Timothy B. Noone. 466-74. Malden, MA: Blackwell, 2002.
Durant, John R. "Scientific Naturalism and Social Reform in the Thought of Alfred Russel Wallace." *British Journal for the History of Science* 12, no. 1 (1979): 31-58.
Earman, John. "The Sap Also Rises: A Critical Examination of the Anthropic Principle." *American Philosophical Quarterly* 24, no. 4 (1987): 307-17.
"Editors' Book Table: The Plurality of Worlds." *The Independent* 6, no. 290 (1854): 200.
"Editors' Table: The History of the Planets. The Elm-Tree." *Godey's Lady's Book and Magazine* 49, no. 173 (1854): 174-76.
Edson, E., and E. Savage-Smith. *Medieval Views of the Cosmos*. Oxford: Bodleian Library, 2004.
Eliade, Mircea. *Myth and Reality*. New York: Harper & Row, 1963.
———. *The Sacred and the Profane: The Nature of Religion*. New York: Harcourt, 1959.
Ellis, George. "Piety in the Sky." *Nature* 371, no. 8 (1994): 115.
Empson, William. *Seven Types of Ambiguity*. London: Chatto and Windus, 1949.
Epstein, Barbara. "Postmodernism and the Left." *New Politics* 6, no. 2 (1997).
Eriksson, Kimmo. "The Nonsense Math Effect." *Judgment and Decision Making* 7, no. 6 (2012): 746–49.
Esser, Doug. "Authors say we're probably alone in the universe." *The Associated Press*, May 15 2000.
Fairweather, Eugene Rathbone. *A Scholastic Miscellany: Anselm to Ockham*. Philadephia: Westminster Press, 1956.

Fara, Patricia. "Heavenly Bodies: Newtonianism, Natural Theology, and the Plurality of Worlds Debate in the Eighteenth Century." *Journal for the History of Astronomy* 35, no. 2 (2004): 143-60.

Ferguson, Andrew. "The Book That Drove Them Crazy: Allan Bloom's 'Closing of the American Mind' 25 Years Later." *The Weekly Standard*, April 9, 2012.

Ferris, Timothy. "I Think, Therefore the Universe Is." *New York Times*, February 16, 1986.

Ferwerda, R. "Democritus and Plato." *Mnemosyne* 25 (1972): 337-78.

Fichman, Martin. "Biology and Politics: Defining the Boundaries." In *Victorian Science in Context*. Edited by Bernard V. Lightman. 94-118. Chicago: University of Chicago Press, 1997.

———. *An Elusive Victorian: The Evolution of Alfred Russel Wallace*. Chicago: University of Chicago Press, 2004.

Fincher, David. "Fight Club." Twentieth Century Fox Film Corporation, 1999.

Findlay, John Niemeyer. *Plato: The Written and Unwritten Doctrines*. New York: Humanities Press, 1974.

Fine, Gail. "Owen's Progress: Logic, Science, and Dialectic: Collected Papers in Greek Philosophy." *Philosophical Review* 97, no. 3 (1988): 373-99.

Fisch, Menachem. "A Philosopher's Coming of Age: A Study in Erotetic Intellectual History." In *William Whewell: A Composite Portrait*. Edited by Menachem Fisch and Simon Schaffer. 31-66. Oxford: Clarendon, 1991.

Fisch, Menachem, and Simon Schaffer, eds. *William Whewell: A Composite Portrait*. Oxford: Claredeon, 1991.

Fontenelle, Bernard le Bovier. *Conversations on the Plurality of Worlds*. Berkeley: University of California Press, 1990.

Fontenelle, Bernard le Bovier. *A Discourse of the Plurality of Worlds written in French by the most Ingenious Author of the Dialogues of the Dead*. Translated by W.D. Knight. Dublin: Andr. Crook and Sam Helsham and William Norman, 1687.

Forrest, Peter. "Aesthetic Understanding." *Philosophy and Phenomenological Research* 51, no. 3 (1991): 525-40.

Foucault, Michel. *Archaeology of Knowledge and Discourse on Language*. Translated by M. Sheridan Smith. New York: Pantheon Books, 1972.

Friedrich, Nietzsche. *Philosophy in the Tragic Age of the Greeks*. Translated by Marianne Cowan. Washington: Regnery, 1998.

Furley, David J. *The Greek Cosmologists*. Cambridge: Cambridge University Press, 1987.

Gairdner, William D. *The Book of Absolutes: A Critique of Relativism and a Defense of Universals*. London: McGill-Queen's University Press, 2008.

"Galileo Was Wrong." http://galileowaswrong.com/galileowaswrong/.

Gardner, Martin. "WAP, SAP, PAP, & FAP." *The New York Review of Books* 33, no. 8 (1986): 22-25.

Gay, Peter. *The Enlightenment, An Interpretation: The Rise of Modern Paganism*. Vol. 1. New York: Norton, 1977.

———. *The Enlightenment, An Interpretation: The Science of Freedom*. Vol. 2. New York: Norton, 1977.

Gelbart, Nina Rattner. "Introduction." In *Conversations on the Plurality of Worlds*. Berkeley: University of California Press, 1990.

George, Marie I. "On the Tenth Anniversary of Barrow and Tipler's Anthropic Cosmological Principle." *American Catholic Philosophical Quarterly* 72, no. 1 (1998): 37-58.

Gerson, Lloyd P. *From Plato to Platonism*. Ithaca, NY: Cornell University Press, 2013.

Gieryn, Thomas F. *Cultural Boundaries of Science: Credibility on the Line*. Chicago: Chicago University Press, 1999.

Gill, Christopher. "The Platonic Dialogues." In *A Companion to Ancient Philosophy*. Edited by Mary Louise Gill and Pierre Pellegrin. 136-50. Malden, MA: Wiley-Blackwell, 2006.

Gillespie, Rosemary G. "The Ecology and Evolution of Hawaiian Spider Communities: The diversification of Hawaiian spiders illustrates universal principles behind community assembly on evolutionary and ecological scales." *American Scientist* 93, no. 2 (2005): 122-31.

Gilson, Étienne. *The Philosophy of St. Thomas Aquinas*. Translated by Edward Bullough. Freeport, NY: Books for Libraries Press, 1971.

Gingerich, Owen. *The Book Nobody Read: Chasing the Revolutions of Nicolaus Copernicus*. New York: Walker, 2004.

———. "The Censorship of Copernicus' *De revolutionibus*." *JASA* 33 (1981): 58-60.

Giorgini, Giovanni. "Radical Plato: John Stuart Mill, George Grote and the Revival of Plato in the Nineteenth-Century England." *History of Political Thought* 30, no. 4 (2009).

Gleiser, Marcelo. *The Dancing Universe: From Creation Myths to the Big Bang* . New York: Dutton Books, 1997.

Glick, Thomas F., Steven John Livesey, and Faith Wallis. *Medieval Science, Technology, and Medicine: An Encyclopedia*. New York: Routledge, 2005.

Glor, Richard E. "Phylogenetic Insights on Adaptive Radiation." *Annual Review of Ecology, Evolution, and Systematics* 41 (2010): 251-70.

Goldman, James A. *Leonardo* 21, no. 3 (1988): 333-34.

Goldstein, William. "The Story behind the Best Seller: Allan Bloom's Closing of the American Mind." *Publishers Weekly*, July 3, 1987.

Gonzalez, Guillermo. "No Other Eden." *Touch Stone Magazine*, http://www.touchstonemag.com/archives/article.php?id=13-05-036-b.

Gonzalez, Guillermo, and Jay W. Richards. *The Privileged Planet: How Our Place in the Cosmos is Designed for Discovery*. Washington, DC: Regnery, 2004.

Goodman, Ellen. "Is *Roe v. Wade* Already Collapsing?" *Boston Globe*, 2006.

Gordon, Margaret Maria. *The Home Life of Sir David Brewster*. Edinburgh: Edmonston and Douglas, 1870.

Gosselin, Edwards A., and Lawrence S. Lerner. "Introduction." In *The Ash Wednesday Supper*. Toronto: University of Toronto Press, 1995.

Gould, Stephen Jay. *Wonderful Life: The Burgess Shale and the Nature of History*. New York: W. W. Norton & Co., 1989.

Gould, Stephen Jay, and Charles E. Morris. "Showdown on the Burgess Shale." *Natural History* 107, no. 10 (1998): 48-55.

Gowanlock, Michael G., David R. Patton, and Sabine M. McConnell. "A Model of Habitability Within the Milky Way Galaxy." *arXiv.org* http://arxiv.org/abs/1107.1286.

Graham, Elaine L. *Representations of the Post/Human: Monsters, Aliens and Others in Popular Culture*. New Brunswick, NJ: Rutgers University Press, 2002.

Grant, Edward. "Aristotelianism and the Longevity of the Medieval World View." *History of Science* 16 (1978): 93-106.

———. *A History of Natural Philosophy: From the Ancient World to the Nineteenth Century*. Cambridge: Cambridge University Press, 2007.

———. "In Defense of the Earth's Centrality and Immobility: Scholastic Reaction to Copernicanism in the Seventeenth Century." *Transactions of the American Philosophical Society* 74, no. 4 (1984): 1-69.

———. "A New Look at Medieval Cosmology, 1200-1697." *Proceedings of the American Philosophical Society* 129, no. 4 (1985): 417-32.

———. *Planets, Stars, and Orbs: Medieval Cosmos, 1200-1687*. Cambridge: Cambridge University, 1994.

"Great Astronomical Discoveries Lately Made By Sir John Herschel, L.L.D, F.R.S, &c. At The Cape of Good Hope." *New York Sun*, August 25, 1835.

"The Great Filter theory suggests humans have already conquered the threat of extinction." *io9*, Dec. 21 2012.

Gregory, Andrew. *Ancient Greek Cosmogony*. London: Duckworth, 2007.

———. "Eudoxus, Callippus and the Astronomy of the Timaeus." *Bulletin of the Institute of Classical Studies* 46, no. S78 (2003): 5–28.

———. *Plato's Philosophy of Science*. London: Duckworth, 2000.

Gregory, Vlastos. "Cornford's Principum Sapientiae." In *Studies in Presocratic Philosophy*. Edited by David J. Furley and R. E. Allen. 42-55. New York: Humanities, 1970.

Gribbin, John. *Alone in the Universe: Why Our Planet Is Unique*. Hoboken, NJ: John Wiley & Sons, 2011.

———. *The Scientists: A History of Science Told Through the Lives of Its Greatest Inventors*. London: Allen Lane, 2002.
Grinspoon, David. *Lonely Planets: The Natural Philosophy of Alien Life*. New York: Harper Collins, 2003.
Griswold, Charles L. "Plato's Metaphilosophy: Why Plato Wrote Dialogues." In *Platonic Writings/Platonic Readings*. Edited by Charles L. Griswold. 143-68. State College, PA: Penn State University Press, 2001.
Grodal, Torben. "Stories for Eye, Ear, and Muscles: Video Games, Media, and Embodied Experiences." In *The Video Game Theory Reader*. Edited by Mark Wolf and Bernard Perron. 129-55. New York: Routledge, 2003.
Gross, Paul R., and Norman Levitt. *Higher Superstition: The Academic Left and Its Quarrels With Science*. Baltimore: John Hopkins University Press, 1994.
Grote, George. *History of Greece*. Vol. 8. New York: Peter Fenelon Collier, 1844.
Grünbaum, Adolf. "Creation as a Pseudo-Explanation in Current Physical Cosmology." *Erkenntnis* 35, no. 1/3 (1991): 233-54.
———. "Narlikar's "Creation" of the Big Bang Universe Was a Mere Origination." *Philosophy of Science* 60, no. 4 (1993): 638-46.
Guthke, Karl S. *The Last Frontier: Imagining Other Worlds, from the Copernican Revolution to Modern Science Fiction*. Translated by Helen Atkins. Ithaca: Cornell University, 1990.
Guthrie, W. K. C. *The Sophists*. London: Cambridge University Press, 1971.
Guthrie, William Keith Chambers. *A History of Greek Philosophy: The Earlier Presocratics and the Pythagoreans*. Cambridge: Cambridge University Press, 2000.
"The Habitability of Worlds." *Christian Review* 80 (1855): 202-19.
Hackforth, R. "Plato's Cosmogony." *Classical Quarterly* 27 (1959): 80-82.
Hall, Andria. "Rare Earth' Ponders Question of Life on Other Planets." CNN, 2000.
Halliday, R. J. "Social Darwinism: A Definition." *Victorian Studies* 14, no. 4 (1971): 389-405.
Hanke, Lewis. *All Mankind is One: A Study of the Disputation Between Bartholomé de Las Casas and Juan Ginés de Sepúlveda in 1550 on the Intellectual and Religious Capacity of the American Indians*. DeKalb: Northern Ilinois University Press, 1974.
Hanna, William. "Memoirs of the Life and Writings of Thomas Chalmers, D. D., KK. D." *The Living Age* 27, no. 333 (1850): 20-23.
Hannaford, Alex. "An Audience with Koko the 'Talking' Gorilla." *Telegraph*, Sep. 17 2011.
Hansen, Suzy. "We've Got Company." *Salon.com*, June 29. http://www.salon.com/2001/06/29/darling/.
Harris, Errol E. *Cosmos and Anthropos: A Philosophical Interpretation of the Anthropic Cosmological Principle*. London: Humanities Press International, 1991.
———. *Cosmos and Theos: Ethical and Theological Implications of the Anthropic Cosmological Principle*. Atlantic Highlands, NJ: Humanities Press, 1992.
Harris, Neil. "*De Revolutionibus* in Bibliography: Analysing the Copernican Census " *The Library* 7, no. 3 (2006): 320-29.
Harris, Sam. *The End of Faith: Religion, Terror, and the Future of Reason*. New York: W. W. Norton, 2004.
Harrison, Mark. "The Extraterrestrial in US Culture." PhD diss., University of Pittsburgh, 2006.
Harrison, Paul. "Post-Structuralist Theories." In *Approaches to Human Geography*. Edited by Stuart Aitken and Gill Valentine. 122-35. London: Sage Publications, 2006.
Hart, Michael H. "An Explanation for the Absence of Extraterrestrial Life on Earth." *Quarterly Journal of the Royal Astronomical Society* 16 (1975): 128-35.
Hart, Stuart N. "Making Sure the Child's Voice Is Heard." *International Review of Education* 48, no. 3/4 (2002): 251-58.
"Hashshamayun; or the Heavens; The Plurality of Worlds." *Christian Observer* 37, no. 23 (1858): 97.
Haskell, Thomas L. "Capitalism and the Origins of the Humanitarian Sensibility: Part 1." *The American Historical Review* 90, no. 2 (1985): 339-61.
———. "Capitalism and the Origins of the Humanitarian Sensibility: Part 2." *The American Historical Review* 90, no. 3 (1985): 547-66.

Hauptman, John. "Rights Are Intact: Vote Turns On Question, 'What is Science?'" *Des Moines Register*, June 2, 2007.
Havel, Vaclav. "From Plato to NATO." *The Herald*, January 31, 1998.
———. "Looking at the World Through Post-modern Eyes." *Palm Beach Post*, July 14, 1994.
———. "The New Measure of Man." *The New York Times*, July 8, 1994.
———. "Our Changing Times, and the New Measure of Man." *Seattle Post-Intelligencer*, July 10, 1994, E1.
Hawhee, Debra, and Christa Olson. "Pan-Historiography: The Challenges of Writing History across Time and Space." In *Theorizing Historiography in Rhetoric*. Edited by Michelle Ballif. 90-105. Carbondale: Southern Illinois University Press, 2013.
Hawkes, Nigel. "Are We Alone? Quite Probably." *The Vancouver Sun*, Jan. 27, 2000.
———. "Earthlings, thank your uniquely lucky stars." *The Australian*, Jan. 31, 2000.
———. "The lonely planet: Maybe we ARE alone in the universe." *The Gazette*, 2000.
———. "Profile of Earth may be quite unique in universe." *Windsor Star*, Feb. 26, 2000.
———. "A Very Lonely Planet: Charmed Place." *Edmonton Journal*, Feb. 13, 2000.
———. "Whose life is it anyway?" *The Times*, Mar. 2, 1998.
———. "Why Earth is a lonely planet." *The Times*, Jan. 26, 2000.
Hebblethwaite, B. L. "Mellor's 'Bridge-Hand' Argument." *Religious Studies* 22, no. 3/4 (1986): 473-79.
Heffernan, William C. "The Singularity of our Inhabited World: William Whewell and A.R. Wallace in Dissent." *Journal of the History of Ideas* 39, no. 1 (1978): 81-100.
Heidegger, Martin. *Introduction to metaphysics*. New Haven: Yale University Press, 2000.
Heller, Michael. *The Review of Metaphysics* 40, no. 3 (1987): 564-65.
Hetherington, Norriss S. "Aristotle's Cosmology." In *Cosmology: Historical, Literacy, Philosophical, Religious, and Scientific Perspective*. Edited by Norriss S. Hetherington. 95-103. New York: Garland Publishing, 1993.
———. "Introduction: The Aristotelian World View." In *Cosmology: Historical, Literacy, Philosophical, Religious, and Scientific Perspective*. Edited by Norriss S. Hetherington. 177-79. New York: Garland Publishing, 1993.
———. "Percival Lowell: Professional Scientist or Interloper?" *Journal of the History of Ideas* 42, no. 1 (1981): 159-61.
Hissette, Roland. *Enquête sur les 219 articles condamnés à Paris le 7 mars, 1277*. Louvain: Publications Universitaire, 1977.
Hitchens, Christopher. *God Is Not Great*. New York: Twelve Books, 2007.
Hodge, Jonathan. "History of the Earth, Life, and Man." In *William Whewell: A Composite Portrait*. Edited by Menachem Fisch and Simon Schaffer. 253-88. Oxford: Clarendon Press, 1991.
Hodgson, Geoffrey M. "Social Darwinism in Anglophone Academic Journals: A Contribution to the History of the Term." *Journal of Historical Sociology* 17, no. 4 (2004): 428-63.
Holder, Rodney D. "Fine-Tuning, Multiple Universes and Theism." *Noûs* 36, no. 2 (2002): 295-312.
Holter, Scott. "Dead Reckoning." *Columns*, 2006 (Sept. 2003), http://www.washington.edu/alumni/columns/sept03/dead02.html.
House Science Committee. *Astrobiology: Search for Biosignatures in Our Solar System and Beyond*, December 4 2013.
Huffman, Carl A. "The Pythagorian Tradition." In *The Cambridge Companion to Early Greek Philosophy*. Edited by A. A. Long. 66-87. Cambridge: Cambridge University Press, 1999.
Hurst, Lynda. "Finally the Unhappy Truth: ET Has no Home to Phone." *The Toronto Star* March 25, 2000.
Hussey, Edward. *The Presocratics*. New York: Scribner, 1995.
———. *The Presocratics*. New York: Charles Scribner's Sons, 1972.
Hyde, Michael J. *The Ethos of Rhetoric*. Columbia: University of South Carolina Press, 2004.
"Intelligent Design vs. Evolution - Stephen Meyer vs. Peter Ward (rematch)." http://www.youtube.com/watch?v=Sakmq5L3IiE&feature=youtube_gdata_player.
Irwin, Terence. *Plato's Moral Theory: The Early and Middle Dialogues*. Oxford: Clarendon Press, 1977.

"Is A Moon Necessary For A Planet To Support Life?" *Talk of the Nation Science Friday*: National Public Radio, 2011.

Israel, Jonathan I. *Enlightenment Contested: Philosophy, Modernity, and the Emancipation of Man, 1670-1752*. Oxford: Oxford University Press, 2006.

Jaeger, Werner Wilhelm. *The Theology of the Early Greek Philosophers*. New York: Oxford University Press, 1947.

Jarratt, Susan C. *Rereading the Sophists: Classical Rhetoric Refigured*. Carbondale, IL: Southern Illinois University, 1991.

Jenkin, Robert. *The Resonableness and Certainty of the Christian Religion*. 3rd ed. Vol. 1. London: Richard Sare, 1708.

Jenner, Ronald A. "Challenging Received Wisdoms: Some Contributions of the New Microscopy to the New Animal Phylogeny." *Integrative and Comparative Biology* 46, no. 2 (2006): 93-103.

Johansen, T. K. *Plato's Natural Philosophy: A Study of the Timaeus-Critias*. Cambridge: Cambridge University Press, 2004.

Jones, Eric. "Where is Everybody." *Physics Today* 38, no. 8 (1985): 11-13.

Kahn, Charles H. *Plato and the Socratic Dialogue: The Philosophical Use of a Literary Form*. New York: Cambridge University Press, 1996.

———. *Pythagoras and the Pythagoreans: A Brief History*. Indianapolis: Hackett, 2001.

Kaufman, Charles. "Argument as Enactment in the Gorgias." *Philosophy and Rhetoric* 12, no. 2 (1979): 114-29.

Kerferd, G. B. *The Sophistic Movement*. Cambridge: Cambridge University Press, 1981.

Khatri, Bhavin S., Tom C. B. McLeish, Richard P. Sear, and José N. Onuchic. "Statistical Mechanics of Convergent Evolution in Spatial Patterning." *Proceedings of the National Academy of Sciences of the United States of America* 106, no. 24 (2009): 9564-69.

King, Barbara J. "When Animals Mourn" *Scientific American* (2013), http://www.scientificamerican.com/article.cfm?id=when-animals-mourn&WT.mc_id=SA_SA_20130618.

Kinsella, William J. "Rhetoric, Action, and Agency in Institutionalized Science." *Technical Communication Quarterly* 14, no. 3 (2005): 303–10.

Kirk, G. S. "Popper on Science and the Presocratics." In *Studies in Presocratic Philosophy*. Edited by David J. Furley and R. E. Allen. 154-77. New York: Humanities Press, 1970.

Klee, Robert. "The Revenge of Pythagoras: How a Mathematical Sharp Practice Undermines the Contemporary Design Argument in Astrophysical Cosmology." *British Journal for the Philosophy of Science* 53, no. 3 (2002): 331-54.

Klein, Joshua. "Joshua Klein: The Intelligence of Crows." Ted.com, 2008.

Kolakowski, Leszek. *Metaphysical Horror*. London: Penguin Group, 2001.

Kottler, Makolm Jay. "Alfred Russel Wallace: The Origin of Man, and Spiritualism." *Isis* 65, no. 2 (1974): 144-92.

Koyré, Alexandre. *From Closed World to the Infinite Universe*. Baltimore: John Hopkins University, 1982.

Kraut, Richard. *Socrates and the State*. Princeton: Princeton University Press, 1984.

Krüger, Lorenz, Lorraine J. Daston, Michael Heidelberger, Gerd Gigerenzer, and Mary S. Morgan, eds. *The Probabilistic Revolution*. Cambridge, MA: MIT Press, 1987.

Kuhn, Thomas. *The Copernican Revolution: Planetary Astronomy in the Development of Western Thought*. Cambridge, MA: Harvard University Press, 1970.

———. *The Structure of Scientific Revolutions*. 3rd ed. Chicago: University of Chicago Press, 1996.

Küpper, Joachim. "The Traditional Cosmos and the New World." *Modern Language Notes* 118, no. 2 (2003): 363-92.

Kyle, W. M. "British Ethical Theories: The Intuitionist Reaction Against Hobbes." In *Thomas Hobbes Critical Assessments*. Edited by Preston King. 12-21. New York: Routledge, 2000.

Laertius, Diogenes. "The Lives of Eminent Philosophers." Edited by Robert Drew Hicks. Translated by Robert Drew Hicks. Cambridge, MA: Harvard University Press, 1925.

Langley, S. P. "The First 'Popular Science Treatise'." *Popular Scientific Monthly* 10 (1877): 718-25.

Larson, Edward J., and Larry Witham. "Leading Scientists Still Reject God." *Nature* 394, no. 6691 (1998).
Lavender, Gemma. "Living in the Galactic Danger Zone." *Space Daily*, Oct. 5, 2011.
Lee, Mi-Kyoung. *Epistemology After Protagoras: Responses to Relativism in Plato, Aristotle, and Democritus*. Oxford: Oxford University Press, 2005.
LeGault, Michael. "Earth to ET: Is Anyone Out There?" *The Globe and Mail*, May 27, 2000.
"Les Pluralite Des Mondes Habites." *Little's Living Age* 83 (1864): 362-66.
Leslie, John. "Anthropic Explanations in Cosmology." *PSA: Proceedings of the Biennial Meeting of the Philosophy of Science Association* 2986 (1986): 87-95.
———. "Anthropic Principle, World Ensemble, Design." *American Philosophical Quarterly* 19, no. 2 (1982): 141-45.
———. *The End of the World: The Science and Ethics of Human Extinction*. New York: Routledge, 1996.
———. "Mackie on Neoplatonism's 'Replacement for God'." *Religious Studies* 22, no. 3/4 (1986): 325-42.
———. "Observership in Cosmology: The Anthropic Principle." *Mind* 92, no. 368 (1983): 573-79.
Lessl, Thomas M. "The Culture of Science and the Rhetoric of Scientism: From Francis Bacon to the Darwin Fish." *Quarterly Journal of Speech* 93, no. 2 (2007): 123-49.
———. *Rhetorical Darwinism: Religion, Evolution, and the Scientific Identity*. Waco, TX: Baylor University Press, 2012.
———. "Science and the Sacred Cosmos: The Ideological Rhetoric of Carl Sagan." *Quarterly Journal of Speech* 71 (1985): 175-87.
Levinton, Jeffrey. "Who Owns the Cambrian Explosion?" *The Quarterly Review of Biology* 74, no. 2 (1999): 201-04.
Lewis, C. S. *The Discarded Image: An Introduction to Medieval and Renaissance Literature*. Cambridge: Cambridge University Press, 2002.
———. *Out of the Silent Planet*. New York: Scribner, 2003.
———. *Perelandra* New York: Scribner, 2003.
———. *That Hideous Strength* New York: Scribner, 2003.
Lewis, George Cornewall. *An Essay on the Influence of Authority in Matters of Opinion*. London: J. Parker, 1849.
Lewis, Jo Ann. "It's Postmodern: And if you don't get it, you don't get it." *The Washington Post*, March 27, 1994.
Lightman, Bernard. "Ideology, Evolution and Late-Victorian Agnostic Popularizers." In *History, Humanity and Evolution: Essays for John C. Greene*. Edited by James Richard Moore. 285-310. Cambridge: Cambridge University Press, 1989.
"Literary and Scientific." *Christian Advocate and Journal* 29, no. 40 (1854): 158.
"Literary Notices." *Zion's Herald and Wesleyan Journal* 30, no. 17 (1859): 66.
Lloyd, G. E. R. *Early Greek Science*. New York: W. W. Norton & Company, 1970.
———. "Plato as a Natural Scientist." *Journal of Hellenic Studies* 88 (1968): 78-92.
Lovejoy, Arthur O. *The Great Chain of Being: A Study of the History of an Idea*. Cambridge: Harvard University Press, 1942.
Lowell, Percival. *Mars and Its Canals*. London: Macmillan, 1906.
Lynch, John M. *'Vestiges' and the Debate before Darwin*. Vol. 6. Bristol, England: Thoemmes, 200.
Lyotard, Jean-François. *The Postmodern Condition: A Report on Knowledge*. Translated by Geoff Bennington and Brian Massumi. Minneapolis, MN: University of Minnesota Press, 1984.
M., T. W. "One Letter From Boston." *American Publishers' Circular and Literary Gazetter* 2, no. 7 (1856): 93-96.
Machor, James L. "The American Reception of Melville's Short Fiction in the 1850s." In *New Directions in American Reception Study*. Edited by Philip Goldstein and James L. Machor. 87-98. New York: Oxford University Press, 2008.

Mailloux, Steven. "Judging and Hoping: Rhetorical Effects of Reading about Reading." In *New Directions in American Reception Study*. Edited by Philip Goldstein and James L. Machor. 23-32. New York: Oxford University Press, 2008.
Malik, Tariq. "Most Americans Believe Alien Life is Possible, Study Shows." *Space.com*, May 31 2005. http://www.space.com/news/050531_alienlife_survey.html
Manovich, Lev. *The Language of New Media*. Boston: MIT Press, 2001.
Manson, Neil A. "Anthropocentrism and the Design Argument." *Religious Studies* 36, no. 2 (2000): 163-76
Marchant, James. *Alfred Russel Wallace: Letters and Reminisences*. New York: Arno Press, 1975.
Marshall, Andrew. "Oh No! Humans Are the Only Intelligent Lifeform." *The Independent*, Feb. 10, 2000, 1.
"The Master of Trinity's Plurality of Worlds." *Littell's Living Age* 12, no. 612 (1856): 403-04.
Matthews, Melvin E. *Hostile Aliens, Hollywood and Today's News: 1950s Science Fiction Films and 9/11*. New York: Algora, 2007.
Maurice, Frederick Denison. *Moral and Metaphysical Philosophy: Fourteenth Century to the French with a Glimpse into the Nineteenth Century*. Vol. 2. London: Macmillan, 1873.
Mazur, Barry. *Harvard Review*, no. 16 (1999): 141-46.
McAleer, Graham James. "Disputing the Unity of the World: The Importance of Res and the Influence of Averroes in Giles of Rome's Critique of Thomas Aquinas concerning the Unity of the World." *Journal of the History of Philosophy* 36, no. 1 (1998): 29-55.
McColley, Grant. "The Seventeenth-Century Doctrine of a Plurality of Worlds." *Annals of Science* 1, no. 4 (1936): 385-430.
McCoy, Marina. *Plato on the Rhetoric of Philosophers and Sophists*. Cambridge: Cambridge University Press, 2008.
McGrath, Alister E. *Darwinism and the Divine: Evolutionary Thought and Natural Theology*. Malden, MA: Wiley-Blackwell, 2011.
McGrew, Timothy, Lydia McGrew, and Eric Vestrup. "Probabilities and the Fine-Tuning Argument: A Sceptical View." *Mind* 110, no. 440 (2001): 1027-37.
McInerny, Ralph. *Aquinas*. Cambridge: Polity Press, 2004.
Mckeon, Richard. *The Basic Works of Aristotle*. New York: Random House, 1941.
McKie, Robin. "Meet the neighbours?" *The Observer*, Feb. 7 2010.
———. "Wake up, it's an alien on line one: British scientists are at the heart of a new project to answer the old question is there anybody out there?" *The Observer*, March 5 2000.
———. "When it comes to intelligent life, we're as good as it gets." *The Observer*, July 16 2000.
McLaughlin, John. *John McLaughlin's "One on One"*: PBS, 2003.
McLean, Murdith. "Residual Natural Evil and Anthropic Reasoning." *Religious Studies* 27, no. 2 (1991): 173-88.
McPhail, Mark Lawrence. *Zen in the Art of Rhetoric: An Inquiry into Coherence*. Albany: State University of New York Press, 1996.
Midgley, Mary. *Science as Salvation: A Modern Myth and Its Meaning*. New York: Routledge, 1992.
Mill, John Stuart. *Utilitarianism*. Chicago: University of Chicago Press, 1906.
———. "Whewell on Moral Philosophy 1852." In *The Collected Works of John Stuart Mill*. Edited by John M. Robson. Toronto: University of Toronto Press, 1985.
Mitchel, Ormsby MacKnight. *Orbs of Heaven*. London: Bradbury and Evans, 1851.
Mitchell, Maria. *Maria Mitchell: Life, Letters, and Journals*. Boston: Lee and Shepard, 1896.
Morris, Simon Conway. "Considering Convergence: Interview with Simon Conway Morris, part 1." Jan. 18. http://www.astrobio.net/index.php?option=com_podcast&task=showall&id=10&msg=&pageNo=2.
———. *The Crucible of Creation: The Burgess Shale and the Rise of Animals*. Oxford: Oxford University Press, 1998.
———. "Darwin Was Right. Up to a Point." *The Guardian*, Feb. 12 2009.
———. "Darwin's Compass: How Evolution Discovers the Song of Creation." http://www.stmarylebow.co.uk/ - /boyle-lecture-2005/4535725162.

———. *Life's Solution: Inevitable Humans in a Lonely Universe*. Cambridge: Cambridge University Press, 2003.

Morrow, Glenn Raymond. *Plato's Cretan City: A Historical Interpretation of the Laws*. Princeton: Princeton University Press, 1993.

Mulhern, J. J. "Treatises, Dialogues, and Interpretation." *The Monist* 53, no. 4 (1969): 631-41.

Myers, P. Z. "The Mediocrity Principle." *The Edge*. http://www.edge.org/q2011/q11_12.html - myerspz.

"Myth." http://www.oxforddictionaries.com/us/definition/american_english/myth.

Nails, Debra. "Mouthpiece Schmouthpiece." In *Who Speaks for Plato?: Studies in Platonic Anonymity*. Edited by Gerald Alan Press. 15-26. Lanham, MD: Rowman & Littlefield, 2000.

"National Book Awards - 1995." *NationalBook.org*, http://www.nationalbook.org/nba1995 - .UmQ7SyQmzyc.

"New Books." *The Albion* 13, no. 27 (1854): 321-22.

"New Titles." *BioScience* 38, no. 1 (1988): 61.

Nichols, Aidan. *Discovering Aquinas: An Introduction to His Life, Work, and Influence*. Grand Rapids, MI.: William B. Eerdman, 2003.

Nicholson, Helen J. *The Crusades*. Westport, CT: Greenwood, 2004.

Nicolson, Marjorie. *Science and Imagination*. Ithaca: Cornell University Press, 1962.

Norman, David. "Curtains Fall on the Freak Show." *The Times Higher Education Supplement*, Feb. 6, 1998.

North, J. D. *Stars, Minds and Fate: Essays in Ancient and Medieval Cosmology*. London: Hambledon, 1989.

North, John D. "One Truth or More? Demarcation in the Universe of Discourse." In *Physics, Cosmology and Astronomy, 1300-1700: Tension and Accommodation*. Edited by Sabetai Unguru. 253-93. Boston: Kluwer, 1991.

Novick, Peter. *That Nobel Dream: The 'Objectivity Question' and the American Historical Profession*. Cambridge: Cambridge University Press, 1988.

Nye, Mary Jo. "Temptations of Theory, Strategies of Evidence: P. M. S. Blackett and the Earth's Magnetism, 1947-52." *The British Journal for the History of Science* 32, no. 1 (1999): 69-92.

O'Driscoll, Michael. "Why life is more than an accident." *Irish Examiner*, Oct. 3, 2007.

O'Meara, Thomas F. "Christian Theology and Extraterrestrial Intelligent Life." *Theological Studies* 60 (1999): 3-30.

O'Reilly, Bill. *The O'Reilly Factor*: Fox News Network, 2000.

Oderberg, David S. "Temporal Parts and the Possibility of Change " *Philosophy and Phenomenological Research* 69, no. 3 (2004): 686-708.

Oresme, Nicole. *Le Livre du Ciel et du Monde*. Translated by Albert D. Menut. Madison: University of Wisconsin Press, 1968.

Osborne, Colin P. , and David J. Beerling. "Nature's Green Revolution: The Remarkable Evolutionary Rise of C4 Plants." *Philosophical Transactions: Biological Sciences* 361, no. 1465 (2009): 173-94.

"Our Mission." *Reasonable Faith*, http://www.reasonablefaith.org/site/News2?page= NewsArticle&id=5299.

Overbye, Dennis. "Hot on Trail of 'Just Right' Far-Off Planet." *The New York Times*, Dec. 3 2011, 1.

———. "Hunting for Life in Specks of Cosmic Dust." *The New York Times*, July 19 2005, 1.

Paglia, Camille. "Ask Camille." *Salon.com*, http://web.archive.org/web/20080411071450/ http://www.salon.com/july97/columnists/paglia2970722.html.

Paine, Thomas. *The Age of Reason*. Vol. 1. London, R. Carlile, London 1826: R. Carlile, 1826.

Paley, William. *Principles of Moral and Political Philosophy*. seventh ed. Philidelphia: Thomas Dobson, 1788.

Partenie, Catalin, ed. *Plato's Myths*. Cambridge: Cambridge University Press, 2009.

Paul, Diana B. "Darwin, Social Darwinism and Eugenics." In *The Cambridge Companion to Darwin*. Edited by Johnathan Hodge and Gregory Radick. Cambridge: Cambridge University Press, 2009.

———. "Wallace, Women, and Eugenics." In *Natural Selection and Beyond*. Edited by Charles H. Smith and George Beccaloni. 263-78. Oxford: Oxford University Press, 2008.

Paul, Erich Robert. *Science, Religion, and Mormon Cosmology*. Urbana: University of Illinois Press, 1992.

Pelikan, Jaroslav. *What Has Athens to do with Jerusalem*. Ann Arbor: University of Michigan Press, 2000.

Peters, John Durham. *Speaking into the Air: A History of the Idea of Communication*. Chicago: University of Chicago Press, 2000.

———. *Speaking into the Air: A History of the Idea of Communication*. Chicago: University of Chicago Press, 1999.

Philo. "De Opificio Mundi." In *The Works of Philo: Complete and Unabridged*. Edited by C. D. Yonge. Translated by C. D. Yonge. Peabody, MA: Henderickson, 2000.

Pinker, Steven. *How the Mind Works*. New York: W. W. Norton & Company, 1997.

Pinker, Steven, Werner Kalow, and Harold Kalant. "Evolutionary Psychology: An Exchange." *The New York Review of Books* (1997), http://www.nybooks.com/articles/archives/1997/oct/09/evolutionary-psychology-an-exchange/?pagination=false.

Plato. "Euthyphro." In *Plato Complete Works*. Edited by John M. Cooper. Translated by G. M. A. Grube. Indianapolis: Hackett, 1997.

———. "Gorgias." In *Plato Complete Works*. Edited by John M. Cooper. Translated by Donald J. Zeyl. Indianapolis: Hackett, 1997.

———. *Greater Hippias*. Edited by John M. Cooper, *Plato Complete Works*. Indianapolis: Hackett, 1997.

———. "Ion." In *Plato Complete Works*. Edited by John M. Cooper. Indianapolis: Hackett, 1997.

———. "Laws." In *Plato Complete Works*. Edited by John M. Cooper Translated by Trevor J. Saunders. Indianapolis: Hackett, 1997.

———. "Letter VII." In *Plato Complete Works*. Edited by John M. Cooper Translated by Glen R. Morrow. Indianapolis: Hackett, 1997.

———. "Meno." In *Plato Complete Works*. Edited by John M. Cooper. Indianapolis: Hackett, 1997.

———. "Phaedo." In *Plato: Complete Works*. Edited by John M. Cooper Translated by G.M.A. Grube. Indianaopolis: Hackett 1997.

———. "Protagoras." In *Plato Complete Works*. Edited by John M. Cooper Translated by Stanley Lombardo and Karen Bell. 746-90. Indianapolis: Hackett, 1997.

———. *Republic*. Edited by John M. Cooper, *Plato Complete Works*. Indianapolis: Hackett, 1997.

———. "Rival Lovers." In *Plato Complete Works*. Edited by John M. Cooper. Indianapolis: Hackett, 1997.

———. "Sophist." In *Plato: Complete Works*. Edited by John M. Cooper Translated by Nicholas P. White. Indianaopolis: Hackett, 1997.

———. "Theaetetus." In *Plato Complete Works*. Edited by John M. Cooper Translated by M. J. Levitt and Myles Burnyeat. 157-234. Indianapolis: Hackett, 1997.

———. *Theaetetus*. Edited by John M. Cooper, Plato Complete Works. Indianapolis: Hackett, 1997.

———. "Timaeus." In *Plato Complete Works*. Edited by John M. Cooper. Indianapolis: Hackett, 1997.

———. *Timaeus*. Translated by Donald J. Zeyl. Edited by John M. Cooper, *Plato Complete Works*. Indianapolis: Hackett, 1997.

"The Anthropic Principle." *The Hamilton Spectator* March 17, 2006.

"The Plurality of Inhabited Worlds." *The Theological and Literary Journal* 7, no. 2 (1854): 276-303.

"Plurality of Worlds." *Circular* 3, no. 150 (1854): 1-2.

"The Plurality of Worlds." *The Eclectic Magazine of Foreign Literature* 37, no. 1 (1856): 25-45.

"The Plurality of Worlds." *Christian Inquirer* 8, no. 47 (1854): 1.

"The Plurality of Worlds." *Christian Watchman and Reflector* 35, no. 49: 194.

"Plurality of Worlds." *Circular* 3, no. 149 (1854): 1-2.
"The Plurality of Worlds." *New Englander* 12, no. 48 (1854): 570-604.
"Plurality of Worlds." *Littell's Living Age* 11, no. 596 (1855): 256.
"The Plurality of Worlds." *Christian Inquirer* 13, no. 39 (1859): 2.
"Poll: U.S. Hiding Knowledge of Aliens." *CNN*, June 15, 1997. http://www.cnn.com/US/9706/15/ufo.poll/.
Popper, Karl. "Back to the Presocratics." In *Studies in Presocratic Philosophy*. Edited by David J. Furley and R. E. Allen. 130-53. New York: Humanities, 1970.
Posner, Richard. *Catastrophe: Risk and Response* Oxford: Oxford University Press, 2005.
Poulakos, John. *Sophistical Rhetoric in Classical Greece*. Columbia: University of South Carolina Press, 1995.
Prelli, Lawrence J. "The Rhetorical Construction of Scientific Ethos " In *Landmark Essays on the Rhetoric of Science: Case Studies* . Edited by Randy Allen Harris. 87-106. Mahway, NJ: Lawrence Erlbaum, 1997.
Press, Gerald A. *Plato: A Guide for the Perplexed*. London: Continuum International, 2007.
Prigogine, I., and J. Geheniau. "Entropy, Matter, and Cosmology." *Proceedings of the National Academy of Sciences of the United States of America* 83, no. 17 (1986): 6245-49.
"Publications." *The Monthly Religious Magazine* 12, no. 6 (1854): 355-40.
Raby, Peter. *Alfred Russel Wallace: A Life*. Princeton: Princeton University Press, 2001.
Ralickas, Vivian. ""Cosmic Horror" and the Question of the Sublime in Lovecraft." *Journal of the Fantastic in the Arts* 19, no. 3 (2008): 364-98.
Raman, Varadaraja V. "Hope from Physics." *Science* 267, no. 5200 (1995): 1042-43.
Randles, W. G. L. *The Unmaking of the Medieval Christian Cosmos, 1500-1760: From Solid Heavens to Boundless Aether*. Brookfield, VT: Ashgate, 1999.
Raub, Cymbre Quincy. "Robert Chambers and William Whewell: A Nineteenth-Century Debate Over the Origin of Language." *Journal of the History of Ideas* 49, no. 2: 287-300.
Raulet, Gérard. "Structuralism and Post-Structuralism: An Interview with Michel Foucault." *Telos*, no. 55 (1983): 195-211.
Raven, J. E. *Plato's Thought in the Making.* Cambridge: Cambridge University Press, 1965.
"Recent Publications." *The British Journal for the Philosophy of Science* 37, no. 3 (1986): 385-93.
"Recent Publications in Philosophy Source." *Noûs* 20, no. 4 (1986): 581-83.
Redbeard, Ragnar, and Darrell W. Conder. *Might is Right or Survival of the Fittest*. New ed. Springfield, MO: Dil Pickle Press, 2005.
Rees, Martin. *Our Final Hour: A Scientist's Warning: How Terror, Error, and Environmental Disaster Threaten Humankind's Future In This Century — On Earth and Beyond*. New York Basic Books, 2003.
Rescher, Nicholas. *Cosmos and Cognition: Studies in Greek Philosophy*. Piscataway, NJ: Transaction, 2005.
———. "Extraterrestrial Science." In *Extraterrestrials: Science and Alien Intelligence*. Edited by Edward Regis Jr. 83-116. Cambridge: Cambridge University Press, 1987.
"Review 0." *The Theological and Literary Journal* 7, no. 1 (1854): 165-67.
"Review: Plurality of Worlds." *Mercersburg Quarterly Review* (1855): 323-24.
Reydams-Schils, Gretchen J. *Demiurge and Providence: Stoic and Platonist Readings of Plato's Timaeus*. Brepols: Turnhout, 1999.
Richards, Guillermo Gonzalez and Jay W. *The Privileged Planet: How Our Place in the Cosmos is Designed for Discovery*. Washington D.C.: Regnery Publishing 2004.
Richardson, Ruth. *The Making of Mr. Gray's Anatomy*. New York: Oxford University Press, 2008.
Ridley, Mark. "Rock of Ages." *The New York Times*, May 10, 1998.
Robinson, Erik. "Message of Earth Day Hits Home." *The Columbian*, April 22, 2000.
Robinson, T. M. "Mind-Body Dualism in Plato." In *Psyche and Soma: Physicians and Metaphysicians on the Mind-Body Problem from Antiquity to Enlightenment*. Edited by John P. Wright and Paul Potter. Oxford: Clarendon, 2000.

Roemer, Kenneth M. "Placing Readers at the Forefront of Nowhere: Reception Studies and Utopian Literature." In *New Directions in American Reception Study*. Edited by Philip Goldstein and James L. Machor. 99-119. New York: Oxford University Press, 2008.
Rogers, James Allen. "Darwinism and Social Darwinism." *Journal of the History of Ideas* 33, no. 2 (1972): 265-80.
Rosen, Frederick. *Classical Utilitarianism from Hume to Mill*. London: Routledge, 2003.
Rosenboom, David. "Propositional Music: On Emergent Properties in Morphogenesis and the Evolution of Music. Part I: Essays, Propositions and Commentaries." *Leonardo* 30, no. 4 (1997): 291-97.
Rosenkranz, K. "Rosenkranz on Hegel's History of Philosophy." *Journal of Speculative Philosophy* 8, no. 1 (1874): 1-13.
Ross, Hugh, and Guillermo Gonzalez. "Live Here or Nowhere." *Reasons.org*, http://www.reasons.org/articles/live-here-or-nowhere.
———. "A Spectrum of Views on ETI." *Reasons.org*. http://www.reasons.org/articles/a-spectrum-of-views-on-eti.
Ross, Hugh Norman. *The Creator and the Cosmos: How the Greatest Scientific Discoveries of the Century Reveal God*. Colorado Springs, CO: NavPress Publishing Group, 1993.
Rossi, Lisa. "Intelligent Design Theory Influenced ISU Tenure Vote." *Des Moines Register*, Dec. 1, 2007.
Roth, Christopher F. "Ufology as Anthropology: Race Extraterrestrials, and the Occult." In *E.T. Culture: Anthropology in Outerspaces*. Edited by Debbora Battaglia. 38-93. Durham, NC: Duke University Press, 2005.
Rotman, Brian. *SubStance* 24, no. 3 (1995): 150-53.
Rowland, Ingrid D. "What Giordano Bruno Left Behind: Rome, 1600." *Common Knowledge* 14, no. 2 (2008): 424-43.
Rubenstein, Richard E. *Aristotle's Children: How Christians, Muslims, and Jews Rediscovered Ancient Wisdom and Illuminated the Dark Ages*. Orlando: Harcourt, 2003.
Ruse, Michael. "Introduction." In *Of the Plurality of Worlds. 1853 A Facsimile of the First Edition with an introduction by Michael Ruse*. Edited by Michael Ruse. 1-31. Chicago: University of Chicago Press, 2001.
———. "The Relationship between Science and Religion in Britain, 1830-1870." *Church History* 44, no. 4 (1975): 505-22.
———. "William Whewell: Omniscientists." In *William Whewell: A Composite Portrait*. Edited by Menachem and Fisch and Simon Schaffer. 87-116. Oxford: Clarendon, 1991.
Ruskin, Steven W. "A Newly-discovered Letter of J.F.W. Herschel Concerning the 'Great Moon Hoax'." *Journal for the History of Astronomy* 33 (2002): 71-74.
Russell, Bertrand. *Mysticism and Logic and Other Essays*. London: Longmans, Green and Co., 1918.
———. *The History of Western Philosophy: And its Connection with Political and Social Circumstances from the Earliest Times to the Present Day*. Great Britain: George Allen and Unwin, 1947.
Russell, Colin A. "Presidential Address: 'Rude and Disgraceful Beginnings': A View of History of Chemistry from the Nineteenth Century." *The British Journal for the History of Science* 21, no. 3 (1988): 273-94.
Sagan, Carl. *The Demon Haunted World*. New York: Random House, 1995.
Samuelson, David N. "Review: Botching the Science in Science Fiction." *Science Fiction Studies* 19, no. 1 (1992): 100-04.
Scade, Paul. "Stoic Cosmological Limits and Their Platonic Background." In *Aristotle and the Stoics Reading Plato*. Edited by V. Harte, M.M. McCabe, R. Sharples and A. Sheppart. 141-81. London: British Institute of Classical Studies, 2011.
Scenters-Zapico, John. "The Case for the Sophists." *Rhetoric Review* 11 (1993): 352–67.
Schiappa, Edward. *Protagoras and Logos: A Study in Greek Philosophy and Rhetoric*. Columbia: University of South Carolina Press, 1991.
Schiffman, Zachary Sayre. *On the Threshold of Modernity*. Baltimore: John Hopkins University Press, 1991.
Schneider, Doug. "Flunking the Tenure Test." *The Washington Times*, April 23, 2008.

Schonfeld, Roger C. *JSTOR: A History*. Princeton, NJ: Princeton University Press, 2003.
Schwartz, Joel S. "Darwin, Wallace, and the "Descent of Man"." *Journal of the History of Biology* 17, no. 2 (1984): 271-89.
“Science.” *Oxford Dictionaries*, http://www.oxforddictionaries.com/us/definition/american_english/science.
Scott, Janny. "Postmodern Gravity Deconstructed, Slyly." *The New York Times*, May 18, 1996.
Secord, James A. *Victorian Sensation: The Extraordinary Publication, Reception, and Secret Authorship of Vestiges of the Natural History of Creation*. Chicago: University of Chicago, 2000.
Sedley, David. "Is Aristotle's Teleology Anthropocentric." *Pronesis* 36, no. 2 (1991): 179-96.
Seely, Ron. "A Setback in the Eternal Hope for Alien Life?" *Wisconsin State Journal*, Feb. 10, 2000.
Segal, Judy, and Alan W. Richardson. "Introduction. Scientific Ethos: Authority, Authorship, and Trust in the Sciences." *Configurations* 11, no. 2 (2003): 137-44.
Sheridan, Mark A. "SETI's Scope: How the Search for Extraterrestrial Intelligence became Disconnected from New Ideas about Extraterrestrials." Dissertation, Drew University, 2009.
Shermer, Michael. *In Darwin's Shadow: The Life and Science of Alfred Russel Wallace*. Oxford: Oxford University Press, 2002.
Shields, Christopher. “Aristotle.” *Stanford Encyclopedia of Philosophy*. http://plato.stanford.edu/entries/aristotle/.
Shklovskii, I.S., and Carl Sagan. *Intelligent Life in the Universe*. Translated by Paula Fern. New York: Dell, 1966.
Shoemaker, Philip S. "Stellar Impact: Ormsby MacKnight Mitchel and Astronomy in Antebellum America." Diss., University of Wisconsin-Madison, 1991.
Shorey, Paul. *The Unity of Plato's Thought* . Whitefish, MT: Kessinger Publishing, 2010.
"The Silence of Scripture." *National Magazine* 13 (1858): 121-28.
Silk, Joseph. "New Series." *Science* 232, no. 4753 (1986): 1036-37.
Simplicus. *On Aristotle's "On the Heavens 2.10-14"*. Translated by Ian Mueller. Ithaca: Cornell University Press, 2005.
Smart, J.J.C. "The Anthropic Cosmological Principle." *The Philosophical Quarterly* 37, no. 149 (1987): 463-66.
Smith, Charles. “The Most Important People in Wallace's Intellectual Life.” *The Alfred Russel Wallace*. http://people.wku.edu/charles.smith/index1.htm.
Snyder, Laura J. "‘Lord Only of the Ruffians and Fiends’? William Whewell and the Plurality of Worlds Debate." *Studies in History and Philosophy of Science* 38 (2007): 584-92.
———. *Reforming Philosophy: A Victorian Debate on Science and Society*. Chicago: University of Chicago, 2006.
———. “William Whewell.” *Stanford Encylopedia of Philosophy*, http://plato.stanford.edu/entries/whewell/.
"Some Regions of the Universe are Not Favorable for Life." *The Daily Galaxy*, Dec. 24, 2012.
Somfai, Anna. "The Eleventh-Century Shift in the Reception of Plato's "Timaeus" and Calcidius's "Commentary'." *Journal of the Warburg and Courtauld Institutes* 65 (2002): 1-22.
Sorabji, Richard. *Time Creation and the Continuum*. Ithaca, NY: Cornell University, 1983.
Spears, Monroe K. "Review: Cosmology and the Common Reader." *Sewanee Review* 99, no. 1 (1991): 113-21.
"Speculators Among the Stars: Part I." *Littell's Living Age* 8, no. 554 (1855): 3-12.
"Speculators Among the Stars: Part II." *Littell's Living Age* 8, no. 555 (1855): 67-91.
"Speculators Among the Stars.—Part I." *Littell's Living Age* 44, no. 554 (1855): 3-12.
Speller, Jules. *Galileo's Inquisition Trial Revisited*. Frankfurt: Peter Lang, 2008.
Spencer, Jones, H. *Life On Other Worlds*. New York: Macmillan, 1940.
Stack, David A. "The First Darwinian Left: Radical and Socalist Responses to Darwin, 1859-1914." *History of Political Thought* 21, no. 4 (2000): 682-710.
Stahl, William A., Robert A. Campbell, Yvonne Petry, and Gary Diver. *Webs of Reality: Social Perspectives on Science and Religion*. Piscataway, NJ: Rutgers University Press, 2002.
Stenger, Victor J. *God the Failed Hypothesis: How Science Shows That God Does Not Exist*. Amherst: Prometheus Books, 2007.

Stephens, Leslie. *The English Utilitarians*. Vol. 2. London: Continuum International Publishing Group, 2005.
———. *The English Utilitarians*. Vol. 3. London: Duckworth, 1900.
Stewart, Albert B. *The Antioch Review* 53, no. 3 (1995): 374-75.
Strauss, Leo. *The City and Man*. Chicago: University of Chicago, 1978.
Sungenis, Robert, and Robert Bennett. *Galileo Was Wrong: The Church Was Right: The Historical Case for Geocentrism*. Vol. 2. State Line PA: Catholic Apologetics, 2008.
———. *Galileo Was Wrong: The Church Was Right: The Scientific Case for Geocentrism*. Vol. 1. State Line PA: Catholic Apologetics, 2008.
"Super-Earth is only 600 light years away; Kepler-22b is the first planet outside our solar system confirmed as possibly habitable." *National Post*, Dec. 6, 2011.
Swirski, Peter. *Between Literature and Science: Poe, Lem, and Explorations in Aesthetics, Cognitive Science, and Literary Knowledge*. Quebec City: McGill-Queen's University Press, 2000.
Taylor, A. E. *A Commentary on Plato's Timaeus*. New York: Garland, 1967.
Taylor, C. C. W. "The Atomists." In *The Cambridge Companion to Early Greek Philosophy*. Edited by A. A. Long. 181-204. Cambridge: Cambridge University Press, 1999.
Tejera, Victorino. *Plato's Dialogues One by One: A Structural Interpretation*. Lanham: University Press of America, 1999.
Temple, Dennis. "Hume's Logical Objection to the Argument from Design Based on the Uniqueness of the Universe." *Religious Studies* 28, no. 1 (1992): 19-30.
Thijssen, Hans. "Condemnation of 1277." *Stanford Encyclopedia of Philosophy*, http://plato.stanford.edu/entries/condemnation/.
Thornton, Brian. "The Moon Hoax: Debates About Ethics in 1835 New York Newspapers." *Journal of Mass Media Ethics* 15, no. 2 (2000): 89-100.
Thucydides. *History*. Translated by P. J. Rhodes. Vol. 2. Warminster: Aris & Phillips, 1988.
Tietge, David J. *Rational Rhetoric: The Role of Science in Popular Discourse*. West Lafayette, IN: Parlor Press, 2008.
Tillyard, E. M. W. *The Elizabethan World Picture*. New York: Macmillan, 1944.
Tipler, Frank. "Postmodern Physics: Colleges Fail to Teach Basics - Even in Physics!" http://www.popecenter.org/commentaries/article.html?id=1843.
———. "TEDx Brussels 2010 - Frank Tipler - The Ultimate Future " YouTube, 2011.
Tipler, Frank J. "Additional Remarks on Extraterrestrial Intelligence." *Quarterly Journal of the Royal Astronomical Society* 22 (1981): 279-92.
———. "The Anthropic Principle: A Primer for Philosophers." *PSA: Proceedings of the Biennial Meeting of the Philosophy of Science Association* 1988, no. 2 (1988): 27-48.
———. "Biography." *Frank J. Tipler Home Page*, http://129.81.170.14/~tipler/biography.htm.
———. "A Brief History of the Extraterrestrial Intelligence Concept." *Quarterly Journal of the Royal Astronomical Society* 22 (1981): 133-45.
———. "Causality Violation in Asymptotically Flat Spacetimes." *Physical Review Letters* 37 (1976): 879-82.
———. "Energy Conditions and Spacetime Singularities." *Physical Review* D17 (1978): 2521-28.
———. "Extraterrestrial Intelligent Beings Do Not Exist." *Physics Today* 34, no. 4 (1981): 9, 70-71.
———. "Extraterrestrial Intelligent Beings Do Not Exist." *Quarterly Journal of the Royal Astronomical Society* 21 (1980): 267-81.
———. "Frank J. Tipler, reply by Martin Gardner." *The New York Review of Books* (1986).
———. "General Relativity and Conjugate Ordinary Differential Equations." *Journal of Differential Equations* 30 (1978): 165-74.
———. "General Relativity, Thermodynamics, and the Poincare Cycle." *Nature* 280 (1979): 203-05.
———. *The Physics of Christianity*. New York: Doubleday Broadway, 2007.
———. *The Physics of Immortality: Modern Cosmology, God, and the Resurrection of the Dead*. New York: Doubleday, 1994.

———. "Rotating Cylinders and the Possibility of Global Causality Violation." *Physical Review* D9, no. 8 (1974): 2203–06.
Todhunter, Isaac. *William Whewell, D.D., Master of Trinity College, Cambridge; An Account of his Writings with Selections from his Literary and Scientific Correspondence*. Vol. 1. New York: Johnson Reprint Corporation, 1970.
———. *William Whewell, D.D., Master of Trinity College, Cambridge; An Account of his Writings with Selections from his Literary and Scientific Correspondence*. Vol. 2. New York: Johnson Reprint Corporation, 1970.
Tooke, John Horne. *Epea Pteroenta or the Diversions of Purley*. London: J. Johnson, 1805.
Topham, Jonathan R. "Beyond the 'Common Context': The Production and Reading of the Bridgewater Treatises " *Isis* 89, no. 2 (1998): 233-62.
Toulmin, Stephen Edelston. "Contemporary Scientific Mythology." In *Metaphysical Beliefs: Three Essays*. Edited by Alasdair MacIntyre. London: SCM Press, 1957.
———. *Cosmopolis: The Hidden Agenda of Modernity* Chicago: University of Chicago Press, 1990.
Townsend, George. "Preliminary Dissertation." In *The Acts and Monuments of John Foxe: A New and Complete Edition*. Edited by Stephen Reed Cattley. 1-533. London: R. B. Seeley and W. Burnside, 1841.
Tressider, Megan. "The Megan Tressider Interview: Meaning of Life Is, Er, God and Omega, Physicist Frank J Tipler, an Atheist, Says He Has Found God." *The Guardian*, 1995.
Trollope, Anthony. *Barchester Towers*. Vol. 1. Leipzig: Bernhard Tauchnitz, 1859.
Tushnet, Mark. "Critical Legal Studies: A Political History." *The Yale Law Journal* 100, no. 5 (1991).
"The Unity of the World, Versus the Plurality of Worlds." *The Albion* 14, no. 20 (1855): 231-32.
"Vast Tides May Explain Life's Origin." *The New Zealand Herald*, April 3, 2013.
Vermeij, Geerat J. "Historical Contingency and the Purported Uniqueness of Evolutionary InnovationsAuthor(s): Geerat J. Vermeij." *Proceedings of the National Academy of Sciences of the United States of America* 103, no. 6 (2006): 1804-09.
Versluis, Arthur. "Antimodernism." *Telos* 137, no. 96 (2006): 96-130.
Vetter, Jeremy. "The Unmaking of an Anthropologist: Wallace Returns from the Field, 1862 –70." *Notes and Records of the Royal Society* 64 (2010): 25-42
———. "Wallace's Other Line: Human Biogeography and Field Practice in the Eastern Colonial Tropics." *Journal of the History of Biology* 39, no. 1 (2006): 89-123.
Vlastos, Gregory. "Cornford's Principium Sapientiae." In *Studies in Presocratic Philosophy*. Edited by David J. Furley and R. E. Allen. 42-55. New York: Humanities, 1970.
———. "Ethics and Physics in Democritus." In *Studies in Presocratic Philosophy*. Edited by R. E. Allen and David J. Furley. 381-408. Atlantic Highlands, NJ: Humanities Press, 1975.
———. *Plato's Universe*. Seattle: University of Washington Press, 1975.
———. "Socrates." In *Socrates: Critical Assessment*. Edited by W. Prior. 136-55. London: Routledge, 1996.
———. *Socrates, Ironist and Moral Philosopher*. Ithaca, NY: Cornell University Press, 1991.
"Volume Information." *American Philosophical Quarterly* 19, no. 1 (1982).
W. "The Plurality of Worlds." *Christian Watchman and Reflector* 35, no. 31 (1854): 1.
Wade, Nicholas. "Scientists Measure the Accuracy of a Racism Claim." *The New York Times*, June 12, 2011.
Wakeford, Tom. "Too Far, Too Fast?" *The Guardian*, Mar. 5, 1997.
Wallace, Alfred Russel. *The Eagle and the Serpent* (1900), http://people.wku.edu/charles.smith/wallace/S580.htm.
———. *Darwinism: An Expositon of the Theory of Natural Selection with Some of its Applications*. London: Macmillan, 1889.
———. "Does Man Exist in Other Worlds "A Reply to my Critics"." *The Daily Mail*, Nov. 12, 1903.
———. *Is Mars Habitable? A Critical Examination of Professor Percival Lowell's Book "Mars and its Canals," with an Alternative Explanation*. London: Macmillan & Co., 1907.

———. *The Malay Archipelago: The Land of the Orang-utan, and the Bird of Paradise*. New York: Harper & Brothers, 1869.
———. "Man's Place in the Universe." *Fortnightly Review* 74, no. 435-411 (1903): 395.
———. *Man's Place in the Universe: A Study of the Results of Scientific Research in Relation to the Unity of Plurality of Worlds*. New York: McClure Phillips & Co., 1903.
———. "Man's Place in the Universe: A Reply to Criticisms." *The Independent* Aug. 27, 1903.
———. *My Life: A Record of Events and Opinions*. Vol. 2. New York: New York, Dodd, Mead, 1905.
———. *The World of Life: A Manifestation of Creative Power, Directive Mind and Ultimate Purpose*. New York: Moffat, Yard and Company, 1911.
Walsh, Lynda. *Scientists as Prophets: A Rhetorical Genealogy*. Oxford: Oxford University Press, 2013.
Wander, Philip C. "The Rhetoric of Science." *Western Speech Communication* 40, no. 4 (1976): 226-35.
Ward, Peter. "No Pace Like Home." *New Scientist*, March 18, 2000, 4646.
Ward, Peter, and Donald Brownlee. *Rare Earth: Why Complex Life is Uncommon in the Universe*. Paperback ed. New York: Copernicus 2003.
Ward, Peter D. "Periodicity of chamber formation in chambered cephalopods: evidence from Nautilus macromphalus and Nautilus pompilius." *Paleobiology* 11, no. 4 (1985): 438-50.
Ward, Peter, and Stephen C. Meyer. "Hear the Great Debate on Intelligent Design!" http://www.discovery.org/a/3456.
———. "Stephen Meyer vs. Peter Ward Transcript." *The Dori Monson Show*, November 16. http://www.discovery.org/a/3421.
Warren, James. *Presocratics*. Berkley: University of California Press, 2007.
Warren, Samuel. "Speculators among the Stars, Part II." *Blackwood's Edinburgh Magazine* 76 (1854): 370-403.
Watters, Ethan. "We Aren't the World." *Pacific Standard* (2013), http://www.psmag.com/magazines/magazine-feature-story-magazines/joe-henrich-weird-ultimatum-game-shaking-up-psychology-economics-53135/.
Wavell, Stuart. "Dissent about Descent." *The Sunday Times*, Jan. 5, 1997.
Weaver, Richard. "Dialectic and Rhetoric at Dayton, Tennessee " In *Landmark Essays on the Rhetoric of Science: Case Studies.* Edited by Randy Allen Harris. 107-26 Mahway, NJ: Lawrence Erlbaum, 1997.
Weeks, Albert L. "Water on Mars? Don't Take a Cup." *Sarasota Herald-Tribune*, July 25, 2000.
Wendt, Alexander, and Raymond Duvall. "Sovereignty and the UFO." *Political Theory* 36 (2008): 607-33
"What If the Earth Had No Moon?". *io9*, May 3, 2013.
Whedbee, Karen E. "Making the Worse Case Appear the Better: British Reception of the Greek Sophists prior to 1850." *Rhetoric & Public Affairs* 11, no. 4 (2008): 603-30.
Whewell, William. *Astronomy and General Physics: Considered with Reference to Natural Theology.* Vol. 3, The Bridge Water Treatises. London: William Pickering, 1833.
———. "A Dialogue on the Plurality of Worlds." In *Of the Plurality of Worlds. 1853 A Facsimile of the First Edition with an introduction by Michael Ruse*. Edited by Michael Ruse. 433-507. Chicago: University of Chicago Press, 2001.
———. *The Elements of Morality including Polity*. Vol. 1. London: John W. Parker, 1845.
———. *The Elements of Morality, Including Polity*. Vol. 1. London: John W. Parker, 1845.
———. *Indications of the Creator: Extracts, Bearing upon Theology, from The History and Philosophy of the Inductive Sciences*. London: John W. Parker, 1845.
———. *Influence of the History of Science upon Intellectual Education: A Lecture Before the Royal Institution of Great Britain.* Boston: Gould and Lincoln, 1854.
———. *Lectures on the History of Moral Philosophy in England.* Cambridge: Cambridge University Press, 1852.
———. *Of Induction: With Especial Reference to Mr. J. Stuart Mill's System of Logic*. London: John W. Parker, 1849.

———. "Of the Plurality of Worlds." In *Of the Plurality of Worlds. 1853 A Facsimile of the First Edition with an introduction by Michael Ruse*. Edited by Michael Ruse. 35-279. Chicago: University of Chicago Press, 2001.
———. *Of the Plurality of Worlds: An Essay Also, A Dialogue on the Same Subject*. London: Longmans, Green, Reader, and Dyer, 1867.
———. *Of the Plurality of Worlds. 1855. A Fascimile of the First Edition with an introduction by Michael Ruse,* 1st ed. Chicago: University of Chicago Press, 1855.
———. *The Philosophy of the Inductive Sciences: Founded Upon Their History*. Vol. 1. London: John W. Parker, West Strand, 1840.
———. *The Philosophy of the Inductive Sciences: Founded Upon their History*. Vol. 2. London: John W. Parker, 1840.
———. *The Philosophy of the Inductive Sciences: Founded Upon their History*. 2 ed. Vol. 1. London: John W. Parker, 1847.
———. *The Platonic Dialogues*. Vol. 2. Cambridge: Macmillan, 1860.
———. *The Platonic Dialogues: For English Readers*. Vol. 3. Cambridge: Macmillan, 1861.
Whewell, William, and Janet Mary Douglas. *The Life and Selections from the Correspondence of William Whewell*. 2 ed. London: Kegan Paul and Trench, 1882.
Whewell, William, Janet Mary Douglas, and Stair Douglas. *The Life and Selections from the Correspondence of William Whewell*. London: C. Kegan Paul, 1881.
Whitaker, M. A. B. "On Hacking's Criticism of the Wheeler Anthropic Principle." *Mind* 97, no. 386 (1988): 259-64.
White, Roger. "Does Origins of Life Research Rest on a Mistake?" *Noûs* 41, no. 3 (2007): 453-77.
Whitehead, Alfred North. *Process and Reality: An Essay in Cosmology*. Corrected ed, Gifford lectures. New York: Free Press, 1978.
Whitney, Barry. "Divine Persuasion and the Anthropic Argument." *The Personalist Forum* 14, no. 2 (1998): 141-69.
"Who Shall Be Leader?". *Circular* 3, no. 127 (1854): 506.
Widtsoe, John Andreas. *Joseph Smith As Scientist: A Contribution to Mormon Philosophy*. Salt Lake City: The General Board Young Men's Mutual Improvement Associations, 1908.
Wiker, Ben. "Billions of Planets, But Only One Earth." National Catholic Register, September 2000.
William, Perry. "Passing on the Torch: Whewell's Philosophy and the Principles of English University Education." In *William Whewell: A Composite Portrait*. Edited by Menachem Fisch and Simon Schaffer. 117-47. Oxford: Clarendon, 1991.
Wilson, E. O. *Consilience: The Unity of Knowledge* New York: Vintage Books, 1999.
Wilson, Patrick A. "What Is the Explanandum of the Anthropic Principle?" *American Philosophical Quarterly* 28, no. 2 (1991): 167-73.
Winkler, E. T. "Religion and Astronomy." *Baptist Quarterly* 5, no. 1 (1871): 58-74.
Woods, Richard, and Christopher Goodwin. "In space, there may be no one there to hear us scream." *The Australian*, 2000.
Woods, Robert, Dominique Schneider, Cynthia L. Winkworth, Margaret A. Rileyand, and Richard E. Lenski. "Tests of Parallel Molecular Evolution in a Long-Term Experiment with *Escherichia coli*." *Proceedings of the National Academy of Sciences of the United States of America* 103, no. 24 (2006): 9107-12.
Wootton, David. *Galileo: Watcher of the Skies*. New Haven: Yale University Press, 2010.
"Worlds in the Sky." *Littell's Living Age* 12, no. 611 (1856): 360-64.
Wright, M. R. *Cosmology in Antiquity, Sciences of antiquity*. London: Routledge, 1995.
Wright, Robert. "Homo Deceptus: Never trust Stephen Jay Gould." *Slate* (1996), http://www.slate.com/articles/news_and_politics/the_earthling/1996/11/homo_deceptus.html.
Yeo, Richard. "Science and Intellectual Authority in Mid-Nineteenth-Century Britain: Robert Chambers and 'Vestiges of the Natural History of Creation'." *Victorian Studies* 28, no. 1 (1984): 5-31.
———. "William Whewell, Natural Theology and the Philosophy of Science in Mid Nineteenth Century Britain." *Annals of Science* 36 (1979): 493-516.

Yeo, Richard R. "Whewell's Philosophy of Knowledge." In *William Whewell: A Composite Portrait*. Edited by Menachem Fisch and Simon Schaffer. 175-99. Oxford: Clarendon Press, 1991.

Zboray, Ronald J., and Mary Saracino Zboray. *Literary Dollars and Social Sense: A People's History of the Mass Market Book*. New York: Routledge, 2005.

Zeyl, Donald J. *Timaeus*. Indianapolis: Hackett, 2000.

Zilioli, Ugo. *Protagoras and the Challenge of Relativism: Plato's Subtlest Enemy*. Burlington: Ashgate, 2007.

Zimmerman, Michael E. "Encountering Alien Otherness." In *The Concept of the Foreign*. Edited by Rebecca Saunders. 153-77. Lanham, MD: Lexington Books, 2002.

Zuckerman, Ben, and Michael H. Hart, eds. *Extraterrestrials: Where Are They?* Cambridge: Cambridge University Press, 1995.

Zycinski, Joseph M. "The Anthropic Principle and Teleological Interpretations of Nature." *The Review of Metaphysics* 41, no. 2 (1987): 317-33.

Index

www.ingramcontent.com/pod-product-compliance
Ingram Content Group UK Ltd.
Pitfield, Milton Keynes, MK11 3LW, UK
UKHW041632031026
15587UKWH00001B/232

9 781498 513012